# Lecture Notes in Geoinformation and Cartography

Publications of the International Cartographic Association (ICA)

Series Editors: William Cartwright, Georg Gartner, Liqiu Meng, Michael P. Peterson

For further volumes:
http://www.springer.com/series/7418

Elri Liebenberg • Imre Josef Demhardt
Editors

# History of Cartography

## International Symposium of the ICA Commission, 2010

Springer

*Editors*
Elri Liebenberg
ICA Commission on the History
of Cartography
Menlo Park
0102 Pretoria
South Africa
elri@worldonline.co.za

Imre Josef Demhardt
University of Texas at Arlington
Department of History
Arlington
USA
demhardt@uta.edu

ISSN 1863-2246 e-ISSN 1863-2351
ISBN 978-3-642-19087-2 e-ISBN 978-3-642-19088-9
DOI 10.1007/978-3-642-19088-9
Springer Heidelberg Dordrecht London New York

Library of Congress Control Number: 2011940791



Printed on acid-free paper

Springer is part of Springer Science+Business Media (www.springer.com)

# Preface

This volume comprises most of the research presented at the 3rd International Symposium of the ICA Commission on the History of Cartography which took place at the University of Texas at Arlington on October 11–12, 2010. Its appearance in published form sets it apart from the proceedings of the 1st International Symposium (2006 Utrecht, NL) and the 2nd International Symposium (2008 Portsmouth, UK) which have been made accessible through the Commission's website only. The publication of this volume was made possible by forging of a new partnership between the International Cartographic Association (ICA) and the international publishing house Springer-Verlag, an agreement which has afforded this Commission the opportunity to present its most recent research findings to a much wider audience. Hoping that this volume would be the first in a series on the history of cartography deriving from future symposia, it might not be inappropriate to also provide some background on the ICA and its Commission on the History of Cartography.

The International Cartographic Association was founded in 1959 and, as the world authoritative body on cartography, it has as its mission the promotion of the discipline and profession of cartography on as wide a scale as possible. The ICA is basically a technical organisation of professional cartographers who are concerned with current aspects of researching, compiling, and producing maps. As historical maps and historical cartographic material are an integral part of any modern cartographic database, the ICA also maintains a keen interest in research on the evolution of modern cartography. It is towards this end that an ICA Working Group on the History of Cartography was formed already in 1972. Since 1979 this Working Group has the status of a full Commission which today encourages the active involvement of all interested researchers and institutions in this field. Further information on the activities of the ICA Commission on the History of Cartography can be obtained from the Commission's website at www.icahistcarto.org

The history of cartography covers a vast field of knowledge and includes virtually all maps and map-like graphics made by humankind since prehistoric times. Map compilation and map use today are, however, seldom dependent on maps which were produced before early modern times. Taking this into account, the ICA Commission decided to concentrate on the history of cartography since the Enlightenment and, more specifically, on cartographic developments during the nineteenth and twentieth centuries.

As already mentioned, this volume emanates from the 3rd International Symposium of the ICA Commission on the History of Cartography. As is customary with symposia of this kind, the 2010 conference in Texas had a general theme which, in this case, was "Charting the Cartography of Chartered Companies". Chartered Companies were decided on as they played an important role in semi-official overseas exploration, trade, expansion, and colonial power. These companies relied upon maps and charts for planning, implementation and operation, and often employed their own surveyors and map making departments. While contributions towards this general theme were encouraged, the symposium was also open to relevant research on other cartographic endeavours than Chartered Companies.

To foster co-operation and also broaden the discussion, the ICA Commission on Maps and Society was invited to join the Symposium in Arlington. This Commission made a contribution from a social sciences perspective as represented in the papers of Peggy Allison, Berenika Byszewski, and the convener of the group, Jörn Seemann. The fact that the Commission on the History of Cartography does not only have a mandate to investigate and put on record the history of paper maps, but also to document the history of Geographic Information Science, manifested in a special exhibition and a paper session at the Symposium presented by the United States Geological Survey. Important contributions by and in association with this institution are the papers by Patrick McHaffie, Lynn Usery, and Dalia Varanka which focus on the twentieth century evolution from aerial to digital cartography.

As is evident from the above this volume contains papers on the general theme of Chartered Companies as well as papers on a wider array of themes within the field of historical cartography. Unfortunately some papers which were presented at the Symposium could not be included. This caveat could, however, be balanced out by the papers of accepted presenters who were unable to present their research personally.

We wish to acknowledge our gratitude to the University of Texas at Arlington, especially Special Collections, for their logistic support, to Prof. Dr. Ferjan Ormeling, Chair of the ICA Publications Committee, and to Mrs. Agata Oelschläger of Springer-Verlag for their kind assistance towards the production of this book.

Pretoria, RSA — Elri Liebenberg
Arlington, Texas, USA — Imre Josef Demhardt

# Contents

## Part VII General Cartography

# Part I
# Early Explorative Cartography

# Chapter 1
# Ferdinand Konšćak – Cartographer of the Compañia de Jesús and his Maps of Baja California

**Mirela Slukan Altić**

**Abstract** Ferdinand Konšćak (Fernando Consag) was a Croatian Jesuit and a missionary who carried out the exploration of Baja California according to the order of Compañia de Jesús. In 1730 he was assigned to the missions in Baja California, where he worked until his death. During his missionary work, he made three major research expeditions (1746, 1751, 1754) and made at least two maps of Baja California. His diaries from the first and second trips, in which he gave descriptions of the terrain and the people of Baja California, were published during his lifetime and have been enjoyed even after his death in several editions and in various languages. His map, "Seño de California y su costa oriental," played a special role after the existence of a California peninsula was officially accepted. Although F. E. Kino and J. de Ugarte had claimed this on their own maps before him, the official position of the Spanish government changed after Konšćak's research.

In this paper we will present the original versions of Konšćak's maps of California and later redactions of the same maps on which Ferdinand Konšćak is listed as the author. With their comparisons with later maps based on Konšćak's template, we will assess his contribution to the cartographic and geographic knowledge of Baja California.

## Ferdinand Konšćak and His Work in San Ignacio

Ferdinand Konšćak (Varaždin/Croatia, 2.12.1703 – San Ignacio/Mexico, 10. 9. 1759), is one of the most famous Croatian missionaries and explorers. He entered the Jesuit Order in 1719. In 1731 he was assigned to the missionaries in Baja California, where from 1732 until his death, he worked in the mission in San Ignacio (he was its principal from 1747). Along with Latin, German, French and Spanish

M.S. Altić (✉)
Institute of Social Sciences, Centre for Urban and Local History, Zagreb, Croatia
e-mail: mirela.altic@zg.htnet.hr

E. Liebenberg and I.J. Demhardt (eds.), *History of Cartography*,
Lecture Notes in Geoinformation and Cartography 6,
DOI 10.1007/978-3-642-19088-9_1, © Springer-Verlag Berlin Heidelberg 2012

languages which he knew before, he quickly learned the language of the Cochimí tribe and those of other neighboring tribes in California, which greatly facilitated not only his missionary work with the native Indians, but also his movements around the land, which he managed over the entire time of this research work.[1] In 1746, he began to work intensively conducting the research that would then lead to a comprehensive description of California and its peoples and drawing geographic maps, which would for the first time conclusively prove that California was a peninsula.

## The Role of Missionaries and Other Explorers in the Geographic and Cartographic Knowledge of California

In the minds of European explorers, California existed as an idea before it was discovered. The earliest known mention of the idea of California was in the 1510 romance novel, "Las Sergas de Esplandián," by Spanish author Garci Rodríguez de Montalvo. The book described the Island of California as being west of the Indies, "very close to the side of the Terrestrial Paradise; and it is populated by black women, without any man among them, for they live in the manner of Amazons."

The first European who reached the present state of Baja California was Fortún Ximénez (?-1533). Sent by Hernán Cortés (1484–1547), he discovered the southern part of Baja California. Cortés himself followed up on the discovery with an expedition to La Paz, but the settlement had to be abandoned soon afterwards. Cortés's limited information on southern Baja California apparently led to the naming of the region after the legend of California and to an initial, but short-lived, assumption that it was a large island. In 1539, Cortés sent the navigator Francisco de Ulloa (?-1540) northwards along the Gulf and Pacific coasts of Baja California. Sailing north from Acapulco, Ulloa reached the head of the Gulf of California which he named the "Sea of Cortés" in honor of his patron. After he sailed the eastern and western coasts of California, it was clear that there was no passage between California and the mainland. An expedition under Hernando de Alarcón ascended the lower Colorado River and confirmed Ulloa's findings. Sebastián Vizcaíno (1548–1624) again surveyed the west coast in 1602, but Father Antonio de la Ascension who traveled with him, still claimed that California was separate from the mainland.[2] According to those explorations, maps published during the sixteenth century, show California as a peninsula. The earliest map of the new land is a hand drawn map for Cortés about 1539.[3] After this map, all the

---

[1]Krmpotić, Davorin. 1923. *Life and Works of the Reverend Ferdinand Konšćak, S.J. 1703–1759 and early missionaries in California*. Boston: The Stratford Company, 1–13.

[2]Suárez, Thomas. 2004. *Early Mapping of the Pacific: The Early Story of Seafarers, Adventurers, and Cartographers Who Mapped the Earth's Greatest Ocean*. Singapore: Periplus Editions, 107.

[3]Kelsey, Harry. 1998. "Spanish Entrada Cartography." In *The Mapping of the Entradas into the Greater Southwest*, eds. Denis Reinhartz and Gerald D. Saxon, 69. Norman: University of Oklahoma Press.

**Fig. 1.1** California as a peninsula on map of Abraham Ortelius, 1571 (Provided by: Barry Lawrence Ruderman Antique Maps)

best known maps of the sixteenth century including those by Gerard Mercator[4] and Abraham Ortelius,[5] correctly show California as a peninsula (Fig. 1.1).

Subsequently, the first known reappearance of the Island of California on a map dates to 1622 on a map by Michiel Colijn from Amsterdam.[6] One contributing

---

[4]America Sive India Nova, 1595.

[5]Americae Sive Novi Orbis Nova Descriptio Antwerp, 1571.

[6]Published as a title page of "Descriptio Indiae Occidentalis per Antonium de Herrera."

**Fig. 1.2** Island of California, Jan Jansson, 1641 (Provided by: Barry Lawrence Ruderman Antique Maps)

factor may have been the second voyage of Juan de Fuca in 1592. Fuca claimed to have explored the western coast of North America and to have found a large opening that possibly connected to the Atlantic Ocean – the legendary Northwest Passage. So after nearly a 100 years of California's appearance on maps as a peninsula, *a* depiction of California as an island revived in the early seventeenth century. Namely, Colijn's representation was immediately used by the most respectable cartographers of the seventeenth century, including Joan Vinckeboons, Nicolas Sanson,[7] Jan Jansson,[8] John Speed,[9] Pieter Goos,[10] Frederick de Wit,[11] Vincenzo Maria Coronelli,[12] and many others (Figs. 1.2 and 1.3).

The Jesuit missionary and cartographer Eusebio Francisco Kino (1645–1711) revived the fact that Baja California is a peninsula. While studying in Europe, Kino had accepted the insularity of California, but when he reached Mexico he began to have doubts. He made a series of overland expeditions from northern Sonora to areas within or near the Colorado River's delta in 1698–1706, in part to provide a practical route between the Jesuits' missions in Sonora and Baja California but also to resolve the geographical question. Kino was satisfied that a land connection must exist, so his maps show California as a peninsula. In this regard, his map "Passage par terre a la Californie" based on his research made between the years 1698 and 1701 is particularly important. Although the map was known to a wide circle of cartographers and church authorities, the official statement of Spain that California was an island, did not change. This stance was followed by European cartographers and the view of California as an island was still held until the middle of the eighteenth century in some places (Heinrich Scherer,[13] Herman Moll,[14] and Matthäuss Seutter[15]). This was especially supported by cartographer Herman Moll who claimed that he had personally spoken with one captain who had circumnavigated the island of California (Fig. 1.4).[16]

## Expeditions of Ferdinand Konšćak

During his missionary work, Ferdinand Konšćak made three expeditions. By order of the Jesuit province of New Spain, in 1746, he sailed the Gulf of California to the mouth of the Colorado River. All the way, Konšćak and his men carried out measurements from which he drew a map that irrefutably proved that Baja California is a peninsula.

[7]Amerique Septentrionale, Paris, 1650.

[8]America Septentrionalis, Amsterdam, 1641.

[9]America, London, 1676.

[10]Paskaerte van Nova Granada et t'Eylandt California, Amsterdam, 1666.

[11]Novissima et Accuratissima Septentrionalis ac Meridionalis Americae Descriptio, Amsterdam, 1690.

[12]L'Amerique Septentrionale ou la Partie Septentrionale du Indes Occidentales, Paris, 1689.

[13]Provinciae Borealis Americae... Munich, 1720.

[14]North America, ca.1720.

[15]Novus Orbis Sive America, Augsburg, 1750.

[16]Goodwin, R. Katherine. 1998. "Entrada: The First Century of Mapping the Greater Southwest." In *The Mapping of the Entradas into the Greater Southwest*, eds. Denis Reinhartz and Gerald D. Saxon, 189. Norman: University of Oklahoma Press.

Fig. 1.3 California as island on the map of Vincenzo Maria Coronelli, 1689 (Provided by: Barry Lawrence Ruderman Antique Maps)

**Fig. 1.4** Californiaan island on the map of Henry Moll, 1720 (Provided by: Barry Lawrence Ruderman Antique Maps)

During the first expedition, Konšćak kept a diary as well. This is a manuscript known as "Derrotero," which will be published many times together with his famous map.[17]

[17] "Derrotero del viaje que en descubrimiento de la costa oriental de California hasta el Río Colorado... hizo el padre Fernando Consag... por orden del padre Cristóbal de Escobar y Llamas, provincial de Nueva España de la Compañía de Jesús; empieza en 9 de junio de 1746."

On his second expedition, in 1751, he crossed the middle of the peninsula in the area of Sierra Madre mountains and went to the Pacific coast of today's Guerrero Negro and continued along the coast north to 30° latitude. The localities identified by this expedition, noted and described in his diary "El diario del viajo que hizo el Padre Fernando Consag en 1751," are not drawn on any version of Konšćak's map. Namely, Konšćak's maps that we know of were completed by 1746. Although Konšćak's diary from this expedition does at many points explicitly mention the field survey, measurements of latitude and compass measuring, cartographic sketches from that expedition have not been found.

Konšćak took his third expedition in 1754. The aim was to find the perfect terrain for a new outpost and to expand missionary activity to the north, around future missionary outpost of San Borja. There is no evidence of Konšćak's cartographic work on that expedition.

Eventually, Konšćak decided to put his experience and knowledge of all three expeditions into literary form in "A comprehensive description of the people and the end of Lower California," known as "Descripción compendiosa."[18] Although "Descripción compendiosa" is not signed, we know that the author of this manuscript is Ferdinand Konšćak. Namely, the handwriting of "Descripción" has an addition under the title "Addiciones a las noticias contendias el la Descripcion compendiosa de lo descubierto y conocido de la California."[19] On two copies of "Addiciones" there is a subsequent note that speaks of Konšćak in the third person, and it explicitly states him as the author of "Descripción compendiosa."

## Konšćak's Hand Drawn Maps of California

Based on the first expedition, Konšćak, in 1746, compiled his first map that shows part of the peninsula which he had visited.[20] He carried out extensive measurements which confirmed the title in his diary, "Report on the trip which was to survey the eastern coast of California to its eventual limits with the Colorado River complied by Father Fernando Consag..." It is a hand drawn map, a copy of which, based on Konšćak's original, was made by California missionary Pedro Maria Nascimben (1703–1754) under the title of "Seno de Californias Y su Costa Orinetal nuevamente descubierta y registrada Desde el Cabo de la Virgines basta su Termino

[18]This manuscript, duplicated in multiple copies, exists today and was published in O'Crouley Pedro Alonso. 1775. *Ydea compendiosa del Reyno de Nueva Espagna*. Ciudad de Mexico.

[19]Unfortunately, that part of the manuscript is preserved in the subsequent transcript from March 1791.

[20]Fascimiles of the Konšćak's maps are published in, Burrus, J. Ernest. 1967. *La Obra Cartograpica de la Provincia Mexicana de la Compañia de Jesus (1567–1967)*. Madrid: Ediciones Jose Porrua Turanzas.

que es Rio Colorado por el P. Fernando Consag de la Compañia de Jesus Missionero de Californias 1746 (Fig. 1.6)."[21]

We can learn something about his work on the maps from the text of his diary. For example, on the way he did not perform measurements of latitude (note dated 25th July), but he knew the exact position of individual sites (on 9th June, states that he started from San Carlos, located at 28° northern latitude). From this we can conclude that he used the map of some of his predecessors, which had a coordinated grid. By all accounts, this was the map of California made by Eusebio Francisco Kino, probably Kino's version from 1701. Namely, the spatial extent and content of Kino's map largely overlaps with Konšćak's hand drawn maps from 1746.[22] However, there are also many differences – Kino's map extends south to 25° latitude and north to 35° and Konšćak's covers the area between 27° and 33° 30′ (Fig. 1.5).

So, Konšćak undoubtedly utilized Kino's template. On this template, he had to insert new toponyms, and correct Kino's errors, especially those related to the contours of the land in the Northern Gulf of California, as well as those pertaining to the islands inside the California Sea (*Mar de Californias*). The scale of Konšćak's map is expressed in Spanish and French miles, the measurements also used by Kino, and the geographic grid is indicated only for latitude. However, there are differences in the latitude of which they both indicate the scale along the left edge of the map. San Ignacio, on Kino's map, was located at approximately 27° northern latitude and on Konšćak's, it was drawn at 28°. So, between their measurements of latitude, there is a difference of about 1°, but Kino's accuracy is better (Konšćak's measurements had a constant error of about 1°).

All toponyms mentioned in Konšćak's diary are, in fact, noted on his map for the first time. Namely, Kino only generally noted the names of peaks in the hinterland for Lower California and the names of only a few missionary outposts. Most of the toponyms on Konšćak's map refer to the names of bays, capes and islands and only a few colony names. He highlighted the localities of sources of drinking water (*aquaje*) with particular attention. As with Kino's, Konšćak's map shows a very schematic method of molehills to show the relief of the land. The coastline on Konšćak's map is very indented and shown with a lot of detail in relation to the generalized view that is found on Kino's template. Konšćak's particularly enhanced representation of the entire northern Gulf of California with the wetlands of the mouth of the Colorado River are, for the first time, accurately mapped on his map. Konšćak also achieved significant progress in the description of the island of Isla Angel de la Guarda and the other islands of the California Sea (his map is the first to note the existence of a rocky island in front of the Bay of San Felipe, which today is named after him, Isla Consag). Many places which, until that time, had not been

[21]Huntington Library, San Marino. One example under the same title is in the British Library.

[22]Mapa del paso por tierra á la California y sus confinantes nuevas Naciones y Misiones nueva de la Compañia de Jesús en la America Septentrional, Descubierto andado y demarcado por el Padre Eusebio Francisco Kino, jesuita, desde el año de 1698, hasta el de 1701.

Fig. 1.5 Map of E. Kino, 1701 (Provided by: Barry Lawrence Ruderman Antique Maps)

marked on maps received names: Purgatorio Bay, San Juan de Pablo, San Pedro y Pablo, and San Felipe de Jesus; whilst many others, which already had names, appeared for the first time on the map.

Besides the aforementioned hand drawn maps of the eastern coast of the peninsula, there are also a few hand drawn copies of Konšćak's maps that show

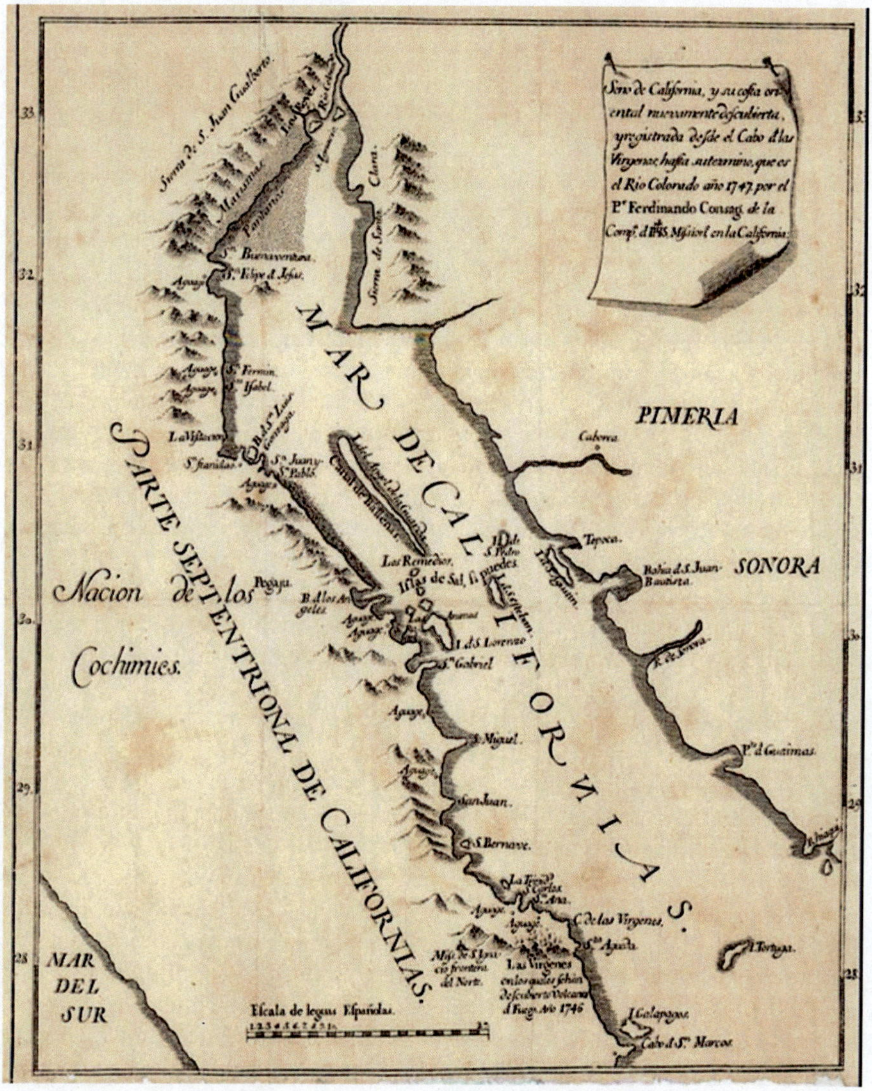


**Fig. 1.6** Map of Ferdinand Konšćak, 1746, published in Venegas-Burriel (Provided by: Barry Lawrence Ruderman Antique Maps)

the whole of the peninsula and which later served as a template for Burriel's printed version of the map. This and all its other variants carry the same title as the previously described map which depicts only the east coast of California. However, since it is clear that the map of the east coast came into existence during the expedition of 1746, with the hand drawn map which shows the entire peninsula, there remains something unknown. Namely, the map is dated in 1746, from which it can be concluded that Konšćak also mapped the southern part of the peninsula by that year, but we have no direct evidence of that. It is known that after arriving in San Ignacio, Konšćak traveled the peninsula conducting visits or going to neighboring missions, especially Loreto. However, for that period of his work between 1732 and 1746, we have no confirmation that could decisively tell us about his cartographic work in that part of the peninsula. His diary describes the southern part of the peninsula, which indicates that this part of California was well known to him, but mapping is not explicitly mentioned. Therefore, it remains to be confirmed whether the representation of the southern part of the peninsula was the result of Konšćak's mapping, or whether it was taken from the maps of his predecessors.

Comparing the maps of his predecessors, especially those of Eusebio Kino, it is immediately visible that Konšćak's map of the southern part of the peninsula has many more details in showing the configuration of the coast and islands through the numerous toponyms. So Kino's map of the southeastern part of the peninsula from 1689[23] or the map "Nuevo Navara" from 1710[24] would have served him only as basic orientations, nevertheless the data on Konšćak's map is obviously the result of detailed personal field observations. This can be confirmed by the comparison of toponyms that Konšćak's "Descripción" contains. Namely that manuscript contains a series of toponyms that Kino did not mention and that all appear on Konšćak's map. In this way, the text of "Descripción" provided the key to the definitive confirmation that Konšćak personally carried the research of the southern part of Baja California and on the basis of which he mapped that part of the peninsula (Figs. 1.7 and 1.8).

It is interesting to note the fact that the copy of the hand drawn maps preserved in Seville feature very few noticeable corrections from its author or copyist (Nascimbien?). In this way, the names of several rivers on the west coast of the peninsula were corrected. It is also notable that, in the far south of the peninsula, the names of several marked missions are missing – alongside the San Jose de Cabo mission with a church symbol is only noted "Mission" and in the position of the Santa Rosa mission is only drawn a church symbol without any inscription. Therefore, the map seems to be an unfinished work. However, every toponym that we come across on Konšćak's map of the east coast of California was also literally assigned on his map of the entire peninsula. This additionally confirms Konšćak's authorship of all of the data on the hand drawn map of Baja California from 1746.

Konšćak's map "Seño de California y su costa oriental" played a special role after which it was officially accepted that a Californian peninsula existed. His map of California was evaluated by the Spanish court and, on that basis, it was officially concluded that California was a peninsula. Although F. E. Kino and J. de Ugarte had

[23]Delineacion de la Nueva Provincia de San Andrés del Puerto de la Paz y de las Islas circunvecinas de las Californias ó Carolinas que al Excmo. Señor Conde de Paredes, Virrey de la Nueva España dedica y consagra la Mission de la Compañia de Jesus de dichas Californias ó Carolinas en 21 de Diciembre dia del Glorioso Apostol de las Indias S. & Thomas, de 1689.

[24]Nuevo Reyno de la Nueva Navara son suos confinantes obros Reynos 1710.

**Fig. 1.7** Map of peninsula by Kino, 1702, in Venegas-Burriel (Source: Burrus, 1967)

**Fig. 1.8** Map of peninsula by Konšćak, published in english version of Venegas-Burriel (Provided by: Barry Lawrence Ruderman Antique Maps)

claimed this finding on their own maps before him, the official position of the Spanish government changed after Konšćak's research and the publication of his maps.

His map, together with his diary known as "Derrotero," was printed in the Venegas-Burriel's book "Noticia de la California" (first edition published in Madrid, 1757). Namely, in order to prepare the work and corresponding maps of California, Burriel employed the help of Pedro Maria Nascimben for the purpose of making multiple copies of both Konšćak's maps. Based on these copies, Burriel gave shape to two copper plates of the maps, one that showed only the east coast and the other showing the whole peninsula. The first plate that shows only part of the peninsula was left with its original title, "Seno de Californias . . .," but this was changed in 1747.[25] The map that showed the whole of the peninsula was given the title "Mapa de la California, su golfo y provincias fronteras en el Continente de Nueva Espagna" and dated 1757. [26] This map, on which he decorated its outer edge with elaborate engravings of the animals and natives of California, is one of the most beautiful maps of the time. Various versions of Konšćak's maps are included in all later editions of Venegas-Burriel's works, as well as many other widely distributed books at the time, such as "Nachrichten von der Amerikanischen Halbinsel Californien," by Jesuit Jacob Bagaert (Manheim, 1772), "Storia dela California" by Mexican Jesuit Francesco Clavigero (Venice, 1789) or Denis Diderot's monumental work "Encyclopedie" (Paris, 1772).

## Authorship over "Addiciones" and Three Sketches of the Southern Part of the Peninsula

As we have noted, in "Addiciones" it explicitly states Konšćak's authorship over "Descripción," but the author of "Addiciones" is clearly not mentioned. Seeing that they are, as in the case of "Descripción" and as in the case of "Addiciones", only preserved transcripts, it is not possible to say whether both of the original manuscripts were written by the same hand or by the same author. "Addiciones" contains valuable information about Californian flora and the customs of the natives and, in this sense, the text continues in "Descripción." At the end of the text, on separate sheets of paper (but written by the same copyist), there are three sketches: one of the tip of the peninsula with Cabo San Lucas, one of the broader belt of the California Bay around the Loreto outpost, and the third of the Pacific coast around the outpost La Purisima. The manuscript had to have originated no later than the start of 1759 because the entire "Addiciones," together with the sketches of 14th

[25]Seno de California, y su costa oriental nuevamente descubierta, y registrada desde el Cabo al las Virgenes, hasta su termino, que es el Rio Colorado año 1747. por el P[e.] Ferdinando Consag de la Comp[a.] de IHS, Mission en la California.

[26]Mapa de la California, su golfo y provincias fronteras en el Continente de Nueva Espagna. I[s.] Peña sculp.1757. It was released in the first volume of Burriel's works.

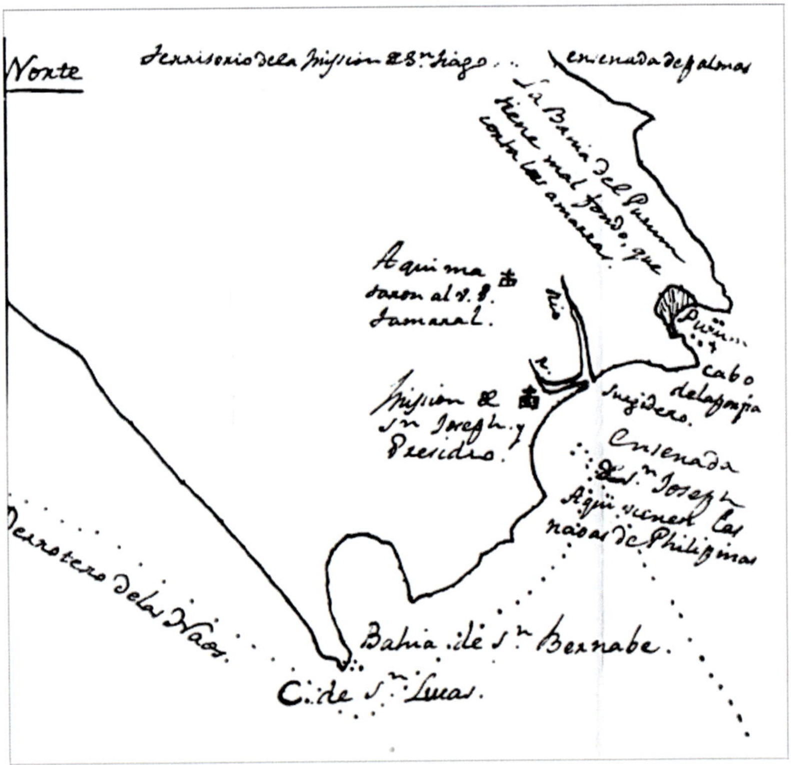


**Fig. 1.9** One of three sketches in "Addiciones" presenting Cabo San Lucas (Source: Burrus, 1967)

April 1759, were sent to Marcos Burriel by the then – Californian visitor Juan de Armesto as a supplement to the new edition of "Noticia de la California." Unfortunately though, the materials arrived too late (Fig. 1.9).[27]

Given the dating of such manuscripts, Aschmann, in his analysis concludes that the author was also probably Konšćak.[28] From the comparison of three sketches that show the same areas as Konšćak's map of peninsula, it is evident that the sketches contain more detail and toponyms than the map of 1746. However, the sketches, together with the manuscripts of "Addiciones," could really have been drawn by Konšćak. Perhaps he was dissatisfied with the accuracy of his map in his later years. In that case, the three sketches of the southern part of the peninsula would be a kind of supplement to his map.

---

[27]The letter is preserved in the Manuscript Collection, Huntington Library, San Marino.

[28]Aschmann, Homer. 1966. *The Natural and Human History of Baja California*. Los Angeles: Dawson's Book Shop, 18–19.

## The Influence of Konšćak's Maps on Later Cartographic Representations of California

Thanks to the spread of Burriel's works, Konšćak's maps quickly became invaluable to foreign cartographers across the world. Those who did not use it literally, used it to complete their own maps. One of the first who evidently used Konšćak's map to complete an old version of Kino's map was Isaak Tirion. On his map of New Mexico and California from 1765,[29] the representation of the peninsula is entirely taken up with Konšćak's map (this is particularly evident in view of the mouth of the Colorado River and in the representation of the Bay of La Paz in the south of the peninsula). In this sense, Tirion's map is one of the most successful compilations of Konšćak's and Kino's research (Fig. 1.10).

The great explorer Alexander von Humboldt also used Konšćak's information. In representing the California peninsula on his map "Carte Generale du Royaume de la Nouvelle Espagne" from 1804,[30] Humboldt also had to use the newest compilation of the areas which were still remaining on Konšćak's map. Namely, although Humboldt mentioned that he created the map according to his observations, it is known that Humboldt during his own Latin American expedition of 1799–1804 did not visit California, so for the description of that area he used someone else's maps.[31] It is the same case with the description of California on Aaron Arrowsmith's map "A Map of America" from 1805, which Humboldt claims was a copy of his map.[32]

Ferdinand Konšćak was one of the most prominent Croatian missionaries and explorers. He spent a full 27 years in California, tirelessly working not only on research but also on improving the living conditions of the local people. Thanks to his modesty, he gained great respect and honor during his life. His diaries and maps were handed to the most eminent historians and travel writers throughout the world, and his work, was published in at least six languages during the eighteenth century. Officially proving the nature of California as a peninsula, his map became one of the most widely reproduced maps of the area and gave him a prominent place as one of the most important cartographers of California.

---

[29]Kaart van het Westelyk Gedeelte van Nieuw Mexico en van California Volgens de laatste Ontdekkingen der Jesuiten en anderen/Isaak Tirion, Amsterdam 1765. The map was included in Tirion's well-regarded *Nieuwe en Beknopte Hand-Atlas*.

[30]Carte Generale du Royaume de la Nouvelle Espagne depuis le Parallele de 16° jusqua'au Parallele de 58° (Latitude Nord) . . .

[31]Humboldt, Alexander von. 1814. *Political Essay on the Kingdom of New Spain*, London: Longman, Hurst, Rees, Orme and Brown and H. Colburn, XXV–XXXIII. He explicitly mentioned that he used the map of California published in Venegas-Burriel.

[32]Humboldt, 174.

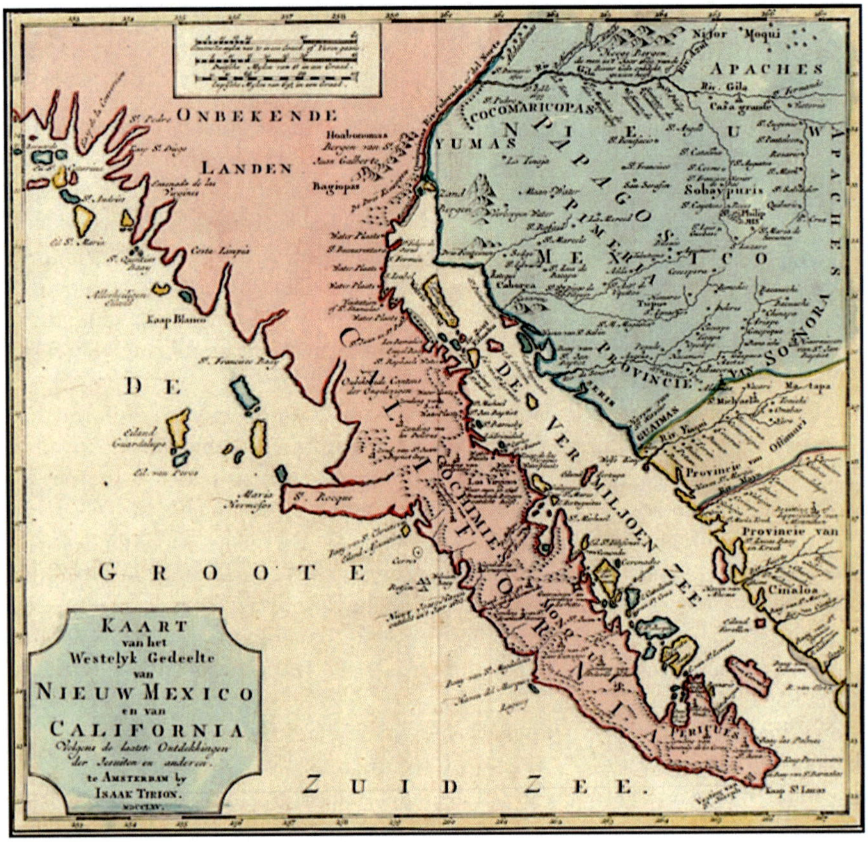


**Fig. 1.10** Tirion's map of California from 1765 based on Konšćak's template (Provided by: Barry Lawrence Ruderman Antique Maps)

## Biographical Note

**Mirela Slukan Altić** is a research advisor in the Institute of Social Sciences Zagreb, Croatia, where she works as a Head of the Centre of Urban and Local History. She specializes in historical geography, historical cartography, urban and local history. At the Department of History of the Faculty of Philosophy, University of Zagreb, Dr. Altić lectures on "Introduction to the Reading of Historic Maps" and "Cartographic Sources for European and Croatian History". She is also participating in education at postgraduate studies at the Department of History. Occasionally, Dr. Altić teaches in universities in foreign countries. She has been the head of Croatian national project, "Historic Towns Atlas," since 2003. She is also the author of 12 books, including five volumes of the Croatian Historical Towns Atlas. In 2004, she won the annual prize for science for her book "Historical Cartography: Cartography Sources in Historical Sciences". Dr. Altić is a member of the ICA Commission on the History of Cartography.

# Chapter 2
# The Caribbean Cartography of Samuel Fahlberg

**Dennis Reinhartz**

**Abstract** Swedish born in Halsingland and educated in Stockholm, Dr. Samuel Fahlberg (1758–1834) came to the Caribbean to the island of St. Barthélemy ("St. Barths") in 1784 as a physician and Government Secretary when it was relinquished by France to Sweden. Two years later, he became the Provincial Medical Officer and the Customs Inspector and Cashier, and in 1803 he also became the Director of Survey of the tiny colony and mapped it extensively for the Swedish West India Company. But because of the ongoing problems between Sweden and France over the island and his too close an association with the island's pro-British faction, Fahlberg eventually was forced to flee under threat to "lose his life, honour, and property" to the neighboring Dutch islands of St. Eustatius (1810–1816 and 1829–1834) and St. Maarten in (1816–1829). He remained in St. Maarten and St. Eustatius ("Statia") as a doctor, surveyor, cartographer, architect, and artist for most of the three remaining decades of his life until his death in St. Eustatius in 1834. During these years, Fahlberg produced several excellent and important maps of these Caribbean islands and their towns, plantations and estates, and fortifications.

## The Life and Times of Samuel Fahlberg[1]

The information regarding Samuel Fahlberg's early life is sketchy. He was born the son of a farmer in the area of Halsingland in Sweden on 5 September 1758. After the death of his father and a twin brother in 1769, he and his mother sailed for

[1] I am indebted to Elsje Bosch, Director of the Sint Maarten National Heritage Foundation and Sint Maarten Museum in Phillipsburg, for her conversations with me about Fahlberg and his work and for her sharing of rare biographical information on him with me.

D. Reinhartz (✉)
The University of Texas at Arlington, P.O. Box 5078, Santa Fe, NM 87502, USA
e-mail: dprein@uta.edu

E. Liebenberg and I.J. Demhardt (eds.), *History of Cartography*,
Lecture Notes in Geoinformation and Cartography 6,
DOI 10.1007/978-3-642-19088-9_2, 

Stockholm to make a living. Along the way they were shipwrecked; his mother drowned, and he barely survived by clinging to some ice.

Upon his arrival in Stockholm, Fahlberg was taken in by Jacob Tjader, Supervisor of Public Baths and Borough Medical Officer, and made his apprentice. Apparently appreciating his intellectual ability, Tjader enrolled Fahlberg in college to study botany and pharmacy under Bergius. He followed this up with an internship in surgery with Olaf of Acrel at Serafimer Hospital where he passed his examination to become a qualified surgeon in 1782.

Fahlberg's first real employment was as a surgeon on a merchant ship. Eventually, he traveled to France where he signed up with the fleet of Jean François de la Perouse bound for North America where he was wounded in a battle with the English in Hudson's Bay. In 1784, he returned to Stockholm to become a deputy surgeon at the Serafimer Hospital.

In 1784, France ceded the West Indian island of St. Barthélemy ("St. Barths") to Sweden in exchange for trading rights at Gothenburg (Sweden sold it back to France in 1878). During the selection process for personnel to administer the new colony, the Swedish government and *Svenska Västindiska Kompaniet* (Swedish West India Company) gave preference to individuals who had some familiarity with the New World. Consequently, in December 1784 Fahlberg was appointed as a government secretary and physician to St. Barthélemy. Two years later he traded in his government secretary position for one as a customs inspector and cashier. At about this time, he also married his first wife Elizabeth Siwars. She died in 1798, and in 1801 he married Elizabeth Evory from the Dutch island of St. Eustatius ("Statia"). Since the island was temporarily (March 1801–June 1802) in British hands, the wedding ceremony was performed by the British governor. In 1805, Fahlberg was reappointed as a government secretary and served in all three positions until he departed the island in 1810.

As a doctor on the island, Fahlberg helped to combat several epidemics, including small pox. He also functioned as an unpaid surveyor and naturalist. By 1803, he was appointed an island engineer with the title of Director of Survey, and eventually, his maps came to the attention of the Swedish government, the Swedish West India Company, and Academy of Sciences in Stockholm.

During this period, Fahlberg also must have wandered the Caribbean and Gulf of Mexico extensively for he collected historical artifacts and natural specimens from Trinidad, Venezuela, and elsewhere for the Academy of Sciences and the University of Uppsala and corresponded with important Swedish scientists and other intellectuals. Some of his articles were published in the Academy's *Lakaren Och Naturforskaren*, and a butterfly, *tortrix Fahlbergiana* was named after him. Fahlberg was granted a doctorate from the University of Uppsala in 1796.

The political situation on St. Barthélemy was quite stressful during the era of the French Revolution and Napoleon. While Fahlberg may have favored the American and French revolutionaries early on, over time his attitudes changed somewhat. St. Bathélemy's population was divided into pro-French and pro-British groups.

Although Fahlberg tried to remain neutral, he was on good terms with the leader of the pro-British faction, John Nordling, who was forced to resign his office in the Swedish West India Company in 1797. Fahlberg later became a commander of a company of Swedish militia in the capital, Gustavia, and word of his attempts to stop the privateering activities of the pro-French faction soon reached the pro-revolutionary Swedish government in Stockholm. Consequently, he was exiled to St. Eustatius in 1810, but continued to work for the Swedish West India Company as an off-island agent for another year. Fahlberg associated with numerous prominent Englishmen throughout the West Indies, and in 1811 in two letters from Antigua to the Swedish government on St. Barthélemy, he advocated that the island be declared neutral under British protection. Thereafter, Fahlberg was tried in absentia for conspiracy against the Swedish government on the island, found guilty, put under a death sentence, and lost all of his positions, titles, and property. Somewhat later, the verdict was confirmed by the Civil Court in Sweden, but altered slightly to allow his property to pass on to his children.

Fahlberg remained in St. Eustatius as a garrison doctor until 1816 when he moved to Dutch Sint Maarten to be with his two married daughters and their families and where he worked as a doctor and surveyor. The closely clustered, privateer-friendly volcanic Leeward Islands of St. Eustatius, St. Maarten, and Saba were under the direct control of the *Geoctroyeerde Westindische Compagnie* ("Chartered West India Company," the Dutch West India Company) from 1621 until 1791, when the Dutch government bought out the failing enterprise. [2] Yet, by Fahlberg's time on the islands, little had changed. In 1829, he returned to St. Eustatius as a garrison doctor and remained there until his death on November 28, 1834. Just before his death, he received a pardon from the Swedish government

[2]Since the American Revolution, there has been an especially close relationship between the United States and St. Eustatius and St. Maarten. Both islands were friendly to the American revolutionaries and opened their ports to their ships and trade. St. Eustatius was known as "Gold Mountain" for its wealth of trade, and in 1776 it was the first foreign territory to officially salute the American flag. In the United States Library of Congress' Geography and Map Division there is a wonderful untitled pencil and wash drawing of Oranjestad, St. Eustatius (177?), probably by an unknown Dutch artist, that shows a fortified harbor surrounded by volcanic peaks, farms, houses, a church, and warehouses. In the harbor, proudly flying their flags, are American ships and those of countries friendly to the American cause – the Netherlands, France, Denmark, and Sweden – engaged in commerce (G5032.S3A35,177-, S., Vault). During World War II, both islands served as bases for hunting U-boats (some of which were being resupplied only a few miles away by Vichy-loyal French St. Martin) in the Atlantic and Caribbean, and today they are major tourist destination for Americans and Canadians, among others. See: Ostindie, Gert. 2005. *Paradise Oversea. The Dutch Caribbean : Colonialism and Its Transatlantic Legacies*. Oxford: Macmillan Education, 2005.

under a general amnesty, but the word did not reach him in time. Today, a main street, the *Rue Fahlberg* in Gustavia, St. Barthélemy is named in his honor.[3]

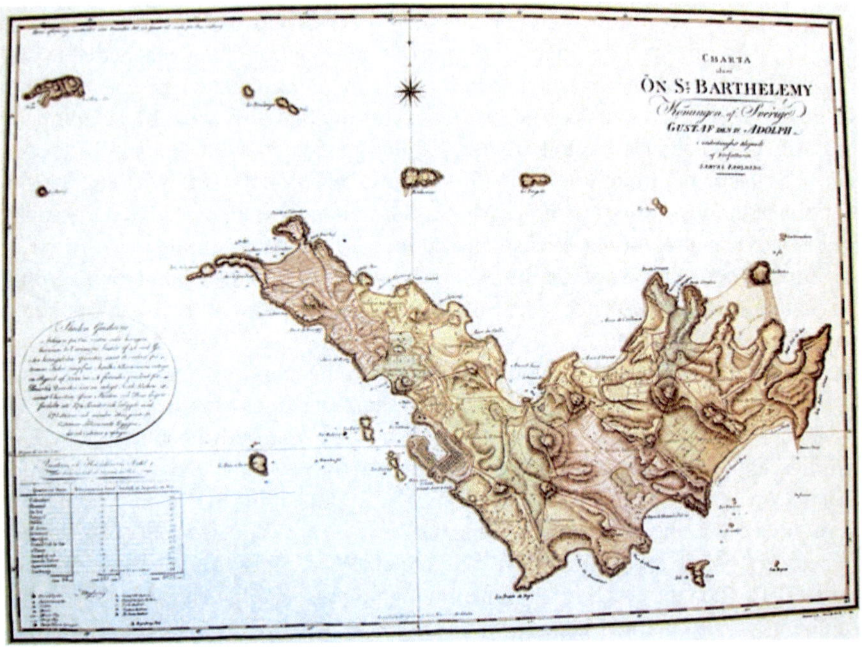


## The Maps of Samuel Fahlberg

Under the ever economically minded Company and colonial administrations, Fahlberg mapped in addition to his other contracted duties; he rarely was paid primarily as a cartographer. The small volcanic 8 square mile island of St. Barthélemy too is in the Leeward Islands very near to St. Maarten and was claimed by France in 1648 at the end of the 30 years War in Europe. Initially, its economy was based primarily on marginal plantation agriculture mainly in sugar for the rum trade. Today, it is a French collectivity and part of the French West Indies along with Guadaloupe, Martinique, and French St. Martin and has international reputation as a center of upscale tourism frequented mostly by visitors from North America and Europe.

[3]Coomans, Henny and Maritza Coomans-Eustatia. 2000. *Sint Maarten in Kaart en Beeld/St. Martin in Maps and Prints*. Bloemendaal: Stichtung Libri Antillani, 16 and 28–31 and Reinhartz, Dennis. 2007. "The Dutch Mapping of Saint Martin." In *Mappae Antiquae: Liber Amicorum Günter Schilder. Essays on the Occasion of his 65th Birthday*, 2 vols., eds. Paula van Gestel van't Schip and Peter van der Krogt. 't Goy Houten: HES & De Graaf Publishers, I, 162–164.

While on St. Barthélemy, Fahlberg drew several maps of the island and its neighboring islands, the most important of which is his *Charta äfver ÖN St. BARTHELEMY*... (Stockholm: 1801), engraved by the Swedish chart maker Erik Åkerland (1754–1832).[4] In fact, it is one of the most significant maps in the cartographic history of the island. It is a scientifically accurate and very aesthetically appealing map, showing the whole island and surrounding close-in smaller ones. The main town Gustavia, named after King Gustav III of Sweden, is at the lower part of the center of the map. It is located at 80° 47′ west of Stockholm and 17° 51′ north latitude and is described in statement in an oval at the center of the left edge of the map. The other villages and *quartiers* also are clearly designated as are the agriculturally productive lands, divided into surveyed French long lots. In the lower left corner of the map, below the Gustavia description and above the guide to the map's symbols there is a list of the *quartiers* (14) indicating the number of habitations (135) and inhabitants (2,072) of each.

The *volcanic* mountains are shown as such, and there are four *salines* (salt ponds) labeled. The numerous bays and respective beaches of the island are identified. There are indications of the depths of some of the bays as there are for the waters off of some of the surrounding smaller islands. The scale is recorded as 1,000 French feet to the inch, and, understandably, the language of much of the map is French.

St. Barthélemy also appears on earlier Fahlberg regional map of 1792, but in nowhere near as refined and detailed manner as in 1801. His *CHARTA öfver Canalerna och Utloppen emellan Öarna frän St. Barthelemy till Dog och Prickle Pear*... (Stockholm: 1792), also engraved by Åkerland, shows the less sophisticated rendition of St. Barthélemy with surrounding islands in the lower right corner.[5] To the northwest is a similarly simplified and undersized depiction of Dutch-French St. Martin and beyond it in the same direction a vastly oversized version of the neighboring British island of Anguilla. Finally, in the upper left corner of the map are the oversized and somewhat misplaced Prickly Pear and Dog islands cited in its title. Inset in the lower left corner is the *Speciale-Charta öfver Inloppet och Hamnen Carenage famt Staden Gustavia* with a guide to its map symbols and again on an indicated scale of 1,000 French feet to the inch. Given the distortions of size and location on the larger part of the map, it is not surprising that no scale is present here.

---

[4]Fahlberg,, Samuel. 1801. *Charta äfver ÖN St. BARTHELEMY Konungen af Sverìge Gustav den IV. Adolph i underdanighet tilegnado af Författaren Samuel Fahlberg*. Stockholm.

[5]Fahlberg, Samuel. 1792. *CHARTA öfver Canalerna och Utloppen emellan Öarna frän St. Barthelemy till Dog och Prickle Pear etter de Gednaste observationer Utlage 1792 och Dedicerad till Kongl. Vetenskaps Academien af Samuel Fahlberg*. Stockholm.

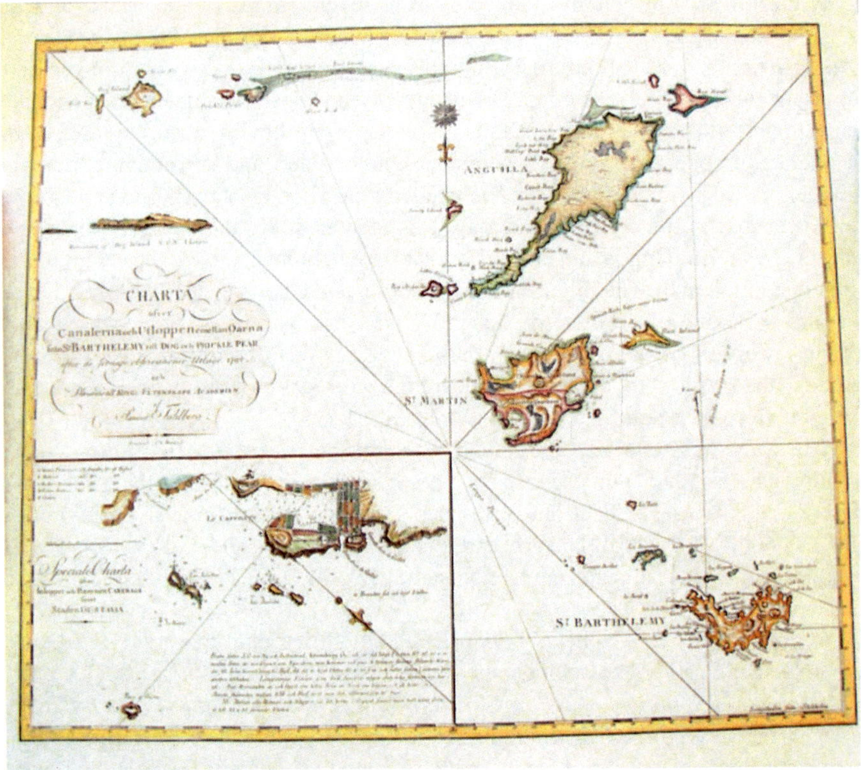


This map covers a larger area and is concerned mainly with the passages and potential routes between islands. Not only St. Barthémey, but all of the islands are relatively generalized. The 1792 and 1801 maps are hand colored. This is still quite an artistically satisfying map, but although better than many of the contemporary maps of the area, the 1792 map is nowhere nearly as scientifically accurate as that of 1801 or other Fahlberg maps, despite whatever its title might profess. It also should be viewed as an earlier example of his cartography. Although rarely cited, his sources for these and others of his maps consisted of previous Spanish, French, Dutch, and English cartography of the islands, local government surveys and maps, and his own careful surveys. All in all, Fahlberg's high-quality mapping of St. Barthélemy and its environs better revealed the island to Sweden and the Swedish West India Company and helped to more firmly establish Swedish possession of it.

In his later years, Fahlberg did a number of exceptional maps of St. Martin. Like St. Eustatius, the 37 square mile St. Martin was first sighted by Europeans on Columbus' second voyage to the New World in 1493. But it had been known to Arawak, Cibonay, and Carib Indians for almost three and a half millennia previously. The Indian name for St. Martin was *Sualouiga* ("Land of Salt"), for its many salt ponds that made it valuable to them and early European interlopers. It was first settled by French and English from nearby St. Christopher (St. Kitts) in 1627 and the Dutch West India Company in 1631 and was officially divided between the French (20.5 sq. mi.) and Dutch (16.5 sq. mi.) in 1648 after the 30 years War, as it remains today. St. Martin's semi-arid climate and soil are poorly suited for plantation agriculture, and cotton, tobacco, sugar, indigo, coffee, and cacao all failed there; its salt remained economically important into the twentieth century. The island began moving toward becoming the tourist mecca that it is today in 1939 when the Dutch and French declared it a duty free area.[6] This status also allows for the ready distillation, importation, blending, and exportation of alcohol, thus creating a developing regional trade in St. Martin flavored rums.[7]

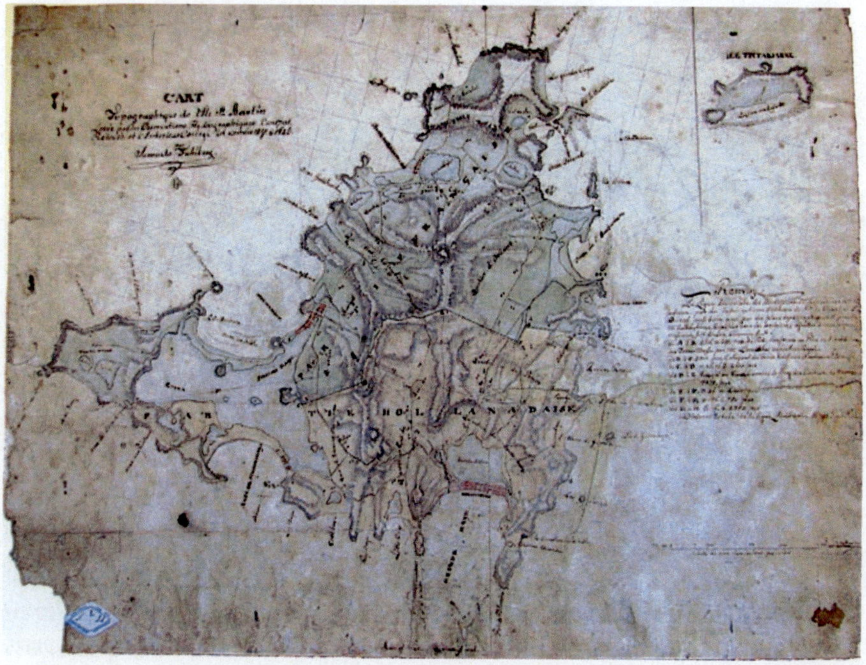


Fahlberg's *Cart Topographique de l'Isle de St. Martin...* (1790–1791, 1817, and 1826?) is on a par with his 1801 St. Barthélemy map. It shows the geographically and politically complex island of St. Martin with its volcanic mountains, salt ponds, external and internal bays, beaches, quarters, towns and villages, and local and national divisions artistically rendered in specific detail. In the upper left corner is the neighboring smaller island of Tintamarre, popular today for its beach and mud bath. Below it is the guide to the maps symbols, and below it, the scale. Perhaps, to

[6]Hartog, J. 1981. *History of Sint Maarten and Saint Martin*. Phillipsburg: St. Maarten Jaycees, 10–31.

[7]Hamilton, Edward. 1997. *Rums of the Eastern Caribbean*. Culebra, PR: Tafia Publishing, 58–65.

emphasize this map's importance and to enhance its broader appeal it is in French rather than Dutch or Swedish. His distinctive 1826? painted state of the map is particularly attractive and exhibits Fahlberg's talents not only as a cartographer, but as an artist as well.[8]

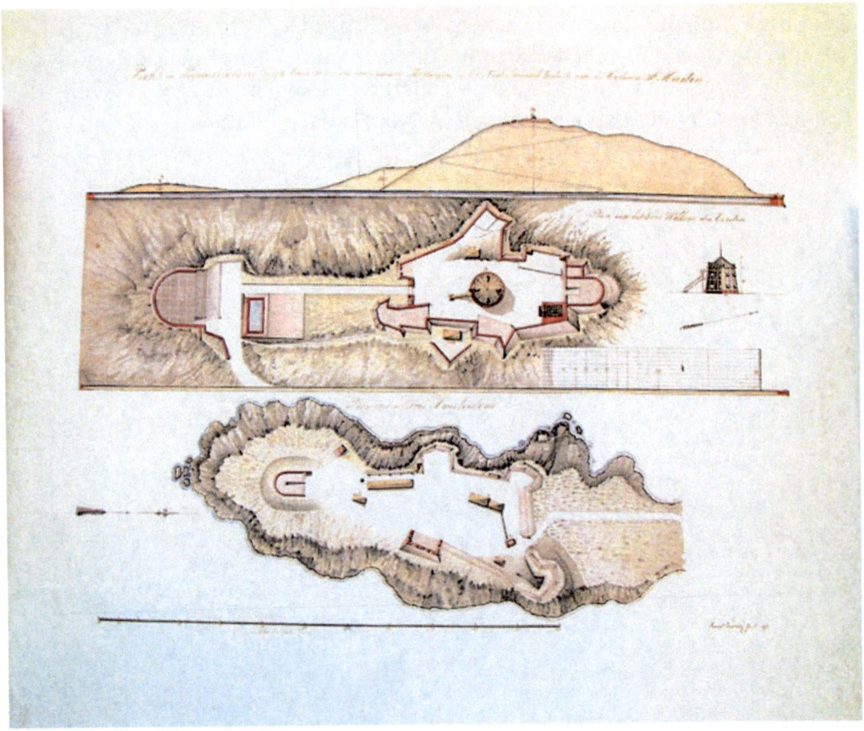

Fahlberg drew, painted, and/or hand colored several other important maps of St. Maarten, especially of the area around Great Bay, including the main town (officially the capital after 1919) of Phillipsburg. In c. 1817, he executed a superb, beautiful watercolor plan of the fortifications protecting the entrance to Great Bay and Phillipsburg, which occupies a strip of land on its north side that separates the bay from the Great Bay Salt Pond, the biggest on the whole island, yet a bit further to the north. It was undoubtedly done for the local Dutch administration. His *Profil en Perpendiculaire hoogte boven de Zee der vornaamste Batterijen in het*

[8]Fahlberg, Samuel. *Cart Topographique de l'Isle de St. Martin Levée par des Observations Hydrographiques l'an 1791. Rélevée e l'IntericurCorrigé les cinnées1817. &1826. par Samuel Fahlberg*, manuscript in the collection of the *Universiteitsbibliotheek, Amsterdam*.

*Nederlandsch van het Eyland St. Martin* is divided into three parts.[9] The top third illustrates the western approach Great Bay from the east, locating its three protective batteries, as acknowledged in the title. The center third, *Plan van het Fort Willem den Ersten*, shows the plan of the western most fort, named after King William I (labeled "A" on the top map) with its principle blockhouse in profile. There is no plan of another smaller gun site located forward of it (labeled "C"). *Plan van het Fort Amsterdam* at the bottom is of the larger eastern battery ("D"), which is separated from Fort Willem by Little Bay. Directional arrows specify the orientations of the plans, their heights above sea level are indicated clearly, and there scales are present. Fahlberg's three extremely accurate maps mirror the eye, ability, presentation, and technique of the skilled surveyor and of the artist.

Fort Amsterdam was constructed in 1631 by the Dutch West India Company to protect its salt extraction on the island. Fort Willem was constructed by the British in 1801 as Fort Trigge to threaten the Dutch and then taken by the Dutch in 1802 and named Fort Gelderland. It fell to the French in 1807 as Fort Louis Napoleon and in 1810 to the British again as Fort Trigge. The Dutch reclaimed it as Fort Willem in 1816 and abandoned it in 1846. Fort Amsterdam has been reconstructed to some extent, in part using Fahlberg's map, as a tourist attraction. It sits just above and south of and is dwarfed by the modern Divi Little Bay Resort. Fort Willem and its outpost on the other hand are in total decay and even difficult to find. They await a commitment of funds from the St. Maarten government from its tourist tax revenues for restoration.

In 1822, Fahlberg painted a charming and notable *View of PHILIPSBURG*.[10] As pointed out in its full title, it looks out upon the town from the estate of one Gerd (Gerhard?) Du Cloux across the Great Bay Salt Pond southward onto Great Bay with the island of Saba in the distance. In front of Saba is Fort Amsterdam and on the heights above it to the left is Fort Willem. It is perhaps Du Cloux's livestock grazing beyond the wall in the foreground. The gatherers of firewood on the near shore of the salt pond and those rowing across it are probably slaves, since slavery was not finally abolished on the French side of St. Martin until 1848 and on the Dutch side until 1863.

---

[9]Fahlberg, Samuel. c. 1817. *Profil en Perpendiculaire* hoogte *boven de Zee der vornaamste Batterijen in het Nederlandsch Gedeelte van het Eyland St. Martin. Samuel Fahlberg...*, manuscript in the collection of the *Universiteitsbibliotheek, Amsterdam*.

[10]Fahlberg, Samuel. *View of PHILIPSBURG. The CAPITAL of the Netherland part of the Island Saint MARTINS. Dedicated to WILLIAM HENRY RINK Esquire the Former Governor of the Islands ST MARTINS and Saba. Taken from the Estate of the Honorble* [sic] *Gerd Du Cloux Esqr. 1822. By His Most Obedient Humble Servant Samuel Fahlberg*, manuscript in the collection of the *Algemeen Rijksarchief*, Division of Maps and Drawings, The Hague. This map also is reproduced on a popular tourist postcard from St. Maarten, *Historical Art of the Caribbean: View of Phillipsburg by Samuel Fahlberg, 1822* (West Falmouth, MA: Caribbean Archives, c. 2005).

Although the town looks quiet, the five larger ships, some of which could be privateers or pirates, anchored or under sail in the bay attest to Phillipsburg's status as a busy West Indian port. Its location at the rear of Great Bay afforded it protection from hurricanes (e.g. 1819) and pirates and other enemies from the sea. And its position between the bay and pond allowed for a type of natural air conditioning against the hot and humid Caribbean climate. Over the years, Fahlberg's *View* has proved to be a rare, valuable, and attractive graphic source on the history of early eighteen century Phillipsburg.

Fahlberg's close acquaintance Willem Hendrik Rink, to whom the *View* was dedicated, was the governor of St. Maarten in 1790–1806. While visiting there in 1816, Fahlberg painted a map of Rink's estate, *THE RETREAT*. . ., out of friendship or perhaps as a commission.[11] It is the picture of a prosperous, fairly good sized estate by St. Maarten standards of the early nineteenth century. There are two great houses, a large barn, out buildings, and slave quarters on a creek surrounded by farm animals, gardens and diversely cropped fields, one being worked by slaves. In the hilly quarter of Cul de Sac, the fields might contain maize and other suitable grains, tobacco, indigo, and even some cotton. Slaves also are engaged at various other activities across the estate, including the opening of the main gate by a slave boy for Rink himself and a retainer approaching on horseback. Coconut and date

[11]Fahlberg, Samuel. *THE RETREAT Situated In The Quarter Cul de Sac; The Property Of WILLIAM HENRY RINK Esquire Formerly Governor Over The Islands ST MARTINS & SABA. To WHOM This VIEW is Humbly Dedicated. By HIS Grateful And Most Obedient Servant Samuel Fahlberg Colonial Surveyor. October The 18th 1816*, collection of the *Algemeen Rijksarchief*, Division of Maps and Drawings, The Hague.

palms and nut and fruit trees are part of the estate as well.[12] In somewhat distorted perspective, it rises up and is all delimited by some of St. Martin's more moderately sized mountains, the higher ones being on the French side.

Like Fahlberg's *View*, *THE RETREAT* is a noteworthy testimonial not only to its creator, but also to the place and time in which he lived. For the historian, it offers not only a scenic portrayal of his friend's property, but useful observations on the operations of an early nineteenth century agricultural holding in the poorer Lesser Antilles (as opposed to the major sugar islands) of the Leeward Islands as well.

12"...'The Retrait' combined with 'The Farm' estate are the most important plantation of the Dutch part at present, a tract of 333 'akkers' numbering 152 slaves...." Teenstra, M.D. 1837. *The Islands of the Netherlands West Indies: St. Marten*. Phillipsburg: Lord & Hunter, 1993, 13.

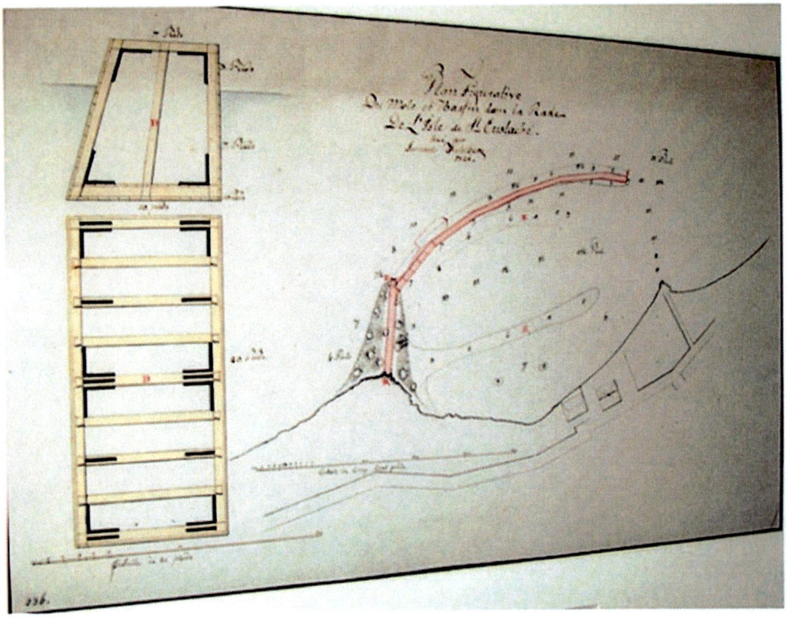

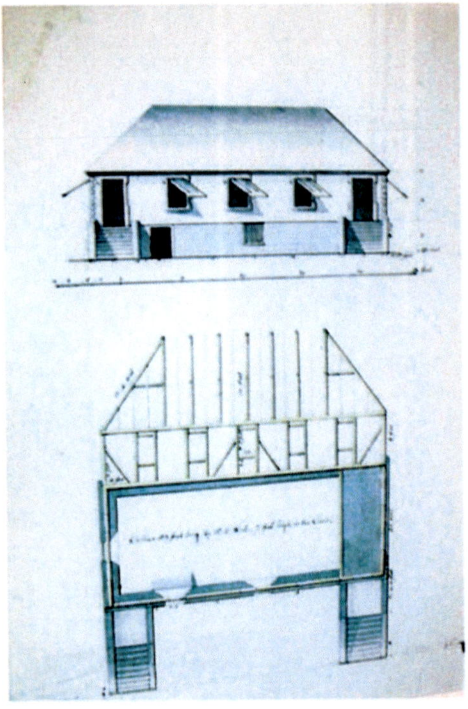

During his final years on St. Eustatius, Fahlberg kept busy in part by working for the local colonial administration in the capital Oranjestad, much as he had done previously in Gustavia and Phillipsburg, producing superb images such as a plan for the erection of a mole (i.e. breakwater) to protect the anchorages of Orange Bay (1828) and an untitled architectural construction drawing for the non-commissioned officers' and soldiers' quarters at Fort Orange (1828).[13] The mole was washed away by a hurricane before it was completed, but the barracks can still be viewed today in a thoroughly restored Fort Orange.

Working for his homeland and the Netherlands, Fahlberg was an "innocuous Swede"[14] whose maps, plans, and other images were inoffensive to but nevertheless of use to the greater powers of Spain, France, and Great Britain in the Caribbean. The quality of his cartography also helped to put the Dutch mapping of St. Martin and its nearby islands at least on a par with, if not over, that of the French and others for much of the remainder of the nineteenth century.[15]

## Biographical Note

**Dennis Reinhartz** is an emeritus professor of history and Russian at The University of Texas at Arlington living with his wife Judy in Santa Fe, New Mexico. He received his A.B. and A.M. degrees from Rutgers University in 1966 and 1967 respectively and his Ph.D. from New York University in 1970. In his almost 44 years at Rutgers, New Jersey Institute of Technology, James Madison University, and UTA, he taught graduate and undergraduate courses in world, Russian, Eastern European, Chinese, and Transatlantic history as well as the history of cartography, discovery and exploration, and historical geography. He also has been a visiting professor at Bridgewater College, University of London, and Oxford University.

[13]Fahlberg, Samuel. "*Plan Figurative Du Mole et Bassin la Rade De L'isle de St. Eustache. levé par Samuel Fahlberg 1828*" and untitled drawing (1828). Copies of both of these images are to be found in the St. Eustatius Historical Foundation Museum in Oranjestad; the originals are in the *Algemeen Rijksarchief*, first section, in The Hague. I am especially thankful to the "guides" at the Museum and to Dorothy F. van Zanten-Pole at the Census Office in Oranjestad for facilitating my research during my brief visit to St. Eustatius.

[14]Engstrand, Iris. "Scientists in New Spain: eighteenth Century Expeditions." Lecture, New Mexico History Museum, Santa Fe, NM, November 12, 2010.

[15]For example, a crucial Dutch map of the later nineteenth century seems to have been the *Kaart van het eiland St. Martin* (Amsterdam and Utrecht: 1883) by Dr. I. Dornseiffen, a copy of which can be found in the *Universiteitsbibliotheek, Amsterdam*. "All present maps are based on this first map which is almost completely correct." Speetjens, Jose. 2002. *St. Martin Yesterday and Today*. Phillipsburg: Foundation History of St. Martin, 2002, 184. Also, see: Reinhartz, 164.

He is the author and editor of 12 books, including *The Mapping of the American Southwest* (1987), *Essays on the History of North American Discovery and Exploration* (1988), *Geography Across the School Curriculum* (1990), *TABULA TERRA NOVA* (1992), *The Cartographer and the Literati: Herman Moll and His Intellectual Circle* (1997), *The Mapping of the Entradas into the Greater Southwest* (1998), *Our World Today: Peoples Places and Issues* (2002), *Mapping of Empire: Soldier Engineers on the Southwest Frontier* (2005), *Transatlantic History* (2006), and *Maps in Detail*, to be published in 2011. He also has published more than 100 book chapters, scholarly articles, and original maps in addition to numerous book reviews and professional presentations.

He is the past president of Society for the History of Discoveries, Arid Lands Studies Association, Western Social Science Association, Western Slavic Studies Association, Southwestern Association for Slavic Studies, Texas Map Society, and Friends of the UTA Libraries. Currently, he is an associate editor for Volume 4 of the *History of Cartography*.

# Part II
# The United States in the Nineteenth Century

# Chapter 3
# Thematic Cartography and Federal Science in Antebellum America

**Susan Schulten**

**Abstract** Among the earliest applications of thematic cartography in the United States was in the study of weather. Inspired by Humboldt, scientists in Federal agencies – primarily the Naval Observatory, the Smithsonian, and the Army – believed that mapping environmental patterns might solve longstanding puzzles such as the causes of storms or the relationship between climate and disease. But this depended on the accumulation of data. Beginning in 1819, the Army began to record rain, wind, and temperature, and disease at its forts and posts. By the 1850s, efforts such as these yielded an enormous amount of information that could only be understood *through* cartographic tools. Analyzing and generalizing from this mass of information was a daunting task that required extensive coordination. Yet the effort was plagued by rivalries when each agency sought to take the lead. For instance, when Smithsonian chief Joseph Henry fired Blodget for insubordination in 1854, the latter took his talents and data directly to Army Surgeon General Thomas Lawson. Lawson jumped at the opportunity to publish Blodget's maps and charts of rainfall and temperature, which in turn enraged Henry, who spent the next few years trying to discredit Blodget's work. Henry knew these maps and charts were not just illustrations of the research, but the product itself, for without them the data had no meaning. Bureaucratic infighting and interagency rivalry fueled cartographic innovation on the eve of the civil war, yet we know little about either the causes or the consequences of this innovation.

The 1850s was a period of significant cartographic creativity in the United States. Scholars have paid close attention to one aspect of this experimentation, namely the "great reconnaissance," when western explorers tremendously amplified cartographic knowledge of the American west. But an equally seismic shift came to

S. Schulten (✉)
Department of History, University of Denver, Denver, CO, USA
e-mail: sschulte@du.edu

E. Liebenberg and I.J. Demhardt (eds.), *History of Cartography*,
Lecture Notes in Geoinformation and Cartography 6,
DOI 10.1007/978-3-642-19088-9_3, 

the *form* of cartographic knowledge during these years, especially with the introduction of maps that focused upon the distribution of information. Though commonly referred to as thematic maps, no such term or category existed in the nineteenth century.

This interest in mapping information was largely inspired by Alexander von Humboldt, one of several who began to use geography to ask and answer questions about environmental relationships and patterns in the nineteenth century. His treatment of the natural world raised the possibility that maps could be used not only to describe the earth's topography and borders with increasing precision, but to identify particular phenomenon in pursuit of specific problems. While Humboldt provided the intellectual inspiration for this new cartography, its execution depended upon two other developments. First, unlike topographic or location-oriented cartography, these maps depended upon large bodies of data, such as temperature or rainfall readings, or the U.S. Census. Without the accumulation of information, cartography simply could not advance much beyond its role as a descriptive and way finding tool.[1]

Finally, this type of cartography grew in the nineteenth century because it suited emergent disciplines, which pursued knowledge by asking questions that could be answered through graphic and cartographic means. Individuals within both Federal agencies and the natural and human sciences began to use thematic cartography for specific purposes: discovering patterns of behavior, visualizing Census data, demonstrating economic growth and settlement patterns, studying disease, or pursuing more efficient agricultural practices. Those physical sciences which most extensively experimented with thematic cartography were geology, topography, and climatology or meteorology; in the human sciences, those who used thematic cartography most aggressively worked in epidemiology, demography, sociology, economics, and other fields exploiting data provided by the expanding Census.

## The Pursuit of Climatology

One of the first areas of cartographic experimentation in this era came in climate and rainfall maps. In part, this resulted from intense competition between Federal agencies over the control of meteorology. The pursuit of meteorology was intense in part because Humboldt made it a priority. Indeed the very concept of meteorology exemplifies his interest in deriving laws and explanations from observation and synthesis. Meteorology also became a kind of Holy Grail in the U.S. by the 1840s because it promised to address longstanding and urgent questions of health and agricultural development. Yet the study of weather demanded – above

[1]Jan Smits, *Petermann's Maps: carto-bibliography of the maps in Peterman's geographische Mitteilungen, 1855–1945* ('t Goy-Houten: Hes & De Graaf), pp. 11, 33–34.

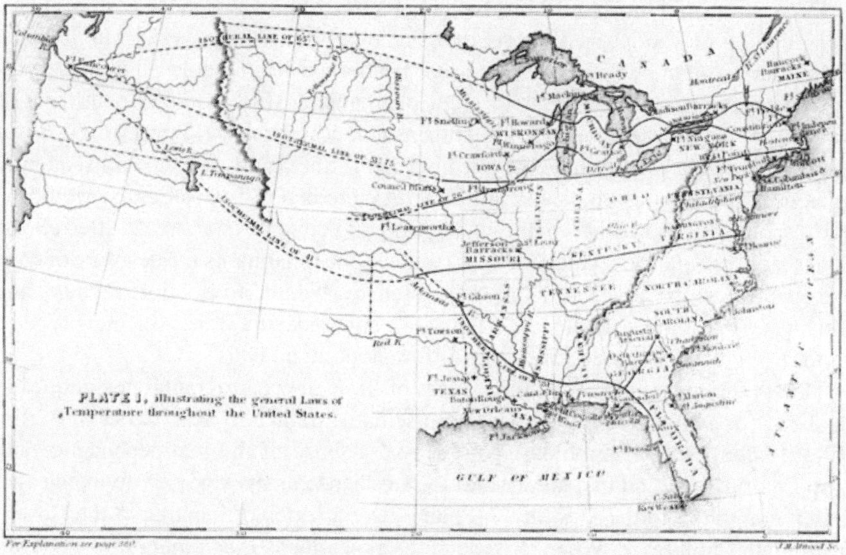


**Fig. 3.1** Samuel Forry's 1842 map of temperature patterns, both known and projected. This map was the first of its kind produced in the United States

all – extensive cooperation, for temperature and rainfall readings needed to be collected over substantial periods of time, and across wide distances. And cooperation was in short supply because these agencies were themselves angling for Federal patronage.

In 1814, the Army requested that soldiers at its disparate posts begin to record temperature, wind, and rainfall on a daily basis as part of its growing concern with the health of soldiers. The Army collected this data through its new Office of Surgeon General, with the goal of ascertaining the environmental causes of disease, particularly the role of climate.[2] This data expanded in the 1820s, when both private individuals and public agencies began to submit readings. After 1848 this information was sent to the newly formed Smithsonian Institution, and kept under the control of Secretary Joseph Henry. These observations formed a vast – if inconsistent – body of raw data, and in 1842 Surgeon General Thomas Lawson asked physician Samuel Forry to compile a map of temperature patterns from this data in order to explore the connection between disease and climate, what he termed "medical geography." In response, Forry compiled one of the first national isotherm maps in the U.S (Fig. 3.1).

[2]The Office of the Surgeon-General was created in 1818, though for several years prior to that medical officers had occupied the post in practice. Stanhope Bayne-Jones, M.D., *The Evolution of Preventive Medicine in the United States Army, 1607–1939* (Washington, D.C.: Office of the Surgeon General, 1968), pp. 79–80.

Forry drew bold lines to show the average temperatures across the eastern states, and dotted lines to project those patterns into the distant west.[3] He invoked Humboldt as his model, and stressed that isothermal lines would allow observers to connect temperature patterns to other phenomena, such as disease. In fact, he argued, one might see in patterns of temperature indications of human progress, for climate "modifies the whole physical organization of man, and consequently influences the progress of civilization." Forry predicted that isotherms would be useful not only to medical men, but to political economists as well. After all, he remarked, "In countries which have no summer, the inhabitants are destitute of taste and genius; whilst in the regions unfavored by winter, true valor, loyalty, and patriotism, are almost unknown."[4] Forry already understood that this new type of cartography could be used (or misused) for geopolitical ends.

One of the most enthusiastic importers of these new cartographic techniques to the U.S. was Arnold Guyot, a Swiss born scientist trained by Karl Ritter in Berlin. Ritter – along with Humboldt – gave Guyot a sense of the interdependence and purpose of nature, and this taught him to see science as the study of divine design. Guyot emigrated to the U.S. in 1848, and gave a series of lectures to the Lowell Institute in Cambridge. When these lectures appeared as *Earth and Man: Physical Geography in Relation to the History of Mankind*, they secured his reputation as one of the nation's preeminent scientists.[5]

Through his work, Guyot publicized Humboldt's ideas in America, particularly the use of geography as a way to study the relationship of humans to their environment. In this framing, maps became not only ways to visualize the environment, but also tools to investigate the laws of the natural world. Temperature maps allowed speculation on the influence of climate over civilization. Maps of vegetation, arable lands, and animal life could be used to analyze patterns of life. These formulations and questions were not merely facilitated by maps, but actually made possible by them, for maps enabled the integration and spatial analysis of information in pursuit of a particular question (Fig. 3.2).

Just a few years after Forry's map of temperature, Guyot published in *Earth and Man* a map of rainfall adapted from the atlas of Heinrich Berghaus. This was among the first thematic maps of the physical world published in the U.S., and it opened questions about the correlation between human progress and the environment. This world map also became a model for Americans as they developed national maps of rainfall.

---

[3]Samuel Forry, *The Climate of the United States and its endemic influences: based chiefly on the records of the Medical department and the adjutant general's office, United States Army* (New York: J. & H.G. Langley, 1842); Fleming, *Meteorology*, p. 68.

[4]Forry, *Climate*, pp. 20–22.

[5]James D. Dana, "Biographical Memoir of Arnold Guyot, 1807–1884," *Biographical Memoirs* v. 2 (1886): 309–348, quote p. 315. Arnold Guyot, *The Earth and Man, or Lectures on Comparative Physical Geography in its Relation to the History of Mankind* (Boston: Gould, Kendall, and Lincoln, 1848); and Guyot's, "IV .—Carl Ritter: An Address to the Society," *Journal of the American Geographical and Statistical Society* (1860): 25–64.

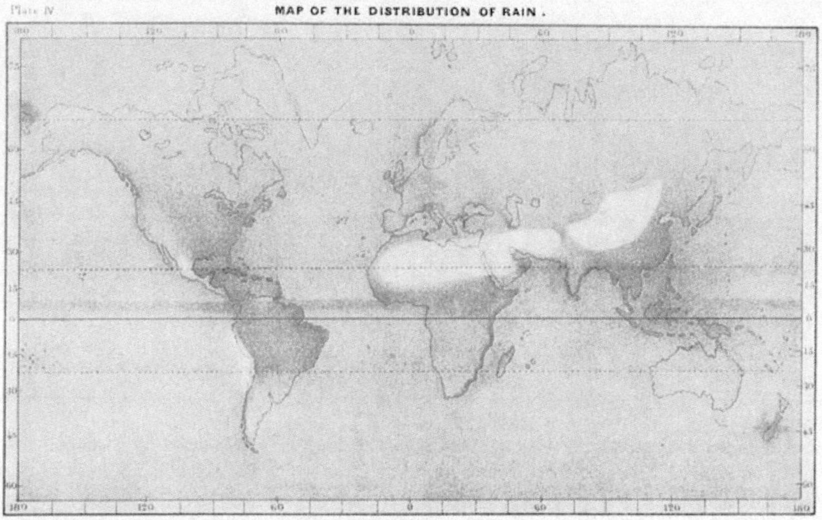


**Fig. 3.2** Arnold Guyot's map of rainfall, from *Earth and Man* (1848) (Image courtesy of Susan Schulten)

## Institutional Rivalry

Soon after his arrival in the U.S., Guyot met Joseph Henry, and the two planned a national system of meteorological observations which became one of the Smithsonian Institution's first major undertakings.[6] Yet the project was hindered by dissent within the Smithsonian, and competition with the Army and the Naval Observatory. In early 1851 Henry hired Lorin Blodget to help Guyot organize the temperature data gathered over the past several decades. Within a few years Guyot was appointed Professor of Physical Geography and Geology at Princeton, which left Blodget to synthesize the data.

Henry's relationship with Blodget began to deteriorate after the latter publicly lectured on climate without sufficiently crediting the Smithsonian. Simultaneously, after a conflict between Henry and Assistant Secretary Charles Jewett over the future of the Institution, the Board dismissed the latter. Henry suspected that Blodget was loyal to Jewett, and asked the Board investigate this perceived insubordination. Part of the conflict concerned the project itself. Henry considered

[6]According to Fleming, the project was initially proposed by James Coffin in 1846 to resolve the debate over the nature and causes of storms. Others give more credit to Henry and Guyot. The first U.S. made map of meteorological data was the Franklin Institute's chart of wind patterns. James Fleming, *Meteorology in America, 1800–1870* (Baltimore: Johns Hopkins University Press, 1990), pp. 59, 75.

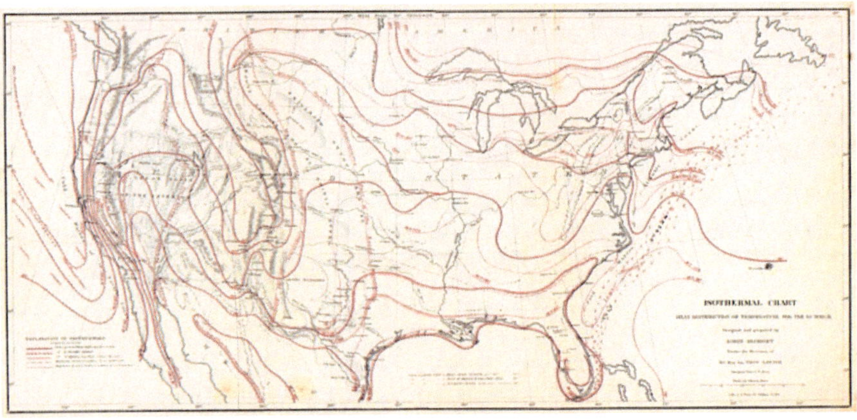

**Fig. 3.3** Lorin Blodget's annual temperature chart created in 1855, one of five comprehensive maps of temperature, and the first of their kind published in the U.S. (Image courtesy of Susan Schulten)

Blodget a clerk hired to disseminate observational data to interested scientists, not someone with special access to the data. Blodget, however, wanted to make a direct contribution to the field of climatology, and was determined to synthesize the data himself. In October 1854 the conflict came to a head when Henry fired Blodget by locking him out of his office.

Blodget was quickly hired by Surgeon General Thomas Lawson to report on climate based on this data. The Army had previously published Forry's 1842 report and map, and in 1855 Lawson updated the Army's *Meteorological Record* with Blodget's lengthy report and ten path breaking maps of temperature and rainfall. Nobody in the U.S. had ever mapped climate and rain patterns so extensively[7] (Figs. 3.3 and 3.4). A year later, Blodget published a widened version of his report for the Army and honored Humboldt as his cartographic inspiration.[8]

The publication of Blodget's report enraged Henry, particularly because the isotherm charts and rain maps were based on data which he considered the province of the Smithsonian. Henry accused Lawson and Blodget of poaching, dishonesty, and shoddy research. Lawson responded with equal irritation, arguing that the Army owned the data, not the Smithsonian. As the Surgeon General, Lawson

[7]"Report on the prominent features of general climate in the United States, as exhibited in the distribution of temperature and of rain, and in explanation of the illustrative charts," in *Army meteorological register: for 12 years, from 1843 to 1854, inclusive* (Washington, D.C: A.O.P Nicholson, 1855); *Papers of Joseph Henry,* vol. 9, introduction, xxvi.

[8]Lorin Blodget, *Climatology of the United States, and of the temperate latitudes of the North American Continent* (Philadelphia: J.B. Lippincott, 1857). Those subscribing to advance copies of Blodget's work came from several different fields, including the disease and medicine (Edward Barton), statistics (the American Geographical and Statistical Society and the Census Office), scientists such as Arnold Guyot and James Dwight Dana, and oceanographer Matthew Maury.

**Fig. 3.4** Lorin Blodget's annual rainfall map, another path breaking effort to comprehensively map the nation's rainfall totals (1855) (Image courtesy of Susan Schulten)

wrote, he never expected to get into the mapmaking business, but hired Blodget to make these charts and maps when the Smithsonian failed to make any cartographic progress. Finally, Lawson reminded Henry that isotherm maps were hardly new. Blodget's maps and charts were just the latest in an established tradition that began with Humboldt's preliminary chart of 1817, followed by Forry's map of 1842. Yet Henry's anger reveals that there was in fact something original, new, and important about Blodget's maps. Though Blodget was not the first to map temperature, his was the first to do so with any kind of certainty and specificity, since Forry's maps had been preliminary and speculative. Furthermore, while Forry used one map to cover the average temperatures for the nation, Blodget devoted one map to each of the seasons in the U.S. The 1855 *Army Meteorological Report* was the first of its kind to include maps, so Lawson must have realized that they would draw attention. Indeed, these maps were not just illustrations of research, but the research itself: without a map, the data was meaningless.[9]

While the Army staked a claim to meteorology, Matthew Maury – Superintendent of the Naval Observatory – announced that the Navy would extend its existing system of weather observation onto land. Maury also publicly championed Blodget's report on climate, which must have doubly irked Joseph Henry.[10] Maury had initially ceded weather observation over land to the

[9]This account is taken largely from *The Papers of Joseph Henry.* See editor's introduction to volume 9 (xiii-xxvii), and letters from Henry to Lawson (February 20 and April 1, 1855), Lawson to Henry (April 5, 1855), Henry to Millard Fillmore, January 25, 1855, and Henry, "Statement of Professor Henry in Reference to Lorin Blodget." See also Fleming, *Meteorology,* pp. 110–115. These maps from the *Army Meteorological Register* were also among the first maps lithographed by Julius Bien, who would become known for his first-rate work with the Geological Surveys and the Census Office after the Civil War.

[10]*Papers of Joseph Henry,* v. 9, xxvii-xxviii.

Smithsonian, but by the 1850s decided that the Naval Observatory would lead American meteorology. In the view of Maury, Lawson, and Blodget, it seemed that Henry was trying to make the Smithsonian the dominant force in the study of American weather.[11] Because of these actions by the Surgeon General's Office and the Naval Observatory, Henry felt substantial pressure to advance or abandon the Smithsonian's Meteorological Project. He convinced Guyot to come back from Princeton to synthesize this data along with Lafayette College Professor James Coffin, who in the late 1840s had pioneered the study of wind patterns in a report before the American Association for the Advancement of Science.[12]

Henry, frustrated by the slow progress of Guyot and Coffin, completed his own study of meteorology in 1856. Noteworthy was his map of agricultural potential, compiled from several types of data (Fig. 3.5). Henry used shading to identify timber, and distinguished between winds "bearing moisture," and those "accompanied by dryness." He also classified arability, separating out "sterile" lands which required irrigation. A map such as this may appear unexceptional today, but Henry was breaking new ground by using cartography to *synthesize* precipitation, winds, and timber. As such, he demonstrated how maps might be used to answer questions and problems, in this case discerning the true potential of the land of the relatively unknown interior. It was, after all, at precisely this time that a battle was raging over the potential usefulness of slavery into the new territory of Kansas.[13]

## The Pressure of Epidemics

Blodget's maps were landmarks in the study of weather as well as cartography. They were spawned by European models, the accumulation of data, and the institutional rivalries within the Federal government. But two urgent conditions

---

[11]Fleming, *Meteorology*, p. 107.

[12]Arnold Guyot, "Biographical Memoir of James Coffin," *Biographical Memoirs*, v. 1(1877): 257–264. James Coffin, "Report on the Winds of the Northern Hemisphere" (AAAS, 1848) and *Winds of the Northern Hemisphere*, in *Smithsonian Contributions to Knowledge* (v.6, 1854). Coffin collected data on wind direction and force, rainfall, and barometric pressure to identify wind patterns in the Northern Hemisphere. After his work on temperature and rainfall Coffin was charged by Joseph Henry with collecting data on winds from 1854 to 1859 from Army posts, published as *Results of Meteorological Observation* (1861).

[13]Henry to James Coffin, April 13, 1855, v. 9 pp. 237–238; Henry to James Coffin, March 14, 1856, v.9 pp. 323; Henry to Arnold Guyot, July 29, 1854, v.9 p. 118; Henry to Sir John Herschel, February 22, 1858, vol. 10, pp. 10–13; Henry to Asa Gray, v.10 p. 20 [all letters in *Papers of Joseph Henry]*. Henry, "Meteorology in its connection with Agriculture," in "Scientific Writings of Joseph Henry," reprinted in *Smithsonian Miscellaneous Publications* (reprinted 1887), maps on pp. 72, 288.

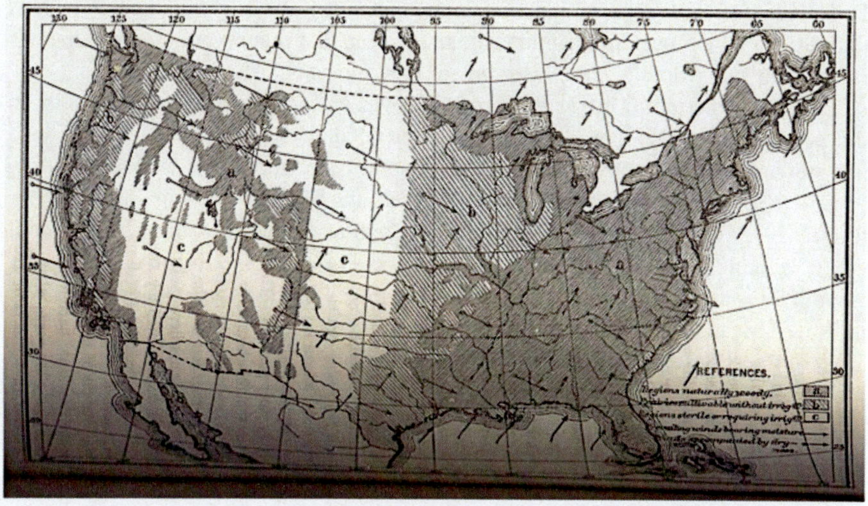


**Fig. 3.5** Joseph Henry's map of meteorology and agriculture (1856). Henry integrated more than one class of information onto the map in order to gauge agricultural potential (Image courtesy of Susan Schulten)

of antebellum life also accelerated the development of environmental mapping in the antebellum era: disease and the politics of expansion. Just as Blodget compiled his maps of rainfall and temperature, medical men were experimenting with maps to explain disease.[14] Best known among these was Daniel Drake, who spent 40 years studying the geography of disease in the Ohio Valley. In 1850 Drake published a monumental study, and his assumption that the environment determined outbreaks of disease led him to treat cartography as "indispensable."

Drake cited the work of Forry and the Army Surgeon General as crucial for the study of disease, and he used maps in several creative ways. Inserted throughout the study were maps of the most growing cities in the Ohio Valley, including St. Louis and Cincinnati, each of which included crops, lowlands, and settlements. But more impressive was a much more modest effort: his small fold-out "hydrographical" map of the interior that removed all civil or political divisions. Nothing else was included on the map, for his goal was to have readers, especially "more gifted medical historians," decide how to think about the geography of disease, and whether to include latitude, altitude, climate, or other environmental variables.

[14] Maps used to identify the incidence of disease exist prior to the nineteenth century, but are rare, and only with the outbreaks of yellow fever and cholera do they become a genre. However, the phrase "medical topography" was used in the 1810s in the Army, suggesting a desire to think about disease in relation to the environment.

Drake's intentionally spare map reveals the quiet shift that allowed thematic maps to flourish in the nineteenth century: the realization that maps were *tools* of analysis rather than self-evident *representations* of the landscape.[15]

Drake's approach to medical topography was enthusiastically read by many in the east, including Blodget.[16] Drake also influenced physician Edward Barton, who had studied the geography of disease in the south since the 1830s.[17] His first study of yellow fever in New Orleans measured monthly mortality rates against temperature ranges, the place of origin of those afflicted, and other variables. But without the new techniques in mapping and graphic knowledge, he could not easily visualize the relationship between these variables. Barton's fear of yellow fever in New Orleans was compounded by a second national outbreak of cholera in 1849. In response, the newly organized American Medical Association asked physicians in several cities to report on public hygiene. Barton's profiled the city's location, situation, layout, temperature and rainfall averages, and wind patterns, and attempted to integrate these with data from "vital statistics" from the census. Barton wondered if New Orleans was unhealthy "by nature," if its geographical situation – proximity to the river and swamps, poor drainage, humidity – invited epidemics. He speculated that yellow fever was miasmatic – conducted and spread through noxious air from poor drainage – and compounded by insufficient trees and intemperance. To demonstrate this hunch he drew a cross section of the bayou and Mississippi River drainage systems.[18]

Within a few years, maps and graphic knowledge became central to Barton's investigations. In 1851 he used graphs to correlate epidemic disease with meteorology and its influence over yellow fever as well as cholera; he isolated as many variables as possible by dividing the population along lines of race, sex, month of mortality, and seasonal changes in wind, temperature, and moisture.[19] The following year, Barton translated the returns of the Seventh Census into "Sanitary Maps" of Louisiana, Mississippi, Arkansas, and Texas, and sent his work to the

[15]Daniel Drake, *A Systematic Treatise, Historical, Etiological, and Practical, on the Principal Diseases of the Interior Valley of North America, as they appear in the Caucasian, African, Indian, and Esquimaux Varieties of Its Population* (Cincinnati: Winthrop B. Smith & Co., Publishers, 1850), p. 705.

[16]Blodget cited Drake's treatise as especially important to him in the final chapters of *Climatology*.

[17]Edward H. Barton, *Account of Epidemic Yellow Fever which prevailed in New-Orleans during the autumn of 1833, with an appendix containing an extensive meteorological account of the year and the mortality; and a chart of the temperature and of the atmosphere* (Philadelphia: Joseph R.A. Skerrett, 1835[1834]).

[18]Edward H. Barton, "Sanitary Report of New Orleans, La.," in *Transactions of the American Medical Association* v.2(1849): 591–610.

[19]E.H. Barton, *Report to the Louisiana State Medical Society, on the meteorology, vital statistics, and hygiene of the state of Louisiana* (New Orleans: Davies, son & co., 1851). Barton approvingly noted Drake's work on the geography of disease on p. 291.

Superintendent of the Census to show possible uses for Census data and to urge the expansion of census schedules for public health.[20]

Despite Barton's efforts, New Orleans experienced its worst outbreak of yellow fever in 1853. Ten thousand, out of a population of 150,000, died from the disease. Panic reigned in the summer of 1853, intensified by the inability of leaders to stem, much less cure, the disease. The course of this epidemic defied previous experience, "leapfrogging" around the city in a way that ignored spatial boundaries and "rendered private spaces public."[21] After the epidemic subsided with the autumn frost, Barton undertook his most ambitious effort yet to catalog the disease and its potential causes. Perhaps it was this sheer geographic unpredictability which prompted Barton's cartographic approach to the crisis (Fig. 3.6).

In his 1854 report to the New Orleans Sanitary Commission, Barton used cartography not just to measure affliction, but to determine how the disease was communicated and how it behaved.[22] He took care not to map the number of deaths, but rather the ratio of those to the population more generally. This reflects an awareness of statistical nuance, which was somewhat lost on John Snow as he mapped the absolute incidence of cholera in London in 1854.[23] Barton also identified any potential causes of yellow fever. Thus, given his interest in drainage and presumption that the disease was miasmic, his attention to canal building and drainage makes sense. Yet, since the 1830s extensive efforts had been taken to properly drain the city of excess moisture, thus the 1853 outbreak raised the possibility of other disease vectors. Mystified at the behavior and scale of the epidemic, Barton used the map to identify several other potentially relevant "nuisances," such as soil disturbances, cemeteries, slaughterhouses, livery stables, manufacturing, unfilled lots, crowded boarding houses, and open markets. Nothing escaped his consideration as a potential threat. Barton concluded that yellow fever was a product of bad air and certain terrain. His conclusions were rooted in place, a product of cartographic thinking. He was mistaken, but his report reveals the

[20]E.H. Barton, letter to J.C.G. Kennedy, reprinted in Kennedy, "The Origin and Progress of Statistics," *Journal of the American Geographical and Statistical Society* (1860); also reprinted in Kennedy, *Review of the Report of the Senate Committee on the Returns of the Seventh Census* (Washington: Gideon and Company, 1852). Barton's "Sanitary Maps" can be found in "Reports on Epidemics," *Transactions of the American Medical Association,* volume 5(1852).

[21]Ari Kelman, *A River and Its City* (Berkeley: University of California Press, 2003), chapter 3, quote pp. 100–1.

[22]*Report of the Sanitary Commission on the Epidemic Yellow Fever of 1853; published by authority of the City Council of New Orleans* (New Orleans: Printed at the Picayune Office, 1854), and reprinted as E.H. Barton, *The Cause and Prevention of Yellow Fever, contained in the Report of the Sanitary Commission of New Orleans* (Philadelphia: Lindsay and Blakiston, 1855). Barton originally published the chart of mortality in his *Report to the Louisiana State Medical Society* (1851).

[23]See Tom Koch, *Disease Maps* (Chicago: University of Chicago Press, 2011), p. 203. As Koch writes, the density of cholera outbreaks on Snow's maps might have been explained by the density of the population.

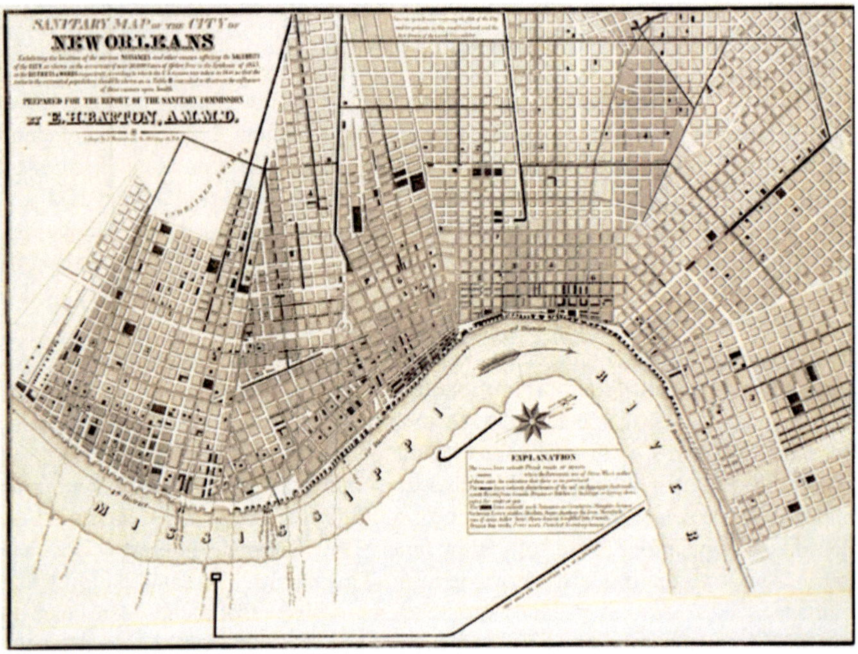


**Fig. 3.6** Barton's map of the 1853 yellow fever outbreak in New Orleans. Notice his extensive inventory of potential influences, including the arrival of ships at the port and several other "nuisances" throughout the city (Image courtesy of the Henry E. Huntington Library, San Marino, California)

commonplace role that maps had taken in public health and sanitation. Barton's urgent reports – and the outbreaks of disease more generally – also motivated Blodget to tackle climate and work toward a comprehensive theory of weather.

Barton's was the most exhaustive of many "sanitary surveys" and "medical geographies" that used maps to identify conditions under which "salubrity" and disease took root. In the mid-nineteenth century, such an open ended problem meant identifying elevation, hydrographic systems, and the placement of roads, hospitals, and sources of drinking water. "Salubrity" and "topography" both implied the importance of place, and maps were uniquely capable of treating the infinite possibilities in both the natural and built environment. Before advances in microbiology identified the precise nature of these diseases, cartography held unique power to explain them.[24]

[24]For a typical example of the use of these isothermal maps in the study of disease, see "Map of the Sacramento and San Joaquin Valleys," in Thomas M. Logan, *Report on the Medical Topography and Epidemics of California, Transactions of the American Medical Association,* v.12(1859): 81–134.

The study of disease directly advanced thematic cartography in two ways. Disease maps were themselves new experiments in thematic mapping, but the urgency around disease also fueled the more general drive toward meteorology. That is, Blodget took Barton and Drake seriously, and epidemics demonstrated the need for a more comprehensive understanding of climate and weather. In turn, once Blodget's maps were published, new questions could be asked about the influence of geographically-specific variables on health and sickness. The American Medical Association, for instance, began to seek cooperation with the Army and Smithsonian Institution regarding meteorological data. The Office of the Surgeon General began to systematically collect data in order to study the geographical distribution of health and disease in the military. And, most impressively, Census Superintendent Francis Amasa Walker used the data amassed by the Ninth Census to create a dazzling array of maps that profiled the population and the environment in new ways that Americans had never seen.[25]

## Expansion and the Politics of Agriculture

Blodget and Henry were constructing their environmental maps just as the nation entered a period of tremendous expansion and intense conflict over territory in the west. The Treaty of Guadalupe Hidalgo extended the country to the Pacific, while the ensuing Compromise of 1850 temporarily settled the status of slavery in the far west. But when Congress passed the controversial Kansas-Nebraska Act in May 1854, the west once again became the central bone of contention between slave and free states. The Act prepared these territories for Anglo settlement, though little was known about the ability of this region to support large-scale agriculture (and by extension slavery). In this context, isotherm maps held great promise to measure the agricultural potential of the American interior.

Indeed, these maps of climate spread quickly through American culture. Malcolm Lewis credits Blodget's *Climatology* for establishing the concept of a "semi-arid" region, which eroded the idea of the "Great American Desert" that had shaped ideas about the interior since the 1820s.[26] With less restraint, western booster William Gilpin used isotherms to "prove" that much of the U.S. lay within the "isothermal zodiac," or the band of climate that hosted great civilizations of the past. Gilpin's work, like Blodget's, eroded generalizations about the American

[25]Richard H. Coolidge, "Report on the Special Committee on Government Meteorological Reports," in *Transactions of the American Medical Association* v. 12(1859): 63–72. For postwar uses of medical cartography, see *Statistics, medical and anthropological, of the Provost-Marshal-General's Bureau* (Washington: GPO, 1875). For postwar Census uses of thematic mapping, see *Statistical Atlas of the United States based on the results of the Ninth Census* (New York: Julius Bien, 1874).

[26]G. Malcolm Lewis, "Regional Ideas and Reality in the Cis-Rocky Mountain West," *Transactions of the Institute of British Geographers*, no. 38(June 1966): 135–150. Blodget was also preoccupied with recent additions of land in the Pacific Northwest, and whether climate was fixed.

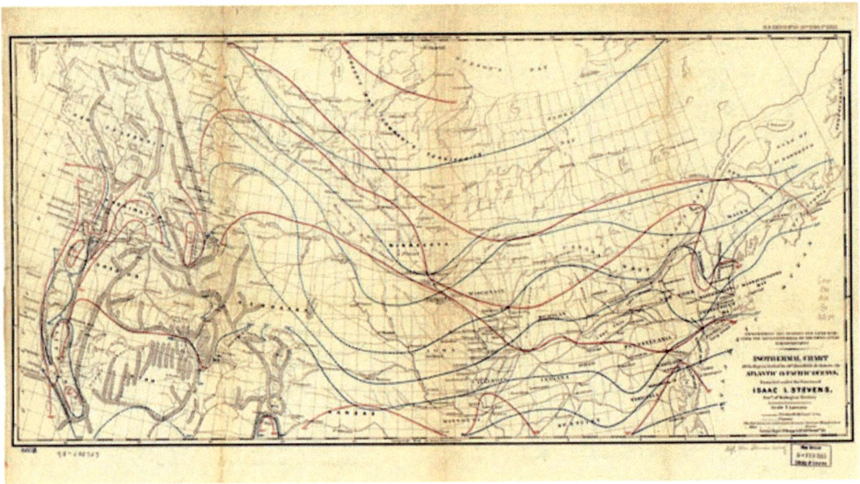

**Fig. 3.7** The War Department adapted this isotherm chart of the west to study the proposed railroad routes to the Pacific in the late 1850s (Image courtesy of the Library of Congress, Geography and Map Division)

interior, but unlike Blodget, Gilpin replaced the "desert" with a pastoral vision that invited development.[27] Isotherms were also used to study the relative advantages of proposed transcontinental railroad routes. In a letter to Secretary of War Jefferson Davis, Henry described his cartographic talents and then offered to map lines of equal temperature. This may have been the basis of the War Department's own map of isotherms, finally published in 1859, used to discuss the competing rail routes to the Pacific (Fig. 3.7).[28] Together with maps of timber and rainfall, isotherms offered new ways to think about the west.

[27]On applying the isotherms to agricultural development see Henry to Sir John Herschel, February 22, 1858, in Rothenberg, ed., *Papers of Joseph Henry*, v.10, p. 12. Even those distant from scientific enterprises were influenced isothermal maps. John W. Draper used Humboldt's isotherms to ground one of the first attempts at writing "scientific history" by using theories about climate to speculate about the course of human history. See Donald Emerson, "Hildreth, Draper, and 'Scientific History,'" in Eric F. Goldman, ed., *Historiography and Urbanization: Essays in American History in Honor of W. Stull Holt*, pp. 139–170 and Donald Fleming, *John William Draper and the Religion of Science* (Philadelphia: University of Pennsylvania Press, 1950).

[28]Henry to Jefferson Davis, August 5, 1858, in *Papers of Joseph Henry*, v. 10. Henry's map was compiled from Army Topographical surveys. See Henry, *Meteorology in its Connection to Agriculture* (Washington, D.C., 1858), p. 479. "Isothermal chart of the region north of the 36$^{th}$ Parallel &c. &c. between the Atlantic & Pacific oceans," compiled under the direction of Isaac I. Stevens, Govnr. of Washington Territory (New York: Sarony Major & Knapp, Lith., 1859). Stevens had worked with Alexander Dallas Bache at the Coast Survey, and acted as commander of the northernmost party of the Pacific Railroad Survey. Robert V. Bruce, *The Launching of American Science, 1846–1876* (New York: Knopf, 1987), p. 204. When Alaska was acquired in 1867, the newly formed Department of Agriculture commissioned Blodget to create an isothermal map of the new territory. "Isothermal Lines of Alaska," by Lorin Blodget (Washington, D.C.: Government Printing Office, 1870).

John Disturnell, whose 1847 map of the west helped to arbitrate the Treaty of Guadalupe Hidalgo, was enthusiastic about these new maps of climate and used them to frame one of the first studies of the effect of climate on human behavior (Fig. 3.8).[29] In the introduction, he excitedly acknowledged the new questions opened by the maps emerging from the Army and the Smithsonian. In the past, he wrote, scientists neglected the study of climate because of the lack of recorded information; now we had access to abundant information, but this data was only useful to the extent that it could be harnessed and made comprehensible through maps.

Disturnell was not a scientist, and made no claim to writing a definitive or original study of climate. Yet his excitement derived entirely from the fact that *cartographic* breakthroughs allowed him to ask new questions about patterns, relationships, and causation, and to connect the human and the natural worlds, just as disease maps had in the 1850s. It was the *synthesis* of information through cartography that was crucial. Ultimately, his conclusions reinforced some standing assumptions about the zones of climate and their effect upon industry: "It is only in the. . .Temperate Climates, ranging from 40° to 70° mean annual temperature, . . .that the human race really thrives, and advances in moral and intellectual culture, and where science and the arts are encouraged."[30] Similarly, during the American Civil War at least two men adapted Blodget's isothermal charts and contemporary geological knowledge to speculate on the future of cotton and slavery. John William Mallet and Edward Atkinson both used charts of climate to speak to political debates over the future of cotton and slavery, indicating that this new environmental intelligence had been assimilated into areas far beyond meteorology.[31]

By the end of the 1850s thematic cartography had been used to account for the distribution of several different phenomena in the nation as a whole. Blodget mapped climate and rainfall by 1855, while Henry mapped climate in relation to agriculture by 1859. Edward Hitchcock developed a map of American geology by 1853, and the naturalist James Graham Cooper mapped vegetation in 1859. Advances were also made in topographical maps in the 1850s, which helped to fill in the blank spaces of the interior. The cumulative effect of this innovation was

---

[29]John Disturnell, *Influence of Climate in North and South America: showing the varied climatic influences operating in the equatorial, tropical, sub-tropical, temperate, cold and frigid regions, extending from the Arctic to the Antarctic circle: accompanied by an agricultural, and isothermal map of North America* (New York: D. Van Nostrand, 1867).

[30]Disturnell, *Influence of Climate,* p. 328.

[31]See the maps included in John William Mallet, *Cotton: the chemical, geological, and meteorological conditions involved in its successful cultivation. With an account of the actual conditions and practice of culture in the southern or cotton states of North America* (London: Chapman and Hall, 1862), and Edward Atkinson, *Report to the Boston Board of Trade on the cotton manufacture of 1862* (Boston, 1863).

**Fig. 3.8** John Disturnell used isotherms and other maps to study the influence of temperature over American settlement in his "Climatological Map of North America" (1867) (Image courtesy of the Denver Public Library)

the ability for scientific and political elites to "see" the country through several different lenses.[32] Joseph Henry even proposed a comprehensive national physical

[32]G.M. Lewis, "Regional Ideas in the Cis-Rocky Mountain West," *Transactions of the Institute of British Geographers,* no. 38(June 1966): 135–150, quote p. 139. G. M. Lewis, "William Gilpin and the Concept of the Great Plains Region," *Annals of the Association of American Geographers* v.56 n.1(March 1966): 33–51.

atlas which, though derailed by the sectional crisis, suggested the enthusiasm and sense of potential attached to thematic cartography in the mid nineteenth century. These maps opened up the possibility for reform, rationalization, and greater control over the environment.[33] They also help to explain why geography offered such promise, and why it dominated American science in the first six to seven decades of the nineteenth century. Conceived in Humboldtian fashion, thematic cartography was uniquely capable of taking both a synthetic and analytical approach to the natural and human world.[34]

These maps also moved the nation toward a more social concept of disease; in fact, the act of mapping disease was an implicit request for more responsive governance. It was no coincidence, then, that this type of mapping was generally undertaken at the behest of the state, either by sanitary commissions or Federal agencies. This challenges the assumption among American historians that the state was virtually non-existent in the nineteenth century. Indeed, in the area of thematic mapping generally, we find not an inert skeletal bureaucracy, but a vigorous one lodged in the Smithsonian, the Army, the Coast Survey, the Census Office, and the Naval Observatory. Indeed, it was the jostling for preeminence among some of these agencies that accounted for the rapid creation of climate maps in the 1850s.

The enthusiasm in Washington for these new maps was palpable. In December 1856, probably at Henry's invitation, Johann Georg Kohl assessed the state of cartography in an address to the Smithsonian. Kohl was then mapping the history of American discovery for the Coast Survey, and used this speech to urge his American colleagues to archive maps to do justice to the history of North America. Kohl's passion for history was matched by a sense of awe at the recent breakthroughs in thematic cartography, which forced him to reconsider the very idea of a map. Maps now identified the depth of the oceans, the pattern of winds, and the structure of geology; they profiled the distribution of crime, disease, manners, languages, and race. While in earlier eras a "map" connoted political boundaries or topographical features, now they organized and divided the world in several new and original ways. The effect was nothing less than an intellectual revolution, Kohl insisted, for it changed the very structure of knowledge. Each new

[33]Fleming, *Meteorology*, p. 133. See also Slotten, *Patronage*, and Daniel Walker Howe, *The Political Culture of the American Whigs* (Chicago: University of Chicago Press, 1979).

[34]Relatedly in the 1850s pressure to make agriculture a higher Federal priority led to a discussion over a separate bureau or cabinet status for a Department of Agriculture. The first attempt to establish a Department of Agriculture failed in 1856, probably due to the sectional conflict. Lincoln created such a Department in 1862, when southern secession gave the Republican Party a freer hand to shape the Federal government. *Papers of Joseph Henry*, v.9, p. 324, fn. 7; A. Hunter Dupree, *Science in the Federal Government: a History of Policies and Activities to 1940* (Cambridge, MA: Harvard University Press, 1957), pp. 113–114. John Jay argued for an expanded Federal commitment to agriculture and an expanded agricultural Census schedule for the 1860 Census in his 1859 address to the AGSS, "Statistical View of American Agriculture,".

map, whether geological, botanical, zoological, magnetic, or ethnographic, "inaugurated a discovery of America in a new sense."[35]

## Biographical Note

**Susan Schulten** is a history professor history at the University of Denver, and the author of *The Geographical Imagination in America, 1880–1950* (University of Chicago Press, 2001). Her recent publications include: "Thematic Cartography and the Study of American History," in Dydia DeLyser, ed., *Envisioning Landscapes, Making Worlds: Geography & the Humanities* (Routledge, 2011); "The Cartography of Slavery and the Authority of Statistics," *Civil War History* (March 2010); "Barack Obama, Abraham Lincoln, and John Dewey," *University of Denver Law Review* 86(2009); and "Get Lost: on the intersection of intellectual and environmental history," *Modern Intellectual History* v.5 n.1(2008).

In 2007 she was a member of the advisory board of the "Maps" exhibit at the Field Museum in Chicago, and contributed a chapter to the companion volume, *Maps: Finding Our Place in History* (University of Chicago Press, 2007). She is a member of the advisory board for the Smith Center for the History of Cartography at the Newberry Library, and volume six of the *History of Cartography* series. In 2008 she was appointed to the Abraham Lincoln Bicentennial Commission in Colorado.

In 2010 she was named a Fellow of the John Simon Guggenheim Memorial Foundation. The award will support her research on the relationship between mapping and the advent of the nation. The project – tentatively titled *A Nation in Maps* – is forthcoming with the University of Chicago Press, and will be accompanied by a website showcasing new forms of cartography that flourished in nineteenth-century America.

Professor Schulten teaches a range of U.S. history courses, including intellectual and cultural history, the Civil War and Reconstruction, the Gilded Age and Progressive Era, the Great Depression, the Cold War, war and the presidency, and the methods and philosophy of history. She also contributes articles on the geographic and cartographic dimension of the Civil War to the *New York Times'* "Disunion" blog, a long-term project commemorating the sesquicentennial of the sectional crisis.

[35]J.G. Kohl, "Substance of a Lecture Delivered at the Smithsonian Institution on a Collection of the Charts and Maps of America," *Annual Report of the Board of Regents of the Smithsonian Institution* (Washington, D.C.: Cornelius Wendell, 1857), quotes p. 124.

## References


Atkinson E (1863) Report to the Boston Board of Trade on the cotton manufacture of 1862. Boston Board of Trade, Boston

Barton EH (1835 [1834]) Account of Epidemic Yellow Fever which prevailed in New-Orleans during the autumn of 1833, with an appendix containing an extensive meteorological account of the year and the mortality; and a chart of the temperature and of the atmosphere. Joseph R.A. Skerrett, Philadelphia

Barton EH (1849) Sanitary report of New Orleans, La. Trans Am Med Assoc 2:591–610

Barton EH (1851) Report to the Louisiana State Medical Society, on the meteorology, vital statistics, and hygiene of the state of Louisiana. Davies, Son & Co, New Orleans

Barton E (1852) Reports on epidemics. Trans Am Med Assoc 5

Baynes-Jones S (1968) The evolution of preventive medicine in the United States Army, 1607–1939. Office of the Surgeon General, Washington, DC

Blodget L (1857) Climatology of the United States, and of the temperate latitudes of the North American Continent. J.B. Lippincott, Philadelphia

Bruce RV (1987) The launching of American Science, 1846–1876. Knopf, New York

City Council of New Orleans (1854) Report of the Sanitary Commission on the epidemic yellow fever of 1853. New Orleans: printed at the Picayune Office; reprinted as E.H. Barton (1855) The cause and prevention of yellow fever, contained in the Report of the Sanitary Commission of New Orleans. Lindsay and Blakiston, Philadelphia

Coffin JH (1848) Report of the Winds of the Northern Hemisphere. paper delivered at the meeting of the American Association for the Advancement of Science

Coffin JH (1853) Winds of the Northern Hemisphere, in *Smithsonian Contributions to Knowledge*, v.6 article 6 (publication 52) (Washington: Smithsonian Institution)

Coolidge RH (1859) Report on the special committee on government meteorological reports. Trans Am Med Assoc 12:63–72

Dana JD (1886) Biographical memoir of Arnold Guyot, 1807–1884. Biogr Mem 2:309–348

Disturnell J (1867) Influence of climate in North and South America: showing the varied climatic influences operating in the equatorial, tropical, sub-tropical, temperate, cold and frigid regions, extending from the Arctic to the Antarctic circle: accompanied by an agricultural, and isothermal map of North America. D. Van Nostrand, New York

Drake D (1850) A systematic treatise, historical, etiological, and practical, on the principal diseases of the interior valley of North America, as they appear in the Caucasian, African, Indian, and Esquimaux varieties of its population. Winthrop B. Smith & Co, Cincinnati

Dupree AH (1957) Science in the Federal Government: a history of policies and activities to 1940. Harvard University Press, Cambridge

Fleming D (1950) John William Draper and the religion of science. University of Pennsylvania Press, Philadelphia

Fleming JR (1990) Meteorology in America, 1800–1870. Johns Hopkins University Press, Baltimore

Forry S (1842) The climate of the United States and its endemic influences: based chiefly on the records of the medical department and the adjutant general's office, United States Army. J. & H.G. Langley, New York

Goldman EF (ed) (1968) Historiography and urbanization: essays in American history in honor of W. Stull Holt. Kennikat Press, Port Washington

Guyot A (1848) The earth and man, or lectures on comparative physical geography in its relation to the history of mankind. Gould, Kendall, and Lincoln, Boston

Guyot A (1860) IV. Carl Ritter: an address to the society. J Am Geogr Stat Soc:25–64

Guyot A (1877) Biographical memoir of James Coffin. Biogr Mem 1:257–264

Henry J (1887) Meteorology in its connection with agriculture. In: Scientific writings of Joseph Henry; reprinted in Smithsonian Miscellaneous Publications

Howe DW (1979) The political culture of the American Whigs. University of Chicago Press, Chicago
Kelman A (2003) A river and its city. University of California Press, Berkeley
Kennedy JCG (1860) The origin and progress of statistics. J Am Geogr Stat Soc 2:92–120
Koch T (2011) Disease maps. University of Chicago Press, Chicago
Kohl JG (1857) Substance of a lecture delivered at the Smithsonian Institution on a collection of the charts and maps of America. Annual report of the Board of Regents of the Smithsonian Institution. Cornelius Wendell, Washington, DC
Lewis GM (1966) Regional ideas and reality in the Cis-Rocky Mountain West. Trans Inst Br Geogr 38:135–150
Logan TM (1859) Report on the medical topography and epidemics of California. Trans Am Med Assoc 12:81–134
Mallet JW (1863) Cotton: the chemical, geological, and meteorological conditions involved in its successful cultivation. With an account of the actual conditions and practice of culture in the southern or cotton states of North America. Chapman and Hall, London
Reingold N, Marc R (1972–2007) The papers of Joseph Henry. Smithsonian Institution Press, Washington, DC
Slotten HR (1994) Patronage, practice, and the culture of American Science: Alexander Dallas Bache and the U.S. Coast Survey. Cambridge University Press, Cambridge
Smits J (2004) Petermann's maps: carto-bibliography of the maps in Petermann's geographische Mitteilungen, 1855–1945. Hes & De Graaf, t Goy-Houten
United States. Surgeon-General of the Army (1855) Army meteorological register: for twelve years, from 1843 to 1854, inclusive. A.O.P. Nicholson, Washington, DC
United States. Office of the Census (1874) Statistical Atlas of the United States based on the results of the Ninth Census 1870 with contributions from many eminent men of science and several departments of the government; Compiled under authority of Congress by Francis A. Walker, M.A. Superintendent of the 9th Census, professor of political economy and history, Sheffield Scientific School of Yale College. Julius Bien, New York
United States. Provost Marshal General's Bureau (1875) Statistics, medical and anthropological, of the Provost-Marshal-General's Bureau: derived from records of the examination for military service in the armies of the United States during the late war of the rebellion of over a million recruits, drafted men, substitutes, and enrolled men. Compiled under direction of the Secretary of War by J.H. Baxter, A.M., M.D., Colonel and chief medical purveyor United States Army, Late Chief medical officer of the Provost-Marshal-General's Bureau. Government Printing Office, Washington, DC

# Chapter 4
# *"An approximation to a bird's eye view, and is intelligible to every eye [...]"*. Friedrich Wilhelm von Egloffstein, the Exploration of the American West, and Its First Relief Shaded Maps

**Imre Josef Demhardt**

**Abstract** The German topographer Friedrich Wilhelm von Egloffstein (1824–1885) migrated in 1849 to the United States where he accompanied three exploration expeditions into the American West: With John C. Frémont (1853/1854) on a winter crossing of the Rocky Mountains from St. Louis to the Great Basin, Edward G. Beckwith (1854) on a railroad reconnaissance from Salt Lake City to California, and Joseph C. Ives (1857/1858) on the Colorado River and across its Southern Plateau. For the last expedition, he developed relief shaded maps of the Grand Canyon, the first such maps of a portion of the United States. After serving in Civil War, he developed an early photomechanical printing method.

By the mid-nineteenth century, the still young United States of America was rapidly expanding westward beyond the banks of the Mississippi, driven by increasing numbers of peasants and tradesmen who had left Europe with their families for the New World, following a mix of economic pressure and wanderlust. Among the immigrant "class of 1846" was a German baronet destined to leave a prominent mark on the pioneering cartography of the American West. Friedrich Wilhelm von Egloffstein was a descendant of a centuries-old noble family in Franconia. Only a few years before his birth on May 18, 1824, in the baronial castle, the tiny but largely autonomous knightdom had been annexed by Bavaria during the Napoleonic wars. He was the youngest of five children of Baron Wilhelm Georg Friedrich Christian Heinrich von Egloffstein (1775–1859) – at times an officer in the Prussian and Bavarian Armies, a chamberlain at the court of Baden, and finally a forestry superintendent in Bavaria – and his (second) wife Caroline von Montperny, who was from an emigrated French noble family. In 1847 Friedrich Wilhelm von Egloffstein surfaced in New Orleans where he left traces as partner in the surveying

I.J. Demhardt (✉)
Department of History, University of Texas at Arlington, Box 19529, Arlington, Texas 76019, USA
e-mail: demhardt@uta.edu

E. Liebenberg and I.J. Demhardt (eds.), *History of Cartography*, Lecture Notes in Geoinformation and Cartography 6,
DOI 10.1007/978-3-642-19088-9_4, 

business of his soon to be brother-in-law. The next year he returned to Germany to marry Irmgard von Kiesewetter, the daughter of a Prussian official. Many parts of Egloffstein's private and business life on both sides of the Atlantic elude detail and call for further research. It seems that he received engineer officer training in the Prussian army before rather to seek his fortune than political exile by migrating back to the United States in 1849. However, he more than once crossed the ocean to visit Germany (Rowan and Szostalo 2005: 22–23; Egloffstein 279, 349, 351; Krygier 1997: 48).

**Fig. 4.1** Friedrich Wilhelm von Egloffstein, 1865 (Huseman 1995: 33)

The adventurous baron emerged in 1850 in St. Louis, Missouri, which had a vibrant community of German immigrants at that time. He earned his living as surveyor, but to this day only a few of his early maps are identified. Among them are a 1852 topographical map of the 2 years before established Bellefontaine Cemetery (scale 1:2,000), and a 1853 rather simple sectional map of Hancock County, Illinois, by J.C. Fonda, in which Egloffstein took part in the lithography (Rowan and Szostalo 2005: 25). In 1852, Egloffstein and surveyor G. Zwanziger published a detailed topographical map of the area around the Maremac River in the St. Louis County, just to the west of the gateway city to the West at the confluence of Missouri and Mississippi Rivers. Adorned at the margin with seven landscape views, this map sheet promoted the Pacific Railroad (later the Missouri Pacific Railroad), which in same year had begun laying its tracks upstream the Maremac and Missouri Rivers towards Kansas City. Slowed down by the Civil War, the railroad reached Kansas City only in 1865. This elaborate map, among the first promotional maps for an American railroad (Rowan and Szostalo 2005: 20–21), which then extended from St. Louis only an half an hour train ride to Sulphur Springs (today Valley Park), St. Louis County, has an eye-pleasing contouring of the surface features and attractive views, recording a landscape today largely lost to the suburbanization of St. Louis. It surely was a strong recommendation for Egloffstein to be invited by John Charles Frémont in September 1853 to join his private rail route reconnaissance expedition as topographer. This effected in Egloffstein's first association with the exploration of the American West.

Since Frémont only published short excerpts from the journal of his fifth and last expedition across the American West in his *Memoirs of My Life* (1887), the 1858 recollections by the expedition's photographer, Solomon Nunes Carvalho, a Sephardic Jew from Charleston, South Carolina, remains the only detailed account of this ill-fated venture (Carvalho 1858 [Reprint Introduction]: V + VII). By the 1850s, the army officer Frémont, nicknamed "The Pathfinder", had achieved a controversial perception by the American public. Having led four major exploration expeditions of the West (1842, 1843, 1845, and 1848), the third expedition resulted in him being court-martialed for mutiny in relation to the U.S. conquest of California, The fourth expedition ended with the tragedy of ten of his men dying of cold and starvation during a winter crossing of the Rocky Mountains. When Frémont, late in the summer of 1853, left St. Louis with a privately funded expedition for yet another winter crossing of the Grand Divide, his leadership abilities where questionable, at best. To restore his reputation, he attempted to compete against the U.S. government-sponsored expeditions to explore and survey a viable transcontinental train route (Carvalho 1858 [Reprint Introduction]: X).

In a short article, which Frémont seemingly prepared before the expedition, but was published only after its commencement in the *American Journal of Science and Arts* in May 1854, he, "*having the natural desire to do something in the finishing up of a great work in which I had been so long engaged*", proposed "*to make a double examination – going out before the snow falls, and returning in mid-winter*" to make him "*acquainted with the depth and prevalence of the snows, and the extent of their impediments to winter travelling*" (Spence 1984: 381). Early in September 1853, Frémont assembled his private expedition in St. Louis, consisting of 22 men, including two Mexicans, ten Delaware Indians, a photographer and "*Mr. Egloffstein, topographical engineer*" (Carvalho 1858: 18–19). Leaving St. Louis on September 8, the expedition proceeded rather slowly up the Kansas River where, by the end of October, they camped at the Saline Fork. Here, the Baron got his desired "*wolf-skin for a saddle-cloth*" by spending a whole day on a tree next to the cadaver of one of the mules in order to shoot the predator (Carvalho 1858: 57–58; Spence 1984: 405). Because of the time consuming daguerreographic procedures, the photographer Carvalho usually stayed behind with the mules of the expedition and his extensive photographic equipment. Here, he was often joined by Egloffstein and his topographic assistant, since their observations also took time, while Frémont and the bulk of the party were far ahead (Carvalho 1858: 76): "*Mr. Egloffstein, Mr. Fuller, and myself were generally at the end of the train, our scientific duties requiring us to stop frequently on the road*" (Carvalho 1858: 117).

By December, the expedition was up in the Rocky Mountains, it snowed, the temperatures dropped down to minus 30°F, and the party increasingly was forced to stop travelling to make warming fires. The provisions had been depleted to such a degree that, by the end of the year, "*horse steaks*" had become a regular diet (Carvalho 1858: 84, 86). By January, Oliver Fuller, Egloffstein's assistant, was the first to lose his horse, forcing him to walk through the snow-covered, rugged terrain. Soon without moccasins too, his feet eventually froze black. Dragged-on for a while by Egloffstein and Carvalho, Fuller finally suffered from paralyzed legs, which forced his two companions to leave him behind. When both finally caught up with the camp, Frémont sent out the strongest men to search for Fuller, who was found and brought into the camp (Carvalho 1858: 117–121). A few days later, however, Fuller succumbed to his horrible frostbites and starvation, just as the expedition met Ute Indians in the vicinity of a Mormon outpost, Parowan, and therefore, was saved from greater losses of life (Carvalho 1858: 129, 134–135).

The expedition reached Parowan, a tiny Mormon settlement in southeastern Utah, on February 8, 1854. Unable to proceed on because of fatigue, illness and starvation, Egloffstein, the topographer, and Carvalho, the photographer, parted from the continuing Frémont-Expedition (Carvalho 1858: 139). Once Frémont had returned from California to the east coast he wrote on August 20, 1854, from New York to George Engelmann, a German born eminent botanist, who in 1832 had emigrated to America where in the 1850s he lived in St. Louis and most likely get to know Egloffstein. Frémont told Engelmann that he had asked a local banker "*to pay to you for Mr. Eggloffstien one hundred dollars which is the amount remaining due him. As you said in yours we are both disappointed. He has many good qualities but is not a good topographer for such service as we are engaged in*" (Spence 1984: 494); a singular complain as future would tell.

While the fifth Frémont-Expedition was just a tragic page in the book of the rail route explorations of the 1850s, the main chapters were written by official expeditions under the command of army engineer officers. The experience of inadequate topographic maps during the War of 1812 against the British prompted the establishment of the U.S. Army Corps of Topographical Engineers in 1813 (although the Topographic Bureau was established only in 1818). It experienced 50 years of independent existence before being merged with the Corps of Engineers in 1863, and was, despite never consisting of more than 36 officers, the major official agent of the Trans-Mississippi explorations (Ehrenberg 2005: 82). With western territorial expansion completed by the treaties with the United Kingdom (1846), the establishment of the southern borders of Canada, and Mexico (1848), ending the U.S.-Mexican War of 1846 with the annexation of almost a third of the southern neighbor's territory, as well as the Gold Rush of 1848 on the Pacific coast, which triggered the influx of hundreds of thousands of fortune seekers and settlers to the newly established state of California (1850), the American public demanded a railroad linking the Mississippi Valley with the Pacific. To investigate "the most practicable and economic route" across the Rocky Mountains, not fewer than six field survey expeditions were sent out between 1853 and 1855 to provide insight into the largely still unknown topographic features of the American West (Ehrenberg 2005: 105–106).

The task of exploring the central proposed route for the railroad, following roughly the 38th and the 39th degree of northern latitude to California, massively lobbied for by the 30 year veteran senator of Missouri, Thomas Hart Benton, also the father-in-law of John Charles Frémont, as well as business circles of St. Louis, was assigned by the Department of War to Captain John W. Gunnison. His expedition left Westport, Missouri, on June 16, 1853, and 4 months later had descended from the Grand Divide into the Great Basin in central Utah. Here, during a clash with a band of Ute Indians at Sevier Lake, eight members of the expedition were killed on October 26, including expedition leader Gunnison, topographer Richard H. Kern, and botanist Friedrich Creuzfeldt. A surviving officer, Lieutenant Edward G. Beckwith, brought the remnants of the expedition to safety by turning about 180 miles northeast to Salt Lake City, which he reached on December 6. There, Beckwith received order to postpone the continuation of the expedition until spring 1854 (Goetzmann 1959: 283, 285; Beckwith 1855a: 113–115).

From his winter quarters in Salt Lake City, Beckwith, acting on the ground of Gunnison's realization that a trans-mountain railroad in the corridor of the 38th and 39th parallels would not be feasible, due to exorbitant costs for tunnels and bridges, petitioned successfully to Washington D.C. to continue the expedition along the considerable more northerly 41st parallel. This had already been conceptualized by his late commander and was also about the latitude of Salt Lake City. This parallel promised, due to reports from earlier explorers, a much more favorable terrain (Goetzmann 1959: 286). Still too weak to ride 2 weeks after their arrival in Parowan, Carvalho and Egloffstein boarded a wagon back into civilization on February 21. "*Mr. Egloffstein* [...] *successfully managed,*" so Carvalho in his account, "*notwithstanding his illness, to make topographical notes all the way to Great Salt Lake City, a distance of three hundred miles, which we accomplished in ten days passing through all the different Mormon settlements on the road* [...]. *We arrived at Great Salt Lake City on the night of the 1st of March 1854 and took lodging's at Blair's hotel; in the morning I learned that Lieut. Beckwith* [...] *with the remnant of Captain Gunnison's expedition, were hibernating in the city. I called Lieut. Beckwith, who invited me and my friend to the mess at their table, at E.T. Benson's, one of the Mormon apostles* [...]. *The arrival of my friend Egloffstein proved very timely; the massacre of the lamented Captain Gunnison and his officers, deprived Lieut. Beckwith of the services of their topographical engineer, to which situation Mr. Egloffstein was immediately appointed*" (Carvalho 1858: 139–140). So fate had delivered Baron Egloffstein at an unlikely spot a second opportunity to still make a cartographic contribution towards the unveiling of the West, since sadly no map ever seemed to have evolved from his observations while accompanying Frémont's last expedition from St. Louis to Parowan in Utah.

On April 4, 1854, Beckwith's expedition left Salt Lake City in a westward direction across the Great Basin, along the Humboldt River in northern Nevada, and to the eastern base of the Sierra Nevada range near Mud Lake, where two favorable passes into California were found. After crossing the Sierra Nevada into California, the expedition terminated on the banks of the Sacramento River at Fort Reading on July 26, having covered a distance of 1,011.71 miles (Beckwith 1855b: 9, 67–69). In his final report, Beckwith stressed the advantages of the 41st parallel between the Great Salt Lake and California: straight course of the railroad, fertile pockets for settlers along the route, and the already existing Mormon settlements to help with the construction and maintenance of the line (Goetzmann 1959: 286). And indeed, the construction of the transcontinental railroad that began in 1863 followed roughly Beckwith's and Egloffstein's survey at least for the section from Salt Lake City to the Humboldt River.

**Fig. 4.2** Friedrich Wilhelm von Egloffstein: Extract from *Map 2: From Great Salt Lake to the Humboldt Mountains* in the scale of 12 miles to 1 in. (1:760,320) with the expedition routes (*broken line*) and the proposed railroad routes (*double line*) (Beckwith 1855b: Map No. 2)

Besides compiling the maps of the portion between Salt Lake City and California, Egloffstein also constructed and drew the maps from the surviving field books of the slain topographer Kern of the first half of the expedition under the late Captain Gunnison from Westport to Sevier Lake. But constructing these maps must have been quite a challenge since Beckwith noted, in a letter from December 1854 that accompanied his final report, that Sheppard Homans, the astronomer, only could observe the latitude because the "*defective instruments not admitting of observation for longitude*", while Egloffstein as "*topographer, to whom I am indebted for superior topographical sketches and the preparation of the accompanying outline map*" still was busy "*making the elaborate maps of the survey*" (Beckwith 1855b: 3).

The sheets of this map were executed in a scale of 12 miles to 1 in. or 1:760,320. Arguably recommended by Egloffstein, the map sheets were supplemented by landscape views "*with no purpose of representing the beauties of the scenery*", so Beckwith, in his explanation of the maps and illustrations, "*but to illustrate its general character* [...]. [...] *Those of the country westward from the Great Salt Lake have, intimately connected as they are with the map of that portion of the country, a still greater value, as the same passes, mountains, and plains which are given on the one are presented in full on the other.*" Furthermore, these views by Egloffstein were taken "*from elevated positions, and consequently partake somewhat of a panoramic character, and being of great extent, the ordinary inequalities of the surfaces of plains and slopes are not perceptible*" (Beckwith 1855b: 9, 125–126). On the map skeleton (light) lines showed information about the countryside obtained only by second hand sources. Astronomical observations were periodically made to determine at least the latitudes. Distances were measured by odometers, and barometric observations were made daily. The topography adjacent to the trail was noted with great care and outline sketches along with magnetic bearings were made of prominent locations. In addition, on an almost daily basis, Egloffstein made short side trips, climbing mountains and elevated ridges to obtain accurate views of distant terrain.

The editing of the final reports on the findings of the reconnaissance and railroad expeditions, not the least because of the extensive cartographical efforts, took some time and was delayed until 1856. In the meantime, it had become a pressing need to have a summary of the cartographical findings of the recent official expeditions into the West. This led to the commissioning of the culmination of the railroad survey expeditions by the Department of War, the huge *Map of the Territory of the United states from the Mississippi to the Pacific Ocean*, 45 by 42 in. in a scale of 1:3 million, on which its author, Lieutenant Gouverneur K. Warren, spent almost 3 years before it was published in 1858 as the first comprehensive topographic compilation of the entire American West based on field surveys since the expedition of Lewis and Clark. The immensely popular, often reprinted and widely copied map, along with a 105 page memoir by Warren summarizing the topographic results of more than 40 expeditions he drew on, provided a terrain presentation by hachures originally inked by Edward Freyhold and Egloffstein (Ehrenberg 2005: 110; Goetzmann 1959:313). Features of the map included existing and projected roads, military installations, settlements, and the hunting grounds of the western Indians. Areas lacking information were left blank and where the data was questionable, features were depicted with faint lines.

The various expeditions of the early 1850s, in search for the viable route for the transcontinental railroad, provided, for many regions, the first scientific insight into the topography of the American West, and naturally triggered the necessity to further investigate the terrain. The biggest "blank" on Warren's map, the middle course of the Colorado River and the plateau to its south, was of prime interest in the later 1850s. Therefore, in 1857, the Bureau of the U.S. Army Corps of

Topographical Engineers commissioned First Lieutenant Joseph C. Ives, the son-in-law of the newly appointed Secretary of War, who had been Whipple's assistant on the 1853 southernmost railroad expedition. Besides a collapsible shallow-draft boat for the ascent up the Colorado River, Ives requested the scientific services of two Germans: Möllhausen and Egloffstein (Goetzmann 1959: 378–380).[1]

The boat used to investigate how far upstream the Colorado River would be navigable was ordered in June 1857 from a Philadelphia shipyard. The 50 ft long iron steamer with a stern paddlewheel was ready in August, taken apart and shipped to Panama, crossed the isthmus by the newly laid out railway, and arrived in San Francisco on October 1. Here, the expedition convened for the first time, including the three scientific officers, with Egloffstein as topographer, John S. Newberry as geologist and physician, and "*a gentleman belonging to the household of Baron von Humboldt*", Balduin Möllhausen, as "*artist and collector in natural history*" (Ives 1861: 21). While most of the expedition members left overland for Fort Yuma on the lower Colorado River, the final meeting point and start of the expedition, Ives, on November 1, with some men sailed on a schooner for a full month with the parcels of the steamer around Baja California, before reaching the mouth of the Colorado River. Here, a provisional yard to assemble the steamer was established (Ives 1861: 22–24). Building the steamer took almost another month, so that the pilot, Captain Robinson, did not arrive too early on December 29. The next day the ship, christened U.S.S. *Explorer*, left with a four-pound howitzer on the small stern cabin upstream for Fort Yuma (Ives 1861: 36).

[1] A striking aspect of the reconnaissance expeditions and transcontinental railroad surveys of the 1840s and 1850s was their cosmopolitan flair with sizeable numbers of Europeans, with the rank and file as well as among the scientific specialists, included in almost each of the field parties. Arguably the largest foreign nationality involved were Germans, with the most notable topographers and map makers being Georg Karl Preuß, known in the United States as George Charles Preuss, who in the 1840s had accompanied Frémont on some of his expeditions to California, and Friedrich Wilhelm von Egloffstein, who also started out with Frémont, and later joined and was responsible for the cartography of the (Gunnison-)Beckwith-Expedition (1853–) 1854 and the Ives-Expedition 1857–1858. Among the other specialists of German origin lending their expertise to these U.S. expeditions, the most colorful character undoubtedly was the Prussian Heinrich Balduin Möllhausen (1825–1905). He came to America in 1849, first gaining frontier experience in 1851 while accompanying Duke Paul of Württemberg's expedition into the Rocky Mountains. This exposure to the rugged wilderness of the West and his art talent made him a valuable member of both the Whipple-Expedition 1853 and the Ives-Expedition 1854, as he made numerous sketches of the topography and drew many illustrations for their final reports. Still in 1854 he returned to Prussia, married the rumored illegitimate daughter of Alexander von Humboldt, and published two accounts of his wanderings, which rank among the best travel books on the West in the 1850s, before embarking on a writing career with about 50 novels on American *sujets* earning him the nickname of "the German Fenimore Cooper" (Goetzmann 1959: 309–310).

**Fig. 4.3** Heinrich Balduin Möllhausen: Drawing of Ives' steamboat U.S.S. *Explorer* on the Colorado River, possibly depicting Friedrich Wilhelm von Egloffstein sketching on the stern casing of the paddle wheel (Huseman 1995: 90)

Mastering the many rapids of the Colorado River in the winter low water again and again slowed the upstream progress of the *Explorer*, while, at the same time, it provided the geologist and the topographer with ample opportunities for their observations. "*Mr. Egloffstein has taken panoramic views of the river and the adjacent country,*" as Ives noted just below the Black Canyon, "*and has now completed a set that extends from Fort Yuma to the present camp. The ascent to a prominent peak on the opposite site of the river (Mount Newberry) has given him a view of the whole Black mountain range*" (Ives 1861: 77–78). On March 5, 1858, the *Explorer*, which just had passed a rapid at the entrance of Black Canyon, hit "*with a stunning crash*" a submerged rock "*so violent that the men near the bow were thrown overboard*", while the steam boiler was thrown out of place and the wheel-house torn away. Although the hull was not seriously damaged and within a few days the boat was provisionally repaired, Ives decided not to proceed on upstream but to wait for the pack-train following the steamer overland (Ives 1861: 81–82).

**Fig. 4.4** Friedrich Wilhelm von Egloffstein: *Rio Colorado of the West, Map No. 1* in four sections in the scale of 6 miles to 1 in. (1:380,160) showing the lower regions from the mouth of the river up to the Black Canyon in relief shading (Ives 1861: Map No. 1)

While they waited for the train of pack mules to arrive, Ives, on a skiff, paddled up into the Black Canyon, the lowest of the gigantic canyon lands along the Colorado River, until a stream entering from the west, the Las Vegas Wash, where he turned around back to the *Explorer*. When the mule-train finally arrived, Ives divided the expedition by sending one half down the river, convinced that he had reached the upper end of the navigability, and proceeded on with the other half, including all scientific officers, eastward overland in search for the Mormon Road leading north along and finally crossing the Colorado River (Goetzmann 1959: 387–388). The overland party consisted, besides Ives and an escort of 24 soldiers, of the three scientific officers, Newberry, Möllhausen, and Egloffstein, as well as "*two cooks, two servants, eleven Mexican and Californian* arrieros [= mule handlers]" (Huseman 1995: 46–47).

By mid-April, the expedition was crossing the plateau south of the Grand Canyon in an easterly direction. On April 10, Ives noted: "*The bluffs are in view, but the intervening country is cut up by side canons and cross ravines, and no place has yet been seen that presents a favorable approach to the gigantic chasm*" (Ives 1861: 104). But 2 days later, the first approach to the south rim was made and, therefore, Ives decided to stay at Camp 73 for a few days to give the mules a rest and to have enough time for a closer investigation of the canyon's edge (see Fig. 4.7). On April 14, Ives, Egloffstein and an exploration party of a dozen men followed an Indian footpath that led through one of the ravines, today's Havasu Creek, with repeated steep drops deeper down into the Grand Canyon. At a drop with a waterfall, the party discussed whether and how to proceed when Egloffstein leaned over the drop and, hidden between the rocks, discovered "*a crazy looking ladder, made of rough sticks bound together with thongs of bark.*" The topographer was eager to be the first to climb down, but his weight was too much for the delicate structure – it broke, Egloffstein fell down and was separated from the party at the bottom of a side canyon which further on opened up, as he explored, to a narrow belt of irrigated farmland with Indian huts before another series of cascades dropping down into the main canyon. The Indians tried to help Egloffstein back to his party, but gave up when they saw the broken ladder. Now his comrades took of the slings of their muskets, knotted them together and finally reached the topographer. "*Whether it would support his weight was a matter of experiment. The general impression was that it would not, but of the two evils – breaking his neck or remaining with the Yampais – he preferred the former, fastened the strap around his shoulders.*" Surprisingly, they were able to lift him up, but only with the help of one of the ladder poles providing at least some support (Ives 1861: 107–108).[2]

A few days later, on April 14, another 30 mile excursion north with barometric readings en-route, which proved the plateau to be about a mile high, brought the exploration party to another portion of the southern rim of the Grand Canyon. "*The formation of the ground was such that the eye could not follow them the whole distance to the bottom, but as far down as the eye traced they appeared almost vertical. A sketch taken upon the spot by Mr. Egloffstein does better justice than any description can do to the marvelous scene*" (Ives 1861: 110). The last northward excursion, which confirmed that there was no passage for the expedition across the gigantic canyon to the plateau to its north, passed through irrigated farmland of the Moquis (= Hopi) Indians where, on May 11, Ives, Newberry and the topographer were invited to visit one of their pueblos. "*While Mr. Egloffstein was making a sketch of the place and its owners, I had a talk with the latter. Spreading out a map of the country we had exploring, I pointed out or route and the places with which I supposed they were familiar. They seemed to comprehend, and the chief designated upon the map the positions of the six Moquis pueblos*" (Ives 1861: 119, 121).

This concluded the investigations of the Ives-Expedition which, on the morning of May 23, reached Fort Defiance on the modern day boundary of Arizona and New Mexico, then a far southwestern outpost of the U.S. Army. While Ives himself returned to Fort Yuma on the lower Colorado River to take care of the *Explorer*, all scientific officers with their collections, notes, and sketches proceeded on via Santa Fe and Fort Leavenworth in Kansas to the east coast (Ives 1861: 130–131). In the Geological Report by John S. Newberry, which was attached to Ives' General Report, the expert provided the first geological cross section of the Grand Canyon: "*For 300 miles the cut edges of the table-lands rise abruptly, often perpendicularly,*

[2]Egloffstein's unbridled enthusiasm repeatedly got the topographer into trouble. One day, as Möllhausen noted in his diary, Egloffstein set out to view the canyon accompanied only by a soldier and an Indian. He selected a difficult route to get on top of one of the bluffs to follow the course of the river farther so that he could make corrections to his draft map. The little party was still out when the sun set and the expedition leader began to worry about the fate of the detachment. Fortunately, the men returned to camp late that night – suffering from thirst, hunger, sore feet – and the loss of "Grizzly", the expedition's beloved pet dog (Huseman 1995: 48).

*from the water's edge, forming walls from 3,000 to 6,000 feet in height. This is the 'Great Cañon of the Colorado', the most magnificent gorge, as well as the grandest geological section, of which we have any knowledge*" (Ives 1861 – Geological Report: 42).

**Fig. 4.5** Friedrich Wilhelm von Egloffstein: Detail of Section 3 of *Rio Colorado of the West, Map No. 1* with the Mojave Range (Ives 1861: Map No. 1)

In the cover letter dated May 1, 1860, Lieutenant Ives explained that, after submitting swiftly a preliminary report on the main objective of the expedition, the navigability of the Colorado River and the exploration of the unknown country along its banks, the (final) general report had "*been delayed until the maps of the region explored should be completed.*" In addition, he mentioned that Egloffstein, besides acting as the expedition's topographer and later drawing its maps, also "*zealously*" assisted Dr. John S. Newberry, the geologist, and finally provided the majority of the views and illustrations in the printed report (Ives 1861: 5). While Ives acknowledged the "*privation and exposure*" to which Egloffstein subjected himself to collect topographical data during the expedition, he remarked about the map drawing technique that Egloffstein had drawn them "*directly upon the plates, which will obviate the ordinary expense for engraving. The style is partly new. The system of light and shade has been frequently adopted; but the application of ruled tints – by which the light sides of the mountains are relieved, and the comparative altitudes of different levels exhibited – is original, I believe, with the artist.*" The printed result was both a "*beautiful and effective representation of the topography*" (Ives 1861: 6).

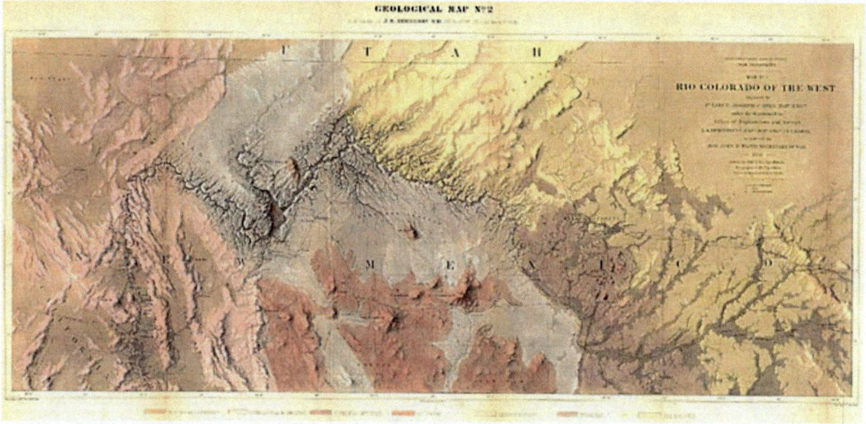


**Fig. 4.6** Friedrich Wilhelm von Egloffstein: *Rio Colorado of the West, Map No. 2* in the scale of 12 miles to 1 in. (1:760,320) showing the Grand Canyon and the plateau to its south, with the route and camps of the Ives-Expedition, in relief shading with an colored overprint as John S. Newberry's geological map (Ives 1861: Geological Map No. 2)

The printed report of the Ives-Expedition concludes with an appendix with remarks upon the construction of the maps, in which Ives, without doubt based on information provided by the map maker, explains Egloffstein's embarkation towards a new printing method. For the maps of the Ives-Report, Egloffstein aimed at combining the direct drawing on the printing medium with an almost three dimensional effect of the printed map. Therefore, he advanced an earlier French technique, "*in which the light is supposed to fall at an oblique angle upon the objects represented*," illuminating some sides of these objects while leaving other sides in the shade. Since, in engraved printing, this method did not show a distinguishable relief on the lit sides of objects like mountains from surrounding plain areas. If printed black and white in lithography upon tinted paper, however, this problem could be solved. Although Ives' description claims that Egloffstein "*conceived the idea* [...] *of endeavoring to give his map the appearance of a small plaster model of the country, with the light falling upon it at a particular angle*", it is rather likely that Egloffstein had been conceptualizing in this direction since the Beckwith-Expedition 1854, and possibly got the initial idea already while assisting the photographer Carvalho during the ill-fated Frémont-Expedition in the winter of 1853/1854. However, Ives continued to explain that in the new style Egloffstein "*constructed from the field-notes directly upon* [steel] *plates*." The lightness or darkness of the tint on the map paper should indicate the elevation or depression of the terrain while "*level plateaus and valleys would have a uniform tint*." For the printing, Egloffstein had "*devised the plan of producing the different tints by means of fine parallel lines, drawn upon the plate with a ruling machine; each part of the ruled portion being brought to the requisite shade by exposing it a longer or shorter time to the action of acid*." While Ives admitted that Egloffstein's experiment would need future improvements, he contended that for a mountainous country at the given medium scale this drawing style was "*in beauty and effectiveness much superior to the old. It possesses the power, which the other does not, of exhibiting the comparative altitudes of plateaus of different elevations*" and therefore, ultimately, "*truer to nature. It is an approximation to a bird's eye view, and is intelligible to every eye.*" And towards the zealous army treasurers, Ives (and Egloffstein) remarked that this new method is economic too, since it saved "*nearly one-half of the most expensive part of map engraving – the hachures upon the mountain sides*" (Ives 1861: Appendix D).

For the mapmaking of Egloffstein, the main difference between the expeditions of Beckwith and Ives was that the first aimed at a linear investigation of a viable railroad route, while the second, after abandoning the navigation upstream the Colorado River, attempted an area exploration. This afforded Egloffstein the chance to develop for the Ives-Report the earliest known maps showing a portion of U.S. territory by use of a shaded relief method for the representation of terrain (Krygier 1997: 27, 32). John B. Krygier suggested that Egloffstein's innovative relief representation maps of the Grand Canyon were already conceptually premeditated in the attempt to simulate nature's varied impression of features in his opulent views and panoramas since accompanying the Beckwith-Expedition (Krygier 1997: 39–42).

**Fig. 4.7** Friedrich Wilhelm von Egloffstein: Detail of uncolored *Rio Colorado of the West, Map No. 2* with relief shading of the Grand Canyon (Ives 1861: Map No. 2)

According to Krygier, the German topographer created a three dimensional plaster model of the terrain for the production of the Colorado River maps. This plaster model was lit from the side and exposed by way of daguerreotype to a steel plate coated with a photographic emulsion. The achieved photographic impression then seemed to be used as guide for creating the engraved terrain shading by line halftones, which by ways of etching the plate with acid achieved to widen the lines for darker printed areas. To further the relief effect, Egloffstein engraved additional hachures for the eastern and southern sides of the mountains, stipples for representing level plateaus and valleys, and crosshatches for an enhanced effect of depth in the canyons (Krygier 1997: 42–43).

The remainder of the life of Baron Friedrich Wilhelm von Eglofstein was not exactly guided by a lucky star. During the American Civil War (1861–1865), Egloffstein, who probably received his topographic education while he was a Prussian military officer, served in the Union forces as colonel commanding the 103rd Regiment New York Infantry, which consisted of many soldiers of German origin. In winter 1862, his regiment was part of the amphibious landing operation in North Carolina under Major-General Ambrose Burnside. In April, the colonel had been successful in an advanced reconnaissance with 200 specially picked soldiers, which Burnside reported to the Secretary of War on April 17 that Egloffstein "*having had several skirmishes with the enemy, in which he captured some 23 prisoners, 80 horses, and quite a quantity of pistols, sabers, &c.*" (Scott 1883: 270–271). But already on April 27, the colonel himself had to report about a skirmish with about 30 Confederates near Haughton's Mill in the vicinity of Pollocksville, about 10 miles west of New Berne, that, while chasing away the enemy, "*Col. Baron v. Egloffstein was wounded in the leg in dismounting, his horse having been killed under him*" (Scott 1883: 332–333). In the report of commanding general to the Secretary of War on May 3, Burnside remarked that Egloffstein "*was shot in the leg just below the knee, making a painful wound, which will probably disable him from service for some three months*" (Scott 1883: 383).[3]

About a decade after the first ever photomechanical printing enterprise had been founded in England, and went bankrupt only 3 years later in 1857, Egloffstein made the second attempt on economically viable heliographic printing, this time in America. Already the maps made for Ives' *Report upon the Colorado River of the West* (1861) had proven that he had been en-route to his own version of heliographic printing, possibly triggered even earlier by his intensive collaboration with the innovative photographer Carvalho during the Frémont-Expedition. Samuel

[3]According to a printed von Egloffstein family history, after the leg injury he returned to active service and participated in the besieging of Petersburg-Richmond in Virginia in 1864–1865 (Egloffstein 1894: 285). As a curiosity, it might be added that the von Egloffsteins obviously had no equestrian luck, since his grandfather Carl Ludwig (1745–1773), a Prussian captain and chamberlain, died from falling off a horse. Similarly, his father, while being captain of a Bavarian cavalry regiment during the Napoleonic Wars in 1814, fell so badly from a horse that he became invalid and had to quit active service (Egloffstein 1894: 278, 347).

Sartain, a noted Philadelphia engraver and long-time friend of the Baron, recalled in 1895 that after the Ives-Expedition, Egloffstein experimented with "*photographically engraving his topographic maps on steel.*" In 1861, while working with Sartain in Philadelphia, the Baron finally succeeded in heliographic engraving two islands for a map on the Amakirima Group of the southern Japanese Okinawa Islands, which had been surveyed in 1855 by the North Pacific Exploring and Surveying Expedition, an around the globe journey of two U.S. Navy vessels in 1853–1856. Then, the Civil War interceded and Egloffstein "*went into the army. He and I finished the plate by hand in 1863, and the following inscription was engraved within the boundaries of the ocean tint: 'In preparing the topography from the curves and views on file, a plaster model was made and photographed. Maikirima and Assa Islands are heliographic etchings. The rest are transfers of photographs by hand. A tint was ruled over the whole to aid the highlights*'" (Hanson 1993: 12–13). According to the printing historian David A. Hanson, the rare Amakirima map was the first (partially) intaglio halftone illustration published in the United States (Hanson 1993: 14).

The only known description of Egloffstein's photomechanical working process, as developed by the summer of 1861, are the recollections of his supportive friend Sartain more than three decades later: "*His method consisted in photographing* [the plaster model] *on a sensitive coating of asphaltum, through the glass screen ruled in one direction only, and also through the glass photographic copy of his subject; then dissolving out the unlit portions, and etching into the steel with acids, to produce the intaglio printing surface.* [...] *I succeeded in making a perfect, evenly-ruled glass plate with the proper proportion of clear and opaque lines which he desired. I also devised a means of preserving the original ruling from the risk of damage attending its use. Soon after my part of the work was accomplished, Baron Egloffstein's stock of benzol, needed for a solvent, was exhausted*" and proved impossible to replace in the needed quality. So all further experiments had to be abandoned, but the process had demonstrated its success (Hanson 1993: 15).

After being severely wounded in North Carolina in April 1862, Egloffstein possibly never returned to active service but was discharged in 1865 as brigadier general. Already in 1864, he took residence in New York City where he opened a "Geographical Institute" at 164 Broadway. In the same year, he provided maps for a book primarily written by the Prussian ambassador in Washington D.C., Friedrich von Gerolt, on his 1830s observations on the mining interest of Mexico.[4] Besides the seemingly not overly successful map business, he pursued the refining of his halftone method of photomechanical printing for which, on November 21, 1865, he received the U.S. Patent 51,103 on "Heliographic and Photographic Spectrum for Producing Line-Engravings" (Hanson 1993: 15). Thanks to the contacts of the Prussian ambassador Baron von Gerolt, the inventor obtained access to wealthy

[4]See F.W. von Egloffstein, F.K.J. von Gerolt, and C. de Berghes. 1864. *Contribution to the Geology, and Physical Geography of Mexico, including a Geological and Topographical Map, with Profiles, of Some of the Principle Mining Districts.* New York: D. Appleton.

capitalists, like senators William Sprague and Oliver Morton, the banker Salmon P. Chase, and the ambassador himself, who were looking for speculation objects. With their backing, Egloffstein was able to establish, in 1867, the Heliographic Engraving and Printing Company on 133–135 West 25th Street in New York City. The start-up business spent more than $ 150,000 to commercialize Egloffstein's patented photomechanical printing method. The production process was a well-guarded secret, as recalled by Stephen Horgan, an expert on halftone printing: The workshops were "*honey-combed with rooms, each devoted to a small detail of the process, but the occupants of one room were not allowed to enter or know what was done in the next room*" (cited in Hanson 1993: 16–17).

Advertising flyers pointed to heliography's ability to deter forgery in hopes of attracting banks as customers for printing their checks. The few surviving products, however, are *The Lord's Prayer in Fifty Four Languages* with map insets for the Catholic Diocese of New York, Beers' *Atlas of New York and Vicinity* (1867), *Genealogical Sketches of the Allen Family of Medfield* (1869), and the pamphlet, *Wooden Ships Superseded by Iron* (1869). Around 1871, economic failure forced Egloffstein to close down the company which, according to Horgan, had a too advanced technological process for the time that, in its infant stage, could not compete with faster, more proven, and established printing methods (Hanson 1993: 17, 19–21). Similar to the disappearance of the printing method with the closure of Heliographic Engraving and Printing Company, the autumn of Friedrich Wilhelm von Egloffstein's life is in the twilight of uncertainty. According to a letter his grandson sent the Smithsonian National Museum of American History in 1952, his grandfather eventually settled in San Francisco where, late in the 1870s, a fire destroyed all of his records. While Horgan, in an 1895 article, claims that Egloffstein died 1885 in New York City (Hanson 1993: 23–24) other authors agree with the year but name Dresden, the capital of Saxony in Germany, as the final resting place of this elusive cartographic pioneer of the American West.

## Biographical Note

**Imre Josef Demhardt** was born and raised in Wiesbaden (Germany) where a Christmas present, the 1896 edition of *Andree's Handatlas*, sparked his lasting interest in the history of cartography. He studied Medieval & Modern History (M.A. 1987) and Geography (M.A. 1991, Ph.D. 1995) at Johann Wolfgang von Goethe University in nearby Frankfort am Main. The research of exploration, colonialism, and mapping brought him into many corners of the globe, but most notably a total of more than three years to sub-Saharan Africa. He has co-curated exhibitions and is the author of four books as well as numerous scholarly papers. In 2008, he took over the Garrett Chair in the History of Cartography and Greater Southwestern Studies at the University of Texas. His current research interest include overseas explorations (from a European perspective) since the

Enlightenment, nineteenth and twentieth century Central European cartography, and (from an American perspective) broader historical studies of "The West".

## References


Beckwith EG (1855a) Report of explorations for a route for the Pacific railroad, Capt. JW Gunnison, Topographical Engineers, near the 38th and 39th Parallels of North Latitude, from the Mouth of the Kansas River, Mo., to the Sevier Lake, in the Great Basin. Government Printing Office, Washington, DC

Beckwith EG (1855b) Report of Explorations for a Route for the Pacific Railroad, of the Line of the Forty-First Parallel of North Latitude. Government Printing Office, Washington, DC

Carvalho SN (1858) Incidents of Travel and Adventure in the Far West with Colonel Fremont's last expedition across the Rocky Mountains: including three months' residence in Utah, and a perilous trip across the Great American Desert to the Pacific. Derby & Jackson, New York Reprinted Lincoln and New York: University of Nebraska Press, 2004. With an introduction by AF Kahn

Egloffstein G von und zu (1894) Chronik der vormaligen Reichsherrn jetzt Grafen und Freiherrn von und zu Egloffstein. Wailandt'sche Druckerei, Aschaffenburg

Ehrenberg RE (2005) U.S. Army Military Mapping of the American Southwest during the Nineteenth Century. In: Reinhartz D, Saxon GD (eds) Mapping and Empire: Soldier-Engineers on the Southwestern Frontier. University of Texas Press, Austin, pp 80–129

Goetzmann WH (1959) Army Exploration in the American West, 1803-1863. Yale University Press, New Haven

Hanson DA (1993) Baron Frederick Wilhelm von Egloffstein: Inventor of the First Commercial Halftone Process in America. Printing History 15(1):12–24

Huseman BW (1995) Wild River, Timeless Canyons: Balduin Möllhausen's Watercolors of the Colorado. Amon Carter Museum, Fort Worth

Ives JC (1861) General report. In: Report upon the Colorado River of the West, explored in 1857 and 1858 by Lieutenant Joseph. C. Ives, Corps of Topographical Engineers [...]. Government Printing Office, Washington, DC

Krygier JB (1997) Envisioning the American West: Maps, the Representational Barrage of 19th Century Expedition Reports, and the Production of Scientific Knowledge. Cartography and Geographic Information Systems 24(1):27–50

Rowan S and Szostalo M (2005) The Baron in the West Country. Friedrich W. von Egloffstein in St. Louis, 1850–1853. In: Gateway: The Quarterly Magazine of the Missouri Historical Society 26 (Summer 2005): 20–29

Scott R (ed) (1883) The War of the Rebellion: A Compilation of the Official Records of the Union and Confederate Armies. Series I, Volume IX. Government Printing Office, Washington, DC

Spence ML (ed) (1984) The expeditions of John Charles Frémont, vol 3, Travels from 1848 to 1854. University of Illinois Press, Urbana and Chicago

# Chapter 5
# Mapping Nationalism: A German Map of German Settlements in the United States of the 1890s

**Paul Rutschmann**

The year 1884 was a pivotal year for the German Reich. It marked the beginning of her entry as a colonial power when Bismarck, after years of hesitation, surprisingly conceded the long-held wishes of colonial agitators and declared a number of territories along western Africa's coastline as protectorates (Fig. 5.1). After decades of colonial debate, the Reich government seemed to be siding with the colonial expansionist agitators, finally making its move towards overseas expansion. For the colonial enthusiasts, Germany appeared to be approaching its "rightful" position alongside its European rivals in the scramble for Africa. Above all else colonial advocates wanted the Reich to coordinate the hitherto uncontrolled mass emigration of Germans overseas through state coordinated programs for settlement. The core of nationalist-imperialist hope was that Germany would become an economical and world-political beneficiary through its colonial acquisitions and settlement programs.[1] It was amid this atmosphere of colonial propaganda and expansionist agitation that a young German cartographer, Paul Langhans, began work on a series of maps of German colonies and settlements in Eastern Europe, Asia, Africa, and North America.

Between the years 1893 and 1897, Langhans delivered a series of maps to the Justus Perthes publishing house, consisting of 30 main maps and 300

[1]Cf. Horst Gründer, *Geschichte der deutschen Kolonien*, (Paderborn: Verlag Ferdinand Schöningh, 1985). Along with the general economic and social imperialist arguments of the colonial discussion in Imperial Germany, Gründer emphasized the important national psychological factors in these debates that underpinned the colonial agitators' desire for expansion overseas and government support in establishing settler colonies. The focal point of these national psychological factors was Janus-faced in character because the demand for colonies expressed, on the one hand, feelings of cultural superiority, while on the other, feelings of inferiority as a nation that was coming up too short in competition with other European powers. For some of the colonial agitators, the colonies could also serve as a destination for the deportation of political agitators and prisoners.

P. Rutschmann (✉)
History dep., University of Texas at Arlington, Arlington, TX, USA
e-mail: prutsch@sta.edu

E. Liebenberg and I.J. Demhardt (eds.), *History of Cartography*,
Lecture Notes in Geoinformation and Cartography 6,
DOI 10.1007/978-3-642-19088-9_5, 

**Fig. 5.1** Langhans: spread of Germandom in North America

supplementary map inserts, later compiled in 1 volume, titled *Deutsche Kolonial-Atlas*. The goal of the Atlas, as Langhans wrote in his preface, was a comprehensive representation of German Protectorates, German settlements in foreign countries, and the spread of German material and spiritual culture on the planet.[2] Langhans was a fierce nationalist, a long time member of the Pan-German League, and later in the 1930s joined the National Socialist Party. As a political cartographer, Langhans was specific in designing his maps as demonstration for and confirmation of Germany's colonial aspirations. The map, "Die Verbreitung des Deutschtums in Nord-Amerika," (The Spread of Germandom in North America), as Tafel 8 in the German edition, is a prime example of a nationalist-imperialist representation that creates the impression of a German sphere of influence, confirming as it were the wishful thinking of German nationalists of the spread of *Deutschthum* (Germandom) in North America. As a closer reading of the map will demonstrate, German settlers appear to have populated America to such an extent that the image conveys overwhelming presence, suggesting the spread of Germandom as an accomplishment to which nationalist spirited Germans of the 1890s could point.

Paul Langhans was born during the year of 1867 in a family of simple means in the port city of Hamburg. During the last years of his basic schooling in the *Realgymnasium*, the young student followed with alacrity the reports of the Reich's

[2]Paul Langhans, *Deutscher Kolonial-Atlas: 30 Karten mit 300 Nebenkarten*, (Gotha: Justus Perthes, 1897).

new overseas acquisitions. In Kiel he began his university studies in geography, natural sciences, and national economy. He went on to study in Leipzig where he was primarily influenced by Friedrich Ratzel, the founder of political geography. Without any formal study in the art and science of map making, and with no more than a layman's comprehension of Germany's acquisitions along the coastlines of Africa, he completed two sketch maps, based on what he read in newspaper articles, over lands acquired by competing nations in their scramble for Africa. He submitted the sketches to the large Justus Geographical Publishing firm whose flagship journal *Petermanns Mitteilungen* surprisingly accepted the work of the young student.[3]

The "Spread of Germandom in North America" was among the last installments that Langhans made to the Justus Perthe Verlag in May of 1897 before it was included in the *Deutsches Kolonial-Atlas*. For the purpose of clarifying the map's distinctness, it will be useful to view it first as broken down into three component parts, consisting of a main map, supplementary inserts, and its title; then to proceed to an interpretive reading of those parts in the map's totality. The main map is drawn on a scale of 1:12.5 million, spread over two pages, measuring approximately 12 by 10 in. included in the upper right hand corner is a color coded legend indicating in descending percentage scale the density of the German population. Spread out across the bottom portion of the two pages, extending from left to right, is a series of 14 supplementary insert maps which are enlargements of selected areas of the United States that had the heaviest concentration of German settlers. The main map is impressive in its detail, at times requiring a magnifying glass to ferret out its richness. In the upper right hand corner is the map's key, a series of colored boxes that specify the various levels of those born in Germany living in the United States. Due to the map's age, the color shaded areas are little faded and difficult to decipher. At best, one can see that the densest areas are colored a reddish-pink, corresponding to concentrations of German-born from 35% to 30%. The pale yellow indicates the areas of Germans populations under 5%. Langhans also included in the legend hyphens and dots to represent the locations of Catholic and German evangelical communities. The key also references German populations in counties and districts represented in the inserts that are based on a population census of 1890/1891.[4] For clarification of Langhans' representation of population density through the use of colored shaded areas, a few figures from the census of 1890 will demonstrate what the map is indicating in image. The foreign-born living in the United States up to 1890 totaled 9,249,547. Of that total, the German population made up 30%, at 2,784,894.[5] But when compared with the total population of the United States, the number of German-born constituted a mere 4.5%.

[3]Imre Demhardt, "Paul Langhans und der Deutsche Kolonial-Atlas 1893–1987" in *Cartographica Helvitica*, Fachzeitschrift für Kartengeschichte, Nr. 40, (Verlag Cartographica Helvitica: Murten, Switzerland, 2009), 20.

[4]Langhans clearly took his numbers from the U.S. Census of 1890, as his shaded areas correspond to the percentage of foreign born provided by that census. 1890 was the first year in which the United States Census undertook an ethnic breakdown of its residents, recording the numbers of all the foreign born according to the countries from which they arrived.

[5]Department of the Interior, Census Office, *Compendium of the Eleventh Census: 1890*, (Washington: Government Printing Office, 1894). These numbers are significant because in relation

The inserts are enlargements of select areas of the main map based on well-known locations of the heaviest concentration of Germans. Langhans selected the New York region, Chicago, Milwaukee, St. Louis, Cincinnati, Detroit, and included Texas and German colonies in Ontario. Langhans' inserts are a cartographic historical overview, and they may be associated with a long history of Germany's early settlement patterns. Together, these inserts are a mosaic of historical events, providing a historical overview of German settlements that begin with the earliest known German colony in East Pennsylvania at the bottom right corner of the map. This particular insert is somewhat unique in that it includes yet another enlargement of the Philadelphia area with its township of Germantown (Germanopolis). This is the only insert that is accompanied with a caption that reads: "the oldest German colony in North America, Germantown next to Philadelphia, founded by Fr. D Pastorius on 24 October, 1689." The insert includes another detail which in proportion to the rest of the cartographic depiction is noticeably large: a printed duplication of the township's original insignia, "Seal of the council-Collegium of Germantown 1691." The enlargement reinforces the idea of German presence since America's earliest days (Fig. 5.2).

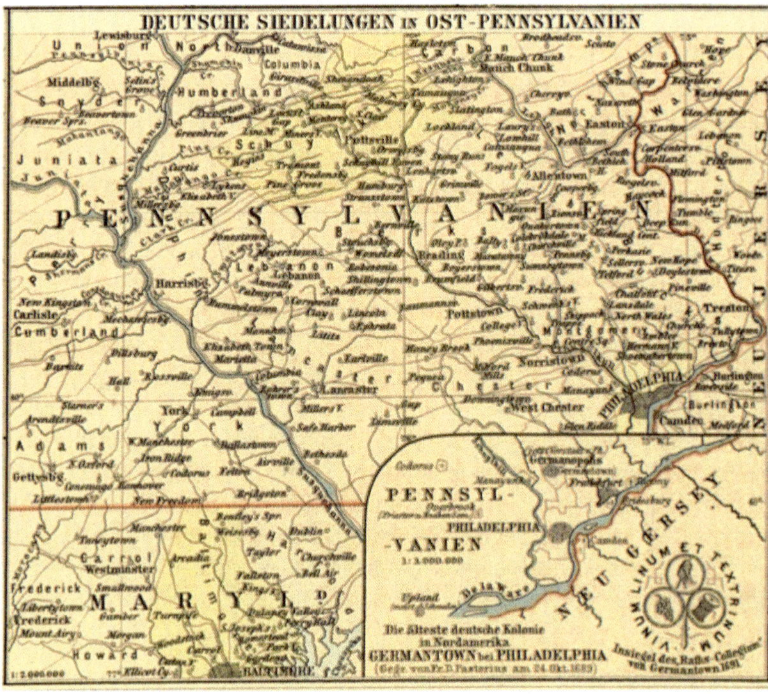


**Fig. 5.2** Langhans: German settlements in East Pennsylvania

to the entire population of the United States, the Germans counted for 25% of that population. In a regional breakdown, of the foreign born population residing in Texas, the Germans totaled 32%, and were second only to the Mexicans; and along the Atlantic seaboard, the Germans made up 12.8%.

Germantown declined in importance as a German cultural center when in 1854 it became an administrative district within the city of Philadelphia, but it remained the symbolic birthplace of German settlement. There were German settlers in the colonies before 1683; but the settlers of Germantown were the first attempt to establish an ethnic enclave. Francis D. Pastorius and his settlers from Crefeld saw their community as Germanopolis, little German city. Though the town, up until the late nineteenth century, possessed a dominant German culture, there was also a sizable mixture of French, British, and Dutch immigrants; and within the German community there was a considerable diversity in dialects and religious practices.[6]

A phenomenon that preoccupied national liberal publicists, expansionist agitators of various patriotic organizations and colonial societies of the 1870s and 1880s was the wave of German emigration overseas that had commenced during the 1830s. Of all Germans who had left their country, a disproportionately large number of them headed for the United States; indeed, throughout the nineteenth century, it is estimated that 85–95% of the emigrants boarded ships for America.[7] Since the Napoleonic wars, German emigrants were leaving their country, and from decade to decade, their numbers were significantly increasing.[8] Between the 1863 and 1913 up to 3.6 million Germans had departed, and between the 1870s and the 1880s this exodus of Germans reached its peak in settlements in the United States.[9] While some argued that the loss of Germans to other lands posed a welcome relief to an ever threatening pauperism after the 1870s, others argued that what was considered a loss in potential labor power was another country's gain. Historian Roger Chickering, in his account of colonial agitation among the patriotic organizations of Imperial Germany, noted that writers of the 1880s were already viewing the emigration "as a direct consequence of economic stagnation and social unrest, and they equated the phenomenon with an irredeemable loss of manpower and talent for the fatherland."[10] Bismarck, like any other German political elites, had witnessed the flood of Germans leaving their homeland. In June of the year in which he had signed off on colonial protectorates, he had given a speech before the Reichstag, condemning those emigrants who had turned their backs on the homeland in stating, "in general I'm not a friend of emigration, and specifically not of this ill-conceived idea of sponsoring emigration which we had in the first years of the

[6]Stephanie Grauman Wolf, "Germantown, Pennsylvania," in Thomas Adam (ed.), *Germany and the Americas: Culture, Politics, and History*, (Santa Barbara: ABC-CLIO Inc., 2005), 443–445.

[7]Beatrice Wedi-Pascha, *Die deutsche Mittelafrika-Politik 1871–1914*, (Pfaffenweiler: Centaurus-Verlagsgesellschaft, 1992), 13.

[8]Hans-Ulrich Wehler, *Bismarck und der Imperialismus*, (Cologne: Deutscher Taschenbuch Verlag, 1969). 155.

[9]Demhardt, "Paul Langhans und der Deutsche Kolonial-Atlas", 17.

[10]Roger Chickering, *We Men Who Feel Most German: A Cultural Study of the Pan-German League*, 1886–1914, (Boston: George Allen & Unwin, 1984), 29.

German Reich. A German who peels off his fatherland as he would an old coat is no longer a German for me. He is no longer of interest to me."[11]

The goal of sponsoring emigration and spreading Germandom has its origins among groups of middle-class men, intellectuals, writers, and journalists who propagated the idea of forming a New Germany. The idea of a New Germany was born in the late 1810s and stemmed mainly from a mixture of "statesmanly paternalism" and "romantic adventurism." The wish for a New Germany in America, as Stefan von Senger und Etterlin writes, was "a projection of the dreams of a united, strong, prosperous, and culturally dominant Germany."[12] A new wave of mass emigration to America during the 1830s was accompanied by public announcements of the need to organize emigration because, as the colonial agitators believed, it was a loss in labor potential and capital, but they could console themselves with the thought that at the very least, "their Germandom should not be lost for the nation."[13]

The agitation for German colony settlements in North America began at the time when the older form of dependent colonies, as it was conducted by the Spanish, British, and French throughout the eighteenth century, was going out of style, giving way to a newer system of colonies that was more informal, based to a larger degree on free exchange, and relieved the mother countries of administrative and financial burdens.[14] The idea of settler colonies in North America was underpinned by the romantic notion that the emigration of people to other lands was an extension of the "fatherland" and that, as von Sengen und Etterlin writes, "if there was to be a larger role for Germany in the world, these emigrants somehow had to become carriers of Germandom. Thus the notion of *colony*, by and large, took on the meaning of *settler colony*."[15]

---

[11]Quoted in Otto Planze, "Germany – Bismarck – America," *Deutschland und der Westen im 19. Und 20. Jahrhundert; Teil 1: Transatlantische Beziehungen*, (Stuttgart: Franz Steiner Verlag, 1993), 73.

[12]Stefan von Senger und Etterlin, "New Germany in North America: Origins, Processes, and Responses, 1815–1860," in *Emigration and Settlement Patterns of German Communities in North America*, (Indiana University: Max Kade German-American Center, 1995), 147.

[13]Von Senger und Etterlin, "New Germany in North America," 149.

[14]See von Senger und Etterlin, "New Germany in North America," for a view of the distinction between older and newer forms of colonies as they were relevant to conditions that existed during the time of Germany's entry among other colonial powers see esp. pages 150–151. The older system of colonies was characterized mainly by rigid control of trade and these politically dependent colonies were settled by subjects of the mother country. Bismarck consented to the protectorate system, following the model of the British charter system because economic development should remain the initiative and responsibility of private entrepreneurs and capital investors.

[15]von Senger und Etterlin, "New Germany in North America," 150. The renowned national political economist Friedrich List, who exemplified pre-imperialist liberal political thought, found resonance among nationalists who pressed for colonial agitation in demanding a national assessment of migration in form of a state directed emigration plan to designated areas of the globe. List argued for the government to intervene and take control of the emigration in order that competing nations would not be strengthened by German labor. See Wedi-Pascha, *Die deutsche Mittelafrika-Politik*, 23.

For German expansionists the purpose and nature of settler colonies was vested in the idea that such a colony would serve as a focal point, as a magnet for further immigration, leading eventually to Germanization of a whole region or even a state. Numerous emigration societies sprang up in Germany throughout early and mid nineteenth century with the stated goal of assisting emigrants. Though their plans were based on the gradual formation of an independent state, they never intended for the formation of such a separate community to become anything other than a member of, rather than a challenge to, the United States. The southwestern regions of America and the Republic of Texas, before it joined the Union, were held to be most suitable for the creation of an independent German colony. The larger societies for emigration and colonization depended entirely on private investors for the necessary capital, and though some of their members were men of influence within their separate state government circles, these governments, avoided active cooperation with these societies.[16] Von Senger und Etterlin cites the example of one prominent emigration society leader who believed that if such a society were well enough organized, the American government, if asked, "would grant them a large piece of land."[17] Also, a number of German-American settlement societies arose in the United States concurrently with their German counterparts. Despite regional differences, they generally shared the common aim of further uniting Germans living in larger American cities to move out west and help populate what should become a New Germany. All the while they were encouraging separation from mainstream American life and the cultivation of German customs and language.[18]

The experiment with Texas during the 1840s, as a site for a future New Germany, is a good example of the expansionist mentality that existed in Germany. This would also help explain why Langhans chose to include an enlargement of the area of Texas near San Antonio in his map. The idea of founding a New Germany in Texas was the guiding principle among two dozen noblemen and high-ranking military officers in and around Hessen in the formation of the *Mainzer Verein deutscher Fürsten und Edelleute zum Schutz deutscher Einwanderer* in Texas (Association of German Princes and Noblemen for the protection of German Immigrants to Texas). Under the direction of the Count of Castell, the association began offering transport and a tract of land to immigrants for a payment of 600 Gulden for families and 300 for unmarried men.

[16]Richard V. Pierard, "The German Colonial Society, 1882–1914," (Ph.D. Dissertation, State University of Iowa, 1964), 21. Prussian Crown Prince Frederick had shown great interest in the colonial society movement but declined the request for active protection on the grounds that the unstable political situation made such endeavors for the Reich inopportune.

[17]von Senger und Etterlin, "New Germany in North America," 152.

[18]von Senger und Etterlin, "New Germany in North America, 162 ff. In almost all the States, at least one such emigration society could be located; but many of them were small and had no effective influence on their countrymen. It is noteworthy that there was the attempt among Germans to influence and organize their fellow compatriots to move further west and establish independent communities where they could maintain their German identity without interference from what appeared to be an ever growing nativist movement.

After a few failed attempts to obtain land concessions from Mexican and Sam Houston's Texas governments, the association succeeded in purchasing a land grant from Henry Fischer, a German living in Houston who had already purchased his tract of land, in what is today known as the Hill Country, from the Texas government.[19] Fischer was appointed as colonial director and in 1845 the first 2,000 immigrants arrived in New Braunfels, northeast of San Antonio. Once larger groups of immigrants began arriving, it became apparent that the Fischer land concession was hardly suitable for the type of settlement that they had envisioned for the founding of a New Germany. From 1845 to 1847 almost half of the 7,380 immigrants died of hunger or disease, and the majority of those who had survived moved to San Antonio or New Braunfels while 1,500 remained in what today are Comal and Gillespie counties.[20] Though the idealistic and romantic motivated venture to found a New Germany in Texas had come to a grinding halt, it is nonetheless noteworthy because of the publicity it received in German newspapers and the vociferous endorsement by the numerous *Burschenschaften*, student organizations characterized by their youthful hotheaded sense of nationalism; it was a publicity that fanned enthusiasm for the spread of Germandom overseas, mirroring the same sentiments that the organizers brought with them when creating the association.

One curious feature of Langhans' enlargement of Texas is the exclusion of half of the region of the Hill Country, just west and northwest of San Antonio, which would be to the left of Langhans' Texas insert map. This region west of San Antonio was also predominantly settled by Germans. It was here that a few groups of emigrants, organized around the idea of forming a utopian/ communist settlement, envisioned a potential site for the fulfillment of their designs. Inspired by the French socialist novelist Étienne Cabet and his *Icarian* model, a group of young men, calling themselves the Darmstädter Forty, signed a contract with the Association for the Protection of German Immigrants, hiding the word communist from the Verein's directors in the description of their projected settlement. They arrived May 1847 in New Braunfels and moved further to the west to begin building their utopian community along the north banks of the Llano River provided them by the Mainzer Verein on the Fischer-Miller land grant. They named their community Bettina in honor of the social visionary writer Bettina von Arnim. Soon thereafter, their experiment failed because there was no agreement between them on distribution

---

[19]Sam Houston had a made generous offer to the Verein, but he made it contingent on payments of taxes of the settlers, which the Verein refused to consider. See von Senger und Etterlin, "New Germany in North America," for a description of the numerous and complex negotiations between the society and Mexicans and Anglo-Americans in Texas for land grants, 165 ff.

[20]Wedi-Pascha, *Die deutsche Mittelafrika-Politik*, 25/5; Gründer, *Geschichte der deutschen Kolonien*, 17; von Senger und Etterlin, "New Germany in North America," 163 ff. Aside from the fact that the failure can largely be attributed to Texas' joining the Union, much of the failure resulted also from a lack of attention to organizational and pragmatic details and inexperience in colonization. The association could not clear its debts until well into the 1880s.

of duties and labor.[21] A similar attempt was made in Sisterdale, which again was a failure. Many of the original Darmstädter, along with larger numbers of Freethinkers who left Germany after the failed revolution of 1848, began settling and building the community of Comfort 10 miles south of Bettina. This region was home to Texas' second largest German community (Figs. 5.3 and 5.4).

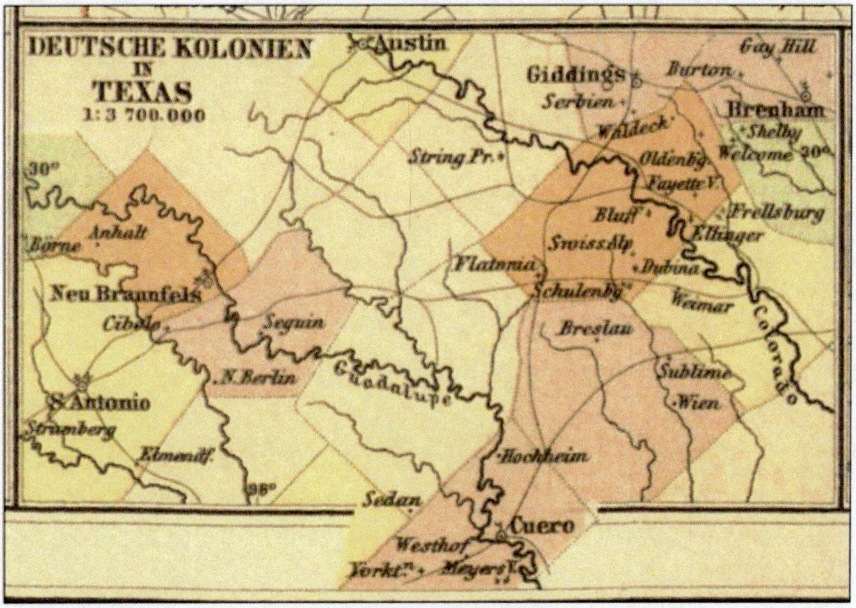


**Fig. 5.3** Langhans: German colonies in Texas

A reading and interpretation of the map ultimately turns on the significance and meaning of the map's third component, its title. The connection between the title of the map and the individual supplementary inserts within the map points to a cartographic literacy in special way. A special domain of knowledge is determined first by the publisher's awareness of Langhan's patriotism and his ability to influence and guide geographic studies in a nationalist direction; and second, by recognizing the map as a text that narrates the spread of Germandom in North America. The map illustrates a certain claim to power-dominance in a social Darwinist tenor by acknowledging the spread of Germandom across Germany's own borders as a commemorative act of citizenship. It was an act of affirming collective identity with a community of its citizens living beyond its own borders. Langhans' cartographic representation of Germandom is a form of calculated visual identity, marking his intention to affirm the extension of homeland identity to fellow Germans overseas while excluding the mix of other communities existent in America who were host to the German settlers.

[21] Andreas Reichstein, "Darmstaedters," in Thomas Adam (ed.) *Germany and the Americas: Culture, Politics, and History*, (Santa Barbara: ABC-CLIO Inc., 2005), 265 ff.

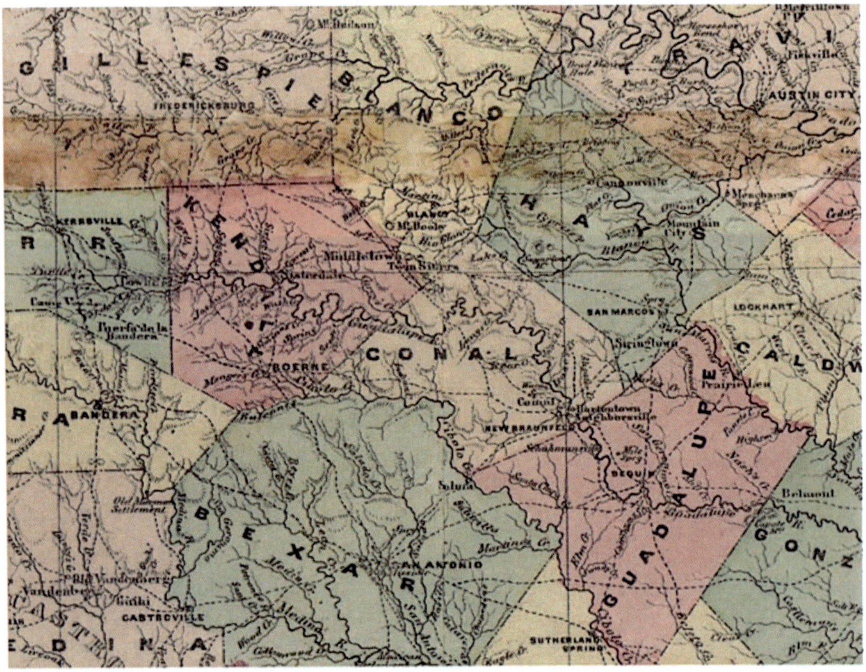


**Fig. 5.4** Charles W. Pressler, *Traveller's Map of the State of Texas*, color photolithograph, 98 × 94 cm (New York: printed by the American PhotoLithographic Co., 1867). UT Arlington Library Special Collections

Langhans *Kolonial-Atlas* was published at a time when cartographic editions were more readily available through improved mechanization of production techniques; thus maps were becoming a mass produced article. Bernard Perthes, Langhans' publisher, was in a bit of a hurry to market the atlas because word got back to him that his primary competitor in Berlin was going to beat him to the press with a comparable atlas. Despite some initial minor criticisms of Langhans' drafts, Perthes decided to rush to the press anyway with the atlas because, as Imre Demhardt noted, "it took for its target group an uncritical mass of people who buy whatever is offered first."[22]

The title in its German original reads "Die Verbreitung des Deutschtums in Nord-Amerika" which can literally be translated as "The Spread of Germandom in North America. The term Deutschtum means the ethno-centric domain of all things German in its spiritual, intellectual, and cultural sense; and more importantly, it implies the commonly held attitude of German intellectuals of a certain sense of cultural superiority. Historian Thomas Nipperdey saw the notion of Germandom in the hands of nineteenth-century German publicists and colonial agitators as the embodiment of nationalist-imperialistic aspirations, an expression of political and

[22]Demhardt, "Paul Langhans und der Deutsche Kolonial-Atlas 1893–1987," 25.

cultural aims regarding their identity as German and their future, a reshaping (*Mitgestaltung*) of the world, and their ability to assert themselves in the contest of national peoples.[23] In turning their attention to the world outside their borders, a world that Langhans and other Pan-German enthusiasts viewed as their spheres of influence, more and more nineteenth-century national-liberal and conservative Germans saw themselves committed to a "type of cultural mission."

The cornerstone of this nationalist-imperialist Weltanschauung was the idea that the German people would be unable to fulfill their mission if they neglected to do their share in spreading German idealism, morals, and culture over other continents. As von Senger und Etterlin described it, the nationalists' credo was the self-conscious assertion that "Germandom was equally worthy of expansion as were other *Culturvölker*."[24] In her account of the Germans' development to a colonial empire, Mary Townsend, writing before the Second World War, argued that "Germany's mission to spread her *Kultur* as a means of maintaining Deutschtum" was in essence the manner in which the country could display before the world her strength as a nation.[25] The aim of spreading *Kultur* became the means to elevate national prestige.

Langhans' map is celebratory, commemorative, and nationalistic. As a Pan-German, Langhans' celebration of Germandom in America is to be seen as a political act. Rather than lament Germany's loss of potential manpower to uncontrolled emigration, Langhans has created by his highly detailed, color coded map a representational reversal of what colonial agitators had previously seen as a loss to the nation. By illustrating instead significant German presence in the United States, Langhans was refashioning and cultivating the nationalist-imperialist discourse by reappropriating the spread of Deutschtum as a cultural *fait accompli*. As can be seen in the map, the significant presence of Germans could be reinterpreted for the readers of the day as a German gain. It was a matter of prestige and a visual confirmation of a new type of social Darwinist conquest by infiltration.

Quite often a map is nothing other than a candid snapshot of an event in the history of mentalities. Langhans' map in particular of the spread of Germandom, is a historical snapshot of a specific Pan-German nationalist-imperialist mentality. Maps have demonstrated usefulness as text especially in cases where they functioned as textual confirmation of acquisition and possession of new territories. Words of books may evoke images in the mind of the reader, but a map supplies a visual, thus fixed image, more forceful in communication as a self-contained, closed system. Here, one should look both inside and outside the map. When looking inside Langhans' map, simply at its individual components, there is an admirable display of detail, conveying a sense of accuracy, thus a correct representation of perceived

[23] Thomas Nipperdey, *Deutsche Geschichte 1866–1918, Bd. II: Machtstaat vor der Demokratie*, (München: Verlag C. H. Beck, 1992), 645.

[24] von Senger und Etterlin, "New Germany in North America", 150.

[25] Mary Townsend, *The Rise and Fall of Germany's Colonial Empire 1884–1918*, (New York: Howard Fertig, Inc., 1930), 79.

reality. When looking outside the map, beyond to the social and intellectual context in which Langhans produced it, there is altogether a different sort of reality.

German emigrants, who since the mid 1830s were part of the larger waves of migration, were made aware of existing German communities in America where they would be welcomed into a familiar cultural enclave devoted to maintaining their language, tradition, and custom. Langhans' selection for enlargement demonstrates the location where these enclaves of Germans were most commonly found: in the cities and their environs of St. Louis, Detroit, Cincinnati, New York, Milwaukee, and Texas. German settlers to these areas were known throughout the nineteenth century to have built schools where instruction in German was conducted, and a variety of clubs, or organizations such as *Gesangverein* (singing club), *Turnerverein*, (quasi-paramilitary gymnastic clubs), *Schiessverein*, (shooting or rifle club), *Literaturverein*, (book club) were established for the sole purpose of preserving distinct forms of German custom and tradition.[26] For example, New York City was host, before the end of the nineteenth century, to an impressive German population. Data from the census of 1890 shows that of all the 639,943 foreign born living in New York City, 210,723 were born in Germany, constituting 1/3 of all foreigners; and the Germans accounted for about 14% of New York City's total population.[27] The city's German community could boast at the time of four daily German language newspapers – more than the cities of Berlin or Leipzig had at the time – and there were eight more publications which appeared weekly, bi-weekly, or monthly.[28]

In looking deeper into its historical context, the map is doubtlessly an exceptional example of an event in the history of a specific mentality. The title page provides a clue as it displayed a drawing of the German eagle spreading its wings while clutching the earth in its talons. In the preface, written in 1892, 5 years prior to its publication, Langhans wrote that "it might appear a daring endeavor to verify a cartographic representation of the complete settlement activity of Germandom with the name Kolonial-Atlas."[29] He then added that "along with the Reich-German Protectorates, those countries have been examined in great detail that have proven themselves most worthy of settlement by German emigrants, and in which Germandom has preserved itself as independent against the influence of foreigners."[30] In unmistakable pan-Germanic fashion, Langhans concluded the

[26]For a more detailed analysis of German societies in America see Jörg Nagler, "*Ubi libertas, ibi patria* – Deutsche Demokraten im Exil," in *Friedrich Hecker in den USA: Eine deutsch-amerikanische Spurensicherung*, (Konstanz: Stadler Verlagsgesellschaft mbH, 1993), 68–70.

[27]Department of the Interior, Census Office, *Compendium of the Eleventh Census: 1890*, (Washington: Government Printing Office, 1894).

[28]Agnes Bretting, "Women's Lives in New York City, 1890–1910: The Women's Pages in the Middle-class *New Yorker Staatszeitung* and the Socialist *New Yorker Volkszeitung*," in *Emigration and Settlement Patterns of German Communities in North America*, (Indiana University: Max Kade German-American Center, 1995), 294.

[29]Langhans, *Kolonial-Atlas*.

[30]Langhans, *Kolonial-Atlas*.

# Deutscher
# Kolonial-Atlas.

30 Karten mit 300 Nebenkarten,

entworfen, bearbeitet und herausgegeben

von

**Paul Langhans.**

**Gotha: Justus Perthes.**

1897.

**Fig. 5.5** Langhans. Title page to Pan-German Colonial Atlas, 1897 preface by declaring that in his Kolonial-Atlas "the spreading out of the spirit of German culture was given its proper due."[31] (Fig. 5.5)

In 1900, just 3 years after the *Kolonial-Atlas* was issued, Justus Perthes Verlag published another atlas researched and prepared by Langhans; this time bearing the title: *Alldeutscher Atlas* (Pan-German Atlas).[32] The title page included a formal acknowledgment of the Atlas' sponsorship by the Pan-German League in which Langhans was an active member. The map of the "Spread of Germandom in North America" was reprinted for this volume and its only difference from the map in the *Kolonial-Atlas* was its title, which reads "Hauptsitze des Deutschtums Übersee," (Centers of Germandom Overseas).

The map of Germandom in America is an important visual documentation of a distinct German nationalist mentality. It is directly associated with the mentality that flourished within the ranks of the politically and socially influential movement of the Pan-German League. Paul Langhans was not only a gifted autodidactic mapmaker, but he was also the editor from 1909 to 1937 of one of Germany's most notable geographic journals, *Petermanns Mitteilungen*, published by J. Perthes.[33] The Pan-German League was one of the most powerful of the many patriotic organizations whose more recent history has raised a complex set of contentious historiographical issues due to its later close ties with the emerging Nazi movement during the 1920s. According to Chickering, too much emphasis on the movement's contribution to the rise of Nazism within these historical debates has tended to disfigure the league's actual significance during the time it emerged.[34] But what is important to note here with respect to the Pan-German movement at the time Langhans was preparing his maps for the *Kolonial-Atlas* and the *Pan-German Atlas* was that, as both Nipperdey and Chickering have argued, more than political parties, this organization and others like them – all grouped under its German non-profit main organizational heading of *Verbände* – "were vehicles of political mobilization in the 'participation revolution' of the late nineteenth century. Many of them had memberships far larger than most of the political parties. The *Verbände* thus quickly found a role in the plebiscitary politics of the German Empire, for

[31]Langhans, *Kolonial-Atlas*.

[32]Paul Langhans, *Alldeutscher Atlas: Statistik der Deutschen und der Reichsbewohner*, (Gotha: Justus Perthes, 1900).

[33]Demhardt, "Paul Langhans und der Deutsche Kolonial-Atlas 1893–1987.

[34]*See Chickering's introduction to We Men Who Feel Most German.* In this introduction, Chickering argues that the Pan-Germans' historical development has unfortunately been obscured because all too often their history has been written in a manner that would support the *Sonderweg* thesis: that the Germans deviated from a normal western path of democratization and thus preserved an authoritarian style of rule that continued on down along an inevitable path to a destructive, dictatorial regime. Against previous arguments maintaining that the Reich government had skillfully manipulated the Verbände for their own political purposes, Chickering argues that the Verbände through their wide range of networks exercised considerable influence over the government by means of its control over public opinion. Members of the Pan-German League were indefatigable writers and publishers of popular nationalist-imperialist inspired literature.

they were in a position to mobilize their formidable constituencies, often with the connivance of the government, in order to exert popular pressure on the Reichstag."[35]

The *Alldeutscher Atlas* deals specifically with statistics of the German population throughout the world and the residents of the Reich. The first eight pages contain charts and tables bearing such titles as "German Major Cities of the Earth," "Distribution of Germans over the Earth," and "United States German Population of Cities." In the table "German Major Cities of the Earth," Langhans has placed, arranged in descending order according to most populous, starting with Berlin, Chicago behind Munich, and in the next column the figure of Chicago's Germans of 407,000 is given, making the Germans 37% of Chicago's total population; similarly with Milwaukee whose German population of 135,000 constituted 66% of Milwaukee's total population. In another table bearing the title "United States," Langhans included a table under the subcategory "German City Population," which provided a more detailed breakdown that distinguished between Germans living in American cities who were born in Germany and Germans born in America to German families. A city like Milwaukee had in 1890 a population of Germans born in Germany that constituted 28% of the city's population, and all Germans living in Milwaukee constituted 66%. Since Milwaukee and Chicago had the highest percentage of Germans compared to other large American cities, Langhans obviously chose to include only two maps inserts of these two cities in the map of the *Alldeutscher Atlas*. The insert maps and the statistical tables point to an overwhelming presence and thus an undeniable spread of Germandom and this apparent spread is confirmed by the map's title: Centers of Germandom.[36] (Fig. 5.6)

Like many Germans of his day, Langhans was a member of other patriotic organizations; the most noteworthy was his membership in the *Deutschbund* (German Fraternity), a formation of loudly anti-Semitic radical right. As *Bundeswart* (Fraternity warden) of the *Deutschbund*, Langhans aligned the organization in 1931 with the NSDAP. The *Deutschbund* had displayed the swastika after the Great War and it may be assumed that Langhans was effectively a member by 1930 of the NSDAP. Lest anyone should assume that the *Petermanns Mitteilungen* thus became a vehicle for Nazi propaganda, Langhans had developed, as Demhardt has pointed out, a few offshoot journals during his tenure as editor for *Petermanns Mitteilungen*. These journals generally served as outlets for the articulation of his private political ideas, thus enabling him to tone down his political leanings in *Petermanns Mitteilungen*.[37]

In preparing his maps of Germandom in the United States, Langhans was renewing the discourse of colonial agitation in the years when in Germany an atmosphere of "colonial weariness" was setting in. As Townsend observed, "a

---

[35]Chickering, *We Men Who Feel Most German*, 25.

[36]Langhans, *Alldeutscher Atlas*.

[37]Demhardt, "Paul Langhans und der Deutsche Kolonial-Atlas 1893–1987," 21.

# Vereinigte Staaten.

**Im deutschen Sprachgebiet Mitteleuropas Geborene*)**
**1890.**

| Staat. | In Deutschland Geborene. | Vom Hundert der Gesamt-Bevölkerung. | Vom Hundert der auswärts Geborenen. |
|---|---|---|---|
| Massachusetts | 31 000 | 1 % | 5 % |
| Connecticut | 31 000 | 4 | 17 |
| Neu-York | 553 000 | 9 | 35 |
| Neu-Jersey | 124 000 | 9 | 37 |
| Pennsilvanien | 262 000 | 5 | 31 |
| Maryland | 54 000 | 5 | 58 |
| Ohio | 254 000 | 7 | 55 |
| Indiana | 91 000 | 4 | 62 |
| Illinois | 366 000 | 10 | 43 |
| Michigan | 173 000 | 8 | 32 |
| Wiskonsin | 283 000 | 17 | 55 |
| Minnesota | 129 000 | 10 | 28 |
| Iowa | 144 000 | 8 | 44 |
| Missouri | 136 000 | 5 | 58 |
| Nebraska | 81 000 | 8 | 39 |
| Kansas | 54 000 | 4 | 37 |
| Kentucky | 35 000 | 2 | 60 |
| Texas | 60 000 | 3 | 39 |
| Kalifornien | 76 000 | 6 | 21 |
| Übrige Staaten | 183 000 | — | — |
| **Vereinigte Staaten** | 2 993 000 | 4,6 % | 32 % |

**In „Germany" geborene Deutsche.**

| Jahr. | Geborene Deutsche. | Vom Hundert der Gesamt-Bevölkerung. | Vom Hundert der auswärts Geborenen. |
|---|---|---|---|
| 1860: | 1 301 136 | 4,06 % | 30,83 % |
| 1870: | 1 690 410 | 4.38 | 30.37 |
| 1880: | 1 966 742 | 3.92 | 29.44 |
| 1890: | 2 784 894 | 4.45 | 30,10 |

*) D. h. den „germanic nations" der nordamerikanischen Statistik Angehörige. Die hier zu viel gerechneten französischen Belgier und Schweizer dürften wieder ausgeglichen werden durch die nicht berücksichtigten Deutschen in den Ländern der „slave-, latin- etc. nations".

**Deutsche Stadtbevölkerung.**

| Stadt. | In Deutschland Geborene. | D. d. Ges.-Bevölkerung. | Deutsche überhaupt. | D. d. Ges.-Bevölkerung. |
|---|---|---|---|---|
| Milwaukee | 57 000 | = 28 % | 135 000 | = 66 % |
| Hoboken | 10 000 | = 24 | 25 000 | = 57 |
| Detroit | 37 000 | = 18 | 88 000 | = 43 |
| Cincinnati | 51 000 | = 17 | 121 000 | = 41 |
| Buffalo | 45 000 | = 18 | 106 000 | = 41 |
| Jersey City | 17 000 | = 11 | 65 000 | = 40 |
| Cleveland | 44 000 | = 17 | 104 000 | = 40 |
| Neu-York (alter Umfang) | 245 000 | = 17 | 583 000 | = 38 |
| Newark | 28 000 | = 15 | 67 000 | = 37 |
| Chikago | 176 000 | = 16 | 407 000 | = 37 |
| St. Louis | 70 000 | = 16 | 168 000 | = 37 |
| Toledo | 13 000 | = 16 | 30 000 | = 37 |
| Alleghany | 14 000 | = 14 | 32 000 | = 32 |
| Rochester | 19 000 | = 14 | 43 000 | = 32 |
| Pittsburg | 27 000 | = 11 | 65 000 | = 27 |
| S. Franzisko | 30 000 | = 10 | 71 000 | = 24 |
| Baltimore | 42 000 | = 10 | 101 000 | = 23 |
| Louisville | 15 000 | = 9 | 35 000 | = 22 |
| Albany | 8 000 | = 8 | 19 000 | = 20 |
| Indianapolis | 8 000 | = 8 | 20 000 | = 19 |
| Philadelphia | 79 000 | = 8 | 189 000 | = 18 |
| Omaha | 9 000 | = 7 | 22 000 | = 16 |
| Minneapolis | 9 000 | = 5 | 21 000 | = 13 |
| Denver | 6 000 | = 6 | 14 000 | = 13 |
| Neu-Orleans | 12 000 | = 5 | 29 000 | = 12 |
| Kansas City | 7 000 | = 7 | 16 000 | = 12 |
| Washington | 6 000 | = 3 | 16 000 | = 8 |
| Boston | 12 000 | = 3 | 27 000 | = 6 |

**Fig. 5.6** Langhans. *Alldeutscher Atlas*

strong reaction from the colonial sentiment of the early eighties set in: then, the people had seen the colonies as supplying a remedy for the national loss from emigration; now, the popular mind of the early nineties regarded them as playfields for the bureaucrats, as battlegrounds for the militarists, and as so many places for the unscrupulous adventurer – in short- possessions not worth the keeping."[38] In an atmosphere in which news of scandals, of mistreatment of indigenous peoples in Africa by German colonial officials, and of the Arab wars, the staunchest supporters of colonial expansion, while public support appeared to be dangerously waning, revived their efforts with renewed vigor and to sway the views of the German public towards Germany's need to expand to benefit her national greatness and international prestige.

After 1893, Germany was able to accommodate its apparent surplus in labor; in fact, Germany had become second only to the United States as host nation for emigrant laborers. Dirk Hoeder observed that between 1894 and 1914, "only between 20,000 and 40,000 German emigrants arrived annually in the United States."[39] Emigrants had departed from regions in which they had been socialized, and thus did not emigrate according to historians' or map makers' categories. The Germans in America brought with them a variety of dialects and differences in the practice of religious ceremonies. As Langhans intended to illustrate, German emigrants did in some measure preserve elements of their Germandom, cultivating their languages, customs, and habits; but, they were not carriers of German nationalism, and certainly not to the extent of planning an extension of the German empire in America. While German emigrants of the nineteenth century were committed to the cultivation of their regional cultures, they openly strove to be good Americans.

Viewed separately, the inserts of Langhans' map are correct in their details, and the title is also correct to a limited degree that the Germans brought their culture with them. But, viewed in its entirety, that is, together with the inserts depicting heavy concentrations of German settlement, without any reference to other ethnicities for comparative purposes, the map as a text in the history of a specific nationalist mentality, more than suggests with its title quite a forceful narrative of the spread of Germandom. It represents a renewed mastery of the nationalist discourse in reappropriating the notion of Germandom to ultimately convey a sense that America had become what it had become due primarily to German influence. No need to lament Germany's loss of citizens to emigration when theses emigrants could be shown to be successfully carriers of Germandom.

[38]Townsend, *Rise and Fall of Germany's Colonial Empire*, 167.

[39]Dirk Hoeder, "North America," an introductory essay to Thomas Adam (ed.) *Germany and the Americas: Culture, Politics, and History*, (Santa Barbara: ABC-CLIO Inc., 2005), 8.

## Biographical Note

**Paul Rutschmann**, completed a B.S. in Economics before turning his attention to history. In 2007 he received his M.A. in History at the University of Texas at Arlington and is currently working on his PhD in the Transatlantic History program at UT Arlington. The focus of his dissertation is a reassessment of the Nazi past through the prism of German and American cinematic docudramas in relation to a historiographical evaluation in a transatlantic perspective. As an intellectual and cultural historian of late 18th and early 19th century Germany, he has broadened his studies to include the history of cartography, focusing mainly on political cartographers and their methods of map making.

## Bibliography

### *Atlas*

Langhans P (1897) Deutscher Kolonial-Atlas: 30 Karten mit 300 Nebenkarten. Justus Perthes, Gotha

Langhans P (1900) Alldeutscher Atlas: Statistik der Deutschen und der Reichsbewohner. Justus Perthes, Gotha

### *Institutional Publication*

Department of the Interior, Census Office (1894) Compendium of the eleventh census: 1890. Government Printing Office, Washington, DC

### *Books*

Chickering R (1984) We men who feel most German: a cultural study of the Pan-German league, 1886–1914. George Allen & Unwin, Boston

Gründer H (1985) Geschichte der deutschen Kolonien. Verlag Ferdinand Schöningh, Paderborn

Nipperdey T (1992) Deutsche Geschichte 1866–1918, Bd. II: Machtstaat vor der Demokratie. Verlag C. H. Beck, München

Townsend M (1930) The rise and fall of Germany's colonial empire 1884–1918. Howard Fertig, Inc., New York

Wedi-Pascha B (1992) Die deutsche Mittelafrika-Politik 1871–1914. Centaurus Verlagsgesellschaft, Pfaffenweiler

Wehler H-U (1969) Bismarck und der Imperialismus. Deutscher Taschenbuch Verlag, Cologne

## Dissertation

Pierard RV (1964) The German colonial society, 1882–1914. Ph.D. dissertation, State University of Iowa

## Articles and Essays

Bretting A (1995) Women's lives in New York city, 1890–1910: the women's pages in the middle-class New Yorker Staatszeitung and the Socialist New Yorker Volkszeitung. In: Emigration and settlement patterns of German communities in North America. Max Kade German-American Center, Indianapolis

Demhardt I (2009) Paul Langhans und der Deutsche Kolonial-Atlas 1893-1987 In: Cartographica Helvitica, Fachzeitschrift für Kartengeschichte, Nr. 40. Verlag Cartographica Helvitica, Murten

Hoeder D (2005) North America. An introductory essay. In: Thomas A (ed) Germany and the Americas: culture, politics, and history. ABC-CLIO Inc., Santa Barbara

Nagler J (1993) *Ubi libertas, ibi patria* – Deutsche Demokraten im Exil. In: Frei AG (ed) Friedrich Hecker in den USA: Eine deutsch-amerikanische Spurensicherung. Stadler Verlagsgesellschaft mbH, Konstanz

Planze O (1993) Germany – Bismarck – America, In: Deutschland und der Westen im 19. Und 20. Jahrhundert; Teil 1: Transatlantische Beziehungen. Franz Steiner Verlag, Stuttgart

Reichstein A (2005) Darmstaedters. In: Thomas A (ed) Germany and the Americas: culture, politics, and history. ABC-CLIO Inc., Santa Barbara

von Senger S. and Etterlin S (1995) New Germany in North America: origins, processes, and responses, 1815–1860. In: Emigration and settlement patterns of German communities in North America. Max Kade German-American Center, Indianapolis

Wolf SG (2005) Germantown, Pennsylvania. In: Thomas Adam (ed) Germany and the Americas: culture, politics, and history. ABC CLIO Inc., Santa Barbara

# Part III
# The United States Geological Survey

# Chapter 6
# The Technology War, the Magical Aeroplane, and the Shift to Photogrammetry in American Public-Sector Mapmaking

**Patrick McHaffie**

**Abstract** This paper examines a shift in U.S. public-sector mapping and its public perception in the first three decades of the twentieth century. In particular the emphasis will be on the first moves toward replacement of tradition-laced systems of apprenticeship and tradesmanship by more technicized approaches to mapping, eventually resulting in a technology-dependent factory system staffed by cadres of skilled and specialized workers. The effects of military financing and R&D through this period became a powerful force shaping mapping technologies both directly (through the funding of civilian mapping agencies during WWI) and indirectly (training personnel who would return to civilian sector mapping following the war). In the minds of the general public at this time, photography from airplanes became synonymous with mapping.

## Introduction

At the start of the twentieth century, systematic topographic mapmaking in the United States was practiced as a tradition-grounded and publicly-funded activity using techniques and technologies that had evolved over centuries. Topographers in U.S. government agencies such as the United States Geological Survey (USGS) worked long months in the field (weather permitting) constructing detailed topographic surveys using plane table, alidade, and spirit level. In winter months they would return to the agency office to complete the drafting of manuscripts sketched during field surveys, assisted by specialized topographic draftsmen and engravers. The maps that were constructed in the slow and labor-intensive early decades of these surveys, intended for the use of natural resource managers, geologists, and engineers, became a central element of geospatial infrastructure, allowing

P. McHaffie (✉)
Department of Geography, De Paul University, Chicago, IL, USA
e-mail: pmchaffi@depaul.edu

E. Liebenberg and I.J. Demhardt (eds.), *History of Cartography*,
Lecture Notes in Geoinformation and Cartography 6,
DOI 10.1007/978-3-642-19088-9_6, © Springer-Verlag Berlin Heidelberg 2012

the opening of the West and the rationalization of land development through the century.

Between 1910 and 1940 this scientific work process was transformed through the development of aerial photography, photogrammetry, and their use in the construction of detailed, systematic mapping programs; the first of three major shifts in the way public-sector maps were made in twentieth century America. During World War I, U.S. military mappers adopted techniques pioneered by French and British military surveyors, combining heavier-than-air aircraft with newly developed photographic technologies to map a dangerous European battlefield. In this paper I borrow a notion of "magical technology" to describe the airplane at this time. At the turn of the twentieth century it joined several other technologies (e.g., telephones, telegraphs, photography, sound recording) that were profoundly redefining experiences and understandings of space and time (Thurschwell 2001).

On returning to their civilian mapping duties, USGS topographers who were conscripted and volunteered to serve in Europe, began to incorporate aerial photography into the labor process, assisted greatly by Air Service flyers who promoted the martial technologies of aerial reconnaissance. Media reports of mapping projects at this time shifted the public's perception of detailed mapping from a slow and tedious field endeavor to a process driven by the modern technologies of aerial photography and photogrammetry. In reality topographic mapping during the period was in transition from field-based triangulation and leveling to a mixed process that gradually incorporated photogrammetry.[1]

## The Apprenticed Government Mapmaker

In the late nineteenth and early twentieth century the craft of mapmaking in the United States was largely practiced in the out of doors for a good portion of the year. This was exemplified in the USGS, whose topographic mappers had begun their detailed quadrangle mapping of the vast American domain (that had been recently wrested from indigenous tribes) during the last 25 years of the nineteenth century. In 1890 the *Los Angeles Times* wrote of the "extremely slow and complicated methods" of USGS topographic surveys, several years after they had begun (Beaman 1890). This situation had not changed appreciably by the early twentieth century as the *New York Times* predicted a "century of work for map makers" – noting that the topographic maps were of great value to the states for transportation and natural resource development. "It is a revelation to lay the eye and mind to look at the work. It is possible to tell the exact nature of every part of the country; the

[1]This is not intended to be a comprehensive history of the USGS topographic mapping operations during this period but rather an examination of a particular facet of their mapping practices. A more comprehensive set of historical resources (Evans and Frye ed. 2009) has been recently released by USGS.

location of the cities and their general appearance, and the towns, the streams, the railroads, and even the common roads." (New York Times 1902)

Longstanding principles of ground based triangulation and plane table surveying underlay the construction of detailed large-scale topographic maps, laid out in systematic quadrangles based on the latitude and longitude grid that was the borrowed product of European mapping – eventually totaling thousands of sheets. The projected time for finishing the map series (even early in the century when the smaller scale maps were the standard) was widely assumed to be at least 100 years, a duration sufficient to guarantee lifetime employment for generations of topographers and effectively remove discussion of increased funding from the public sphere. Mappers were viewed by the public as something of a curiosity, much like social insects working on an inconceivably difficult and immense task.

> Every year with the coming of the open-field season survey parties hurry away from Washington to the four quarters of the United States and Alaska. The result of their season's work is the topographic and geologic mapping of tens of thousands of square miles of all sorts of country, ranging from the most forbidding swamps and morasses to the loftiest of the glacier-covered mountains of the Rockies and high Sierras. (Washington Post 1911)

The purpose of this mapping was largely to assist the exploitation of natural resources, water power, timber extraction, the creation of transportation networks to support these activities, and to allow the rapid expansion of agriculture and land development in the West.

In the USGS mappers passed through a semi-formal system of apprenticeship that had become normal during the period of heroic surveys of the Western U.S. This entailed being hired as junior topographers or topographic aides, progressing to assistant topographer, and then to topographer. Chief administrators were classed as geographers or topographic engineers. "Practically every officer having large administrative responsibilities connected with its field operations has been connected with the service for at least a score of years and has worked in every grade of field position" wrote G.O. Smith in his Director's Report (USGS 1913 p. 8). Topographers initially did all of the work to produce the topographic maps up to the point of lithography, which were made principally to serve as a base for geologic mapping. It was typical for the employees of the Topographic Branch, when in the field or the office, to work an extraordinary number of hours overtime, in fact "of the scientific members of the organization, several will be found working at their desks every evening, during the winter, often until midnight" and "in the clerical force, where there is the same devotion to duty but possibly not the same personal interest in the work, the record is no less creditable in the amount of voluntary and, indeed, volunteered service." In the field it was little different "and even after darkness has fallen there may be several hours of accounting and clerical work before the chief of party or his principal assistants turn in for the night." The astounding conclusion derived from this overtime was that "the motive of public service is often as effective in these kinds of government work as the incentive of personal gain in private enterprise." (USGS 1913 p. 9)

Topographers entered government service at USGS at pay levels that reflected the amount of education and experience they brought to the job. In 1889 "topographic aids" could become qualified by taking an examination in Washington. The subjects included "algebra, geometry, plain trigonometry, surveying, geodesy, astronomy, topographic drawing". Salaries for these junior staff ranged from $600 to $840 per annum (approximately $12,900–$18,000 in 2006 dollars). (Washington Post 1889) In 1897 topographers working in the Pacific Section were earning $1,600 per annum (approximately $38,200 in 2006 dollars) and their "camp men" were employed at $2/day (letter dated June 16, 1897). This would allow one of these locally-hired workers to earn nearly $312 in a 6 month field season (approximately $7,600 in 2006 dollars). New junior-level mappers would enter the profession from a variety of educational and experiential backgrounds. In the decades through the turn of the twentieth century the engineering and surveying professions were becoming more organized. The American Society of Civil Engineers (ASCE) was the first engineering professional society in the U.S. (1852) and the professionalism that emerged stressed at various times business – managerial leadership, status as an employee of a large organization, or keeper and practitioner of esoteric scientific knowledge and technique (Noble 1977). These different forms of professionalism and their deployment at USGS created tensions through the early formative decades of the mapping agency, as employees of unequal and differentiated status organized themselves into a working enterprise. Topographers may have aspired to professional status as they gained experience and authority in the USGS, or they occasionally entered the mapping agency with engineering credentials. This status, that of a professional engineer, would become a *de facto* credential for advancement as the agency became larger and more complex. As mapping became more technicized in the twentieth century, the differentiation of employees between various skill levels and professional statuses became more determinative and necessary to a long-term drive for efficiency and economy.

The way of life for topographers in the field during this period prior to 1920 was rugged. They worked in a raw and sparsely settled land, finding themselves in frontier outposts, remote from the Eastern and Pacific Coast areas (that had been mapped first) as settlement progressed during the nineteenth century. Transport in the field was most often accomplished with the aid of horses and mules, and camp life in close quarters with many animals was the common experience. As late as 1917, Rodmen were reported as being seriously injured by being kicked or trampled by horses (Memorandum 1917). It was common practice for party chiefs to hire residents from the areas being mapped as assistants for various assignments related to the work of mapping or the day-to-day activities of the surveyors, giving the party chief a local status as employer and man of authority. These temporary workers were employed as "camp men", working as cooks, mule skinners, or later mechanics and drivers, as cars and trucks came to be used in mapping. Local workers who showed potential as mappers were also hired to help directly with the tasks of mapping, often taking the title of Rodman, a reference to the graduated rod held vertically at a forward or rear location while the topographer used transit and level to fix its position in topographic space. The technical staff of a typical party would include the topographer, a junior topographer, two rod men, and a recorder. The camp men and mapping assistants were hired at a rate that remained relatively stable around $2/day through the early twentieth century These wages paid in the deflationary period of 1897 remained fixed at the same rate in 1915, even though inflation had removed over 17% of its buying power. Runaway inflation during WWI would require a reevaluation of all wage rates in government service. (USGS 1897, 1915)

There was an informal process of acculturation to the routines of field and office work for lower-grade mappers, and progress was judged from year to year as they advanced in rank. As mentioned above, it was not uncommon for a locally hired worker who had distinguished himself through outstanding work as an aide to be promoted to junior topographer and enter the Topographic Branch's informal system of apprenticeship. While the credentials of technical training as an engineer or surveyor were appreciated there was a strong culture of accepting workers who showed themselves to be capable and bright into the fraternity of topographers. A list of requirements for promotion to junior topographer, the entry level professional mapping position in the mapping agency, sought "men" who were a "graduate of a technical institution or with equivalent experience" and "1 year of practical field experience in topographic mapping." Personal characteristics that would be evaluated included "honesty, industry, and administrative ability" (Memorandum for Colonel Marshall 1919). Rate of pay for this position at that time was $80/month ($922/month 2006 dollars). There were many anecdotal instances of men who had started as field aides and worked themselves to positions of high rank within the agency. This practice of de facto apprenticeship, later supplemented with internal engineering training and support for employee attendance in external technical schools lasted well into the middle of the 20th century.

## World War I: The Technology War

World War I was the first major war of the twentieth century (notwithstanding the Russo-Japanese conflict 10 years earlier) and during this war there was an unprecedented reliance on new technologies as direct or indirect tools in the massive bloodletting that took place. Neither side was more or less reliant on new innovations from transportation (gasoline engines, aircraft, lorries, tanks etc.), electronics (electric signaling gear, lighting, etc.), chemical industries (poisonous gas, explosives, photography, etc.), and many other related areas. In 1915 the Naval Consulting Board was established to mobilize American scientists to study war-related problems, and Thomas Edison was named to head the board. Also invited to join were Orville Wright, Alexander Graham Bell, and Henry Ford (Chicago Daily Tribune 1915; 1916).

There was substantial speculation about the value of airships as reconnaissance platforms in the years before the war. Major General Baden-Powell, British colonial military commander, was unequivocal: "An officer going up in it can spy out the

**Fig. 6.1** Goddard1: A military aerial photography unit training at the U.S.A School of Aerial photography at cornell University in early 1918. Thousands of soldiers were trained at these schools in early techniques of photo interpretation and mapping for the Army Air Service and Signal Corps

enemy's land and see where his strength and weakness lie and report to headquarters. This undoubtedly will be the use to which the flying machine will be put in the next war" (Chicago Daily Tribune 1906). By 1911 the U.S. military, who had established their first aviation training field at College Park, Maryland, were carrying out experiments in aerial photography and mapping. Photographs were "taken from various altitudes and at different speeds, while the sketch and map work (was) done with pencil and pad by the aviator" (New York Times 1911).

No technology in World War I changed mapping more than the use of aerial photography by both sides. In the U.S., preparation for this eventuality started during the late 19th century and continued for more than a decade with experiments in kite photography. French and British military kite units were also mustered during the early twentieth century. In 1910 a Boston Daily Globe headline blared "WAR KITES – Man carrying Flyers of French Army – Act best in Winds that are bad for Balloons – Mapmaker Lifted Half a Mile in the Air." (Boston Daily Globe 1910) These great heights, achieved by British military kite-flyers, were also approached by the French, who developed a sophisticated photographic system for reconnaissance work. For military purposes the greatest needs initially were oriented towards particular targets rather than systematic survey – it would not be a great leap for civilian mappers, however, to adapt a system developed for martial reconnaissance to systematic survey.

As the text for strategic planning, detailed maps were a crucial component of the infrastructure necessary to the forces on both sides, allowing them to accurately and rapidly capture the spatial character of the battlefield. Mappers entered the war in a state of transition, with the infant technologies of aerial photography and

**Fig. 6.2** In World War I the branding of mass-marketed products such as Kodak cameras and film was enhanced by the company's association with and contributions to mapping and surveillance technologies used by the Army Air Service

photogrammetry offering distinct advantages to the older approach of risky ground-based field survey or sketching on maps while in the aircraft. Regular army officers had been assigned to the aerial photography units at the start of the war but, after disappointing results the photographic section was reorganized and "nonfighting men" with prior photographic experience were sought for the photo service. The Signal Corps put out a call for 1,000 photographers, especially "men to fit the finished prints into their proper places, in the photographic reproduction of the German front, including men familiar with map compilation, map reading and interpretation, topographical science and drafting" (Boston Daily Globe 1918).

In the U.S., schools for aerial photographers and mappers were established quickly at various locations including Rochester, Cornell University, and the Great Lakes Naval Training Station in Waukegan, Illinois. Hundreds of soldiers were trained in these schools very quickly (800 students in the first class at Rochester alone). In several instances well-known photographers enlisted and were assigned as teachers in these schools or were commissioned as officers, including Joseph Toloff, the best known photographer in Evanston, and Walton Reeves, a well-known photographer from Atlanta. Much of the initial training was done by English and French specialists. Just before shipping to Europe, Reeves reported "the course (at Cornell) has been given by English officers, who are expert in this new phase of photography... The map work and interpretation of the aerial photographs has been taught by French experts and is undoubtedly the most interesting work of any branch of the army" (Chicago Daily Tribune 1918; Atlanta Constitution 1918). At the Rochester school "all of the facilities of a large factory" were converted to the task of training the aerial photographers. "Officers from the Allied armies, who have seen actual service in the flying zone, (were) detailed to the school to instruct those whose task (was) to operate and maintain the photographic devices of the scouting airmen" (Los Angeles Times 1918).

The USGS was involved early supporting the military by mapping aerodromes, military bases, naval facilities, and other areas deemed necessary to the defense of the nation, even before the formal declaration of war. In spring of 1912 the War Department requested a substantial supplemental appropriation for the Geological Survey "on account of the urgent importance of securing accurate topographic sheets showing the military features required by the War Department for immediate use in case of emergency" (House of Representatives 1912). Later supplemental requests were made right until the entry of the U.S. into the war. In 1916 the supplemental request bore an ominous tone, "Since the last annual estimates were submitted conditions at home and abroad have so changed as to make the topographical mapping of our frontiers an imperative necessity in connection with the national defense." (House of Representatives 1916) In addition to the worsening European conflict, the northern Mexican insurgency led by Doroteo Aranga Arambula (Pancho Villa) and the U.S. response to related border unrest created a great demand for accurate topographic mapping along the southern border (Chicago Daily Tribune 1916).

Shortly after the U.S. entered the war in April 1917, the War Department requested more than 15 times the previous supplemental request and asked for the assistance of the Geological Survey, the Coast and Geodetic Survey, and "other mapping agencies of the Government." The purpose of this assistance was to complete detailed mapping of "the entire Atlantic and Gulf coasts between Maine and Texas… for a distance back from the shore line of not less than 100 miles at any point." The appropriation was "deemed imperatively necessary on account of existing emergency conditions, which could not be foreseen at the time the regular estimates were submitted." (U.S. Senate 1917) For their part the Topographic Branch was stretched too thin to be able to effectively complete this request. In the years before the war, USGS Director G.O. Smith had complained that the agency was "undermanned and underequipped", adding that many workers were leaving the Survey to enter the private sector, drawing salaries "two and one-half times the salaries paid them by the Geological Survey." (op. cit. 1913)

Because of this shortage of trained mappers the U.S. entered the war somewhat behind the national surveys of other belligerent nations, with some catching up to do in the use and implementation of aerial photography and photogrammetry. As mentioned above, the immediate response was to begin an accelerated training program for topographers, aides, and other mapping specialists at numerous sites around the country. Some of the schools were operated by the military through various Engineer units, others were created at institutions of higher learning, and still others were created and managed by the various civilian mapping agencies. Through these schools hundreds of soldiers, nearly exclusively male, were trained in traditional techniques of topographic surveying. The standard textbook was the U.S.G.S. *Manual of Topographic Methods*, by Henry Gannett, that had been originally published in 1892 and was revised and republished in 1906. This manual dealt strictly with traditional field-based topographic mapping. This was supplemented for many by a publication of the Geological Survey (Bagley 1917), *The Use of the Panoramic Camera in Topographic Surveying with notes on the Application of Photogrammetry to Aerial Surveys* by James Bagley. Bagley had spent several years working in the Alaska territory, and had developed new methods and instruments that allowed a combination of field plane table surveying with topographic mapping based on the use of photographs taken with a ground-based panoramic camera. His work was based on pioneering work in the use of ground-based photography in topographic mapping by the French and Germans during the nineteenth century, both mainly spurred by military surveys in the Franco-Prussian War. Bagley was very excited about the possibilities of aerial survey however, as in the opening pages he stated, "Aerial photography will in the future not only have a prominent part in military reconnaissances in this country, but quite certainly will also have a direct and important bearing on the practices of civil engineering." (p. 15) His section on aerial photography was substantial, amounting to one fourth of the 86 pages in the report, drawn mainly from the earlier French work, presenting cases where aerial photography could be used in district reconnaissance and route survey. Other texts used at the schools were quickly republished by the Department of War in 1917.

Not many documented narratives or oral histories of mappers from this period exist, however we can draw some understanding of the experience from William Tufts, a USGS topographer from New York (Tufts 1916–1919). By January of 1918 Tufts was in France, and was working at the Army Engineer School, Topographic Section, of the American Expeditionary Force. The French were the principal instructors in courses related to airphoto interpretation and use at the school. As the courses progressed through this period Tufts also instructed regarding aerial photographic use, but the French instructors seemed to teach the most technically advanced courses, for example on February 23 "Lecture and study on photographs by Capt. Pepin, a.m., Constructing maps from photographs, p.m.". Quite often Tufts would lecture to very large groups. For February 27, "Capt. Tufts lectured on airplane photographs at Chatillon, 2nd Corps School to 320 Officers." In one of the courses taught by Pepin and assisted by Tufts on airplane photographs "Capt. Patton and 20 students from the tank school attended." Late in this period 2 days of instruction were given on "practical restitution of photographs on maps" and on quizzes students were asked to define "restitution of photographs" – a technical photogrammetric procedure.

The story of Tufts is illustrative of the experience of an elite group of mappers in the war, conscripted as officers and serving as trainers. The experience of enlisted men who served in some form of mapping capacity can be drawn from the letters home of Albert Kleinecke, a surveyor who served in France. Kleinecke, enlisted in fall 1917 at the age of 28, and served until the summer of 1919 in the 29th Engineers (the same unit as Tufts), achieving the rank of Sergeant First Class upon his discharge. Between April 1918 and May 1919, Kleinecke frequently wrote to his family to tell them of his experiences and his well-being. He seems to have spent much of his time well back from the front, although he commented early on "we are near enough to the front to meet many right from the trenches and get their experiences... We are in Zone No. 1, and after 6 months in this locality will wear a yellow emblem on our sleeves signifying '6 months in the danger zone.'"

(Veterans History Project, Kleinecke letters, p. 2) He later states the nearest he had been to the front was about 25 miles. Later in April 1918 writing to his mother he said

> War is not very hard on me. Indeed, it is rather soft. I am busy all the time. I was on a plane table survey, and am now studying aerial photography, which I think will be extremely interesting. My work will be that of assisting in the production of military maps from information gathered from these sources: surveys made in the field when possible, aerial photos of the dangerous sections, and French and Belgium maps of the war zone. The photos are snapped by members of the signal corps or aviators, and shaped into maps by our outfit. (Veterans History Project, Kleinecke letters, pp. 7–8)

His trainers sound familiar – he appears to have been trained by Tufts and others from USGS. In letters home Kleinecke describes his work in some detail:

> We go up near the front for the purpose of mapping areas that the Huns are likely to occupy so that the allies can more easily get their range and locate their batteries, supply stations, etc. As yet, they have never occupied a section newly mapped, but that is not our fault, is it? (Veterans History Project, Kleinecke letters, p. 35)

After the armistice Kleinecke worked with teams who were mapping the Hindenburg Line, the Argonne, Verdun, and other locations to determine sites where Allied soldiers had been buried during hostilities.

The experience of the war was life-changing for everyone who was part of the effort. Certainly, government topographers and the thousands of others who entered the military to map the dangerous battlefield were no different in many respects from any other member of the American Expeditionary Force. American soldiers returned to the U.S. psychologically and ideologically shaped by their experiences. For their part mappers had become firm believers in the new technologies of aerial photography and photogrammetry, as well as the centrality of the fledgling air-arm to future military power. They believed that the Allies would not have succeeded without the intelligence that was derived from aerial mapping during the war, and they were determined to bring these techniques back into the civilian world by applying them to the gigantic task of mapping America. Another lesson of the war was that of preparedness – in the early stages of war the time required to requisition necessary equipment and supplies, train personnel, and develop new techniques as adaptations to circumstances was seen as being dangerously slow and poorly planned (perhaps unavoidably so). This sense that the nation could not afford such slow preparation for war in the future led to the entrepreneurial penetration of civilian mapping agencies and state and local government entities by the Air Services of the army and navy in the years immediately after the conflict. The military service of the principal administrators of these civilian agencies would make what was to become a strong cooperative relationship between military aviators and aerial photographic mappers and civilian mapping agencies much easier.

## Cameras into Plowshares: Aerial Mappers Return to Federal Service

After their European service the USGS topographers who enlisted or were conscripted into the Engineers Officers Corps returned to the U.S. having seen and used techniques for air photo mapping in collaboration with Allied forces. Upon returning to service in the USGS, former military mappers, heading virtually every division in the Topographic Branch, began implementing new ways of map-making driven by the new technologies of photogrammetry. This resulted in the establishment of a Photographic Mapping unit within the Topographic Branch early in the 1920s. Most of the actual photography would be supplied by military aviators and photographers in several high-profile projects. By the end of the decade the Topographic Instructions were rewritten to include a large section on photogrammetry.

In the decade immediately following the war there was general fascination with the airplane and the possibilities of social and economic change presented by the new technology. Following the war thousands of airplanes, associated *materiel*, and technical equipment was scrapped or sold for pennies on the dollar as the Air Service was slashed to a small fraction of its wartime size. Much of this surplus was snapped up by enterprising investors interested in establishing the first commercial flying services in the U.S. Although it would be several years before a viable aircraft services industry would evolve, many journalists eagerly promoted these new enterprises. New companies organized shortly after the war that included Army Air Service personnel in their organizing groups included the Aero Service Corporation and the Pioneer Aerial Surveying Company. Lieutenant Sydney Bonnick, who was a principal aerial photographer for Aero Service at this time, served as an aerial photographer in the British, Canadian, and American air services from 1914 till the end of the conflict. (Atlanta Constitution 1921; Los Angeles Times 1921)

There was particular interest in the area of "aerial mapping" and aerial photography in the science community as well as the general public, although it was still very clear that most progress that had taken place was under the impetus of military necessity. At the Second Pan American Aeronautic Congress Lt. M.C. Lawrence claimed

> The map of the future is the air mosaic... Whether it be the military map, the automobile road map, or a map of air routes, the only absolutely accurate information of the terrain, the roads, the elevation of towns, cities, and all topographic features, is the minute mosaic, the map produced by detail pictures snapped from the airplane of any given territory. The Government already has begun using aviators in making a map that is to cover every inch of territory in the United States. This is a stupendous undertaking. It probably will require years of hard work and application, but its *military* and *commercial* value, once completed, will be unlimited. (emphasis added, New York Times 1919)

By late 1921 the secretary of the War Department claimed that the Army Air Service had developed aerial photography to a degree where the "entire United States

**Fig. 6.3** Goddard2: George Goddard, a central figure in the early adoption of aerial photography in public sector mapping, preparing to fly in the early 1920's. Goddard championed the adoption of aerial photography by civilian mapping agencies, notably the U.S. Geological Survey, in the decade following World War I

could be photographed in a comparable short time" (Washington Post 1921). Army air service personnel working at Langley Field in Virginia and Arcadia, California "have perfected automatic timing of exposures to cover the proper ground area and at the same time secure the necessary overlap for the accurate fitting together of individual negatives... Lenses have been discovered especially adapted for this type of work..." (Washington Post 1921). Beginning as early as 1921 army photographers began acquiring aerial photography that was used for "city planning, traffic control, street widening, and fire protection work" in several cities across the country. In Los Angeles cooperation between the Chamber of Commerce, city government, USGS, and army officers from Fort MacArthur allowed the initiation of 1:24,000 topographic mapping covering 6-min sheets, with 5-ft contour intervals for much of the city during 1922 (Los Angeles Times 1922). Earlier (during 1920) Bibb County, Georgia (in Atlanta) had been photographed in a similar fashion by the Army Air Service at the request of the National Geographic Society (Atlanta Constitution 1920).

In the first few years after the war there was a sense that the gains made during the conflict (particularly in the solution of problems associated with the quality of photography and the restitution of photographic distortion) in the application of photogrammetry to planimetric mapping by both the Allies and the Germans were a sign that other problems would soon be solved. In particular mappers felt that technical solutions would unlock the potential of stereo photography for contour or terrain mapping using photogrammetric techniques. It was widely known of the importance of stereometry in aerial photography during the war and where it had come from:

> The old-fashioned stereoscope through which Uncle Elmer, Aunt Hetty and little Chester – Sunday visitors – used to gaze so breathlessly upon marvelous views of the Mammoth Cave, Lincoln's funeral and sardine fishers off Brittany, this quaint semi-ridiculous old instrument of Sabbath entertainment carried roaring, smashing Hell to the Huns of Wilhelm, the 'scourge of the world'. (Boston Daily Globe 1919)

Experiments by the British Ordinance Survey were reported in 1920 where cameras were fixed to either end of the wing of a large aircraft securing a separation of 97 ft and allowing plotting using a Zeiss instrument – no strong method for using the photographs to produce contours had yet been found however (Newcombe 1920). By 1922 Bagley, USGS' expert on ground-based photogrammetric mapping, was writing of the complexity of determining elevations from aerial photographs sufficient for placing contours on maps (Bagley 1922). Whatever the U.S. and British mappers were doing in the immediate postwar years in this regards, it is clear that they were playing catch-up with German and Austrian technology in optical and mechanical stereoplotter development (c.f. Hinks 1922; McLeod, 1922; Jack et al. 1922).

The USGS Topographic Branch entered the postwar years with newly recognized status as a critical provider of national infrastructure, essential to national security and economic development, and poised to leverage the rapidly evolving technology of photogrammetry to accomplish long-held goals of detailed seamless mapping. Unfortunately, immediate postwar fiscal support for topographic mapping didn't match requested levels, in part due to the withdrawal of War Department supplementary appropriations as well as political considerations in the wake of the Teapot Dome scandal. Funds allotted to the Topographic Branch had soared to an all-time high of over $1 million in 1918 (over half of that from the Department of War); by 1922 total funding had fallen back to $564,000, even with a substantial increase in state support through cooperative programs. In 1922 the Department of War contribution to domestic topographic mapping was only 2.5% of the wartime peak. Added to this retrenchment was the efficiency program in the Department of the Interior forced onto the Branch by Secretary Albert Bacon Fall (who would later be humiliated and convicted while in office as the central figure in the Teapot Dome scandal). (Washington Post 1922) USGS Director Smith, perhaps in anticipation of postwar belt-tightening and a fiscally conservative and newly-elected Republican administration, in his 1921 annual report proposed a new program of efficiency and economy:

> This type of economy, which can be obtained only through increased efficiency, is not new; it is the engineering ideal set up in every productive industry; but in the Government bureau, where service is the product, such economy and efficiency are difficult to measure and to appraise... Scientific work can be organized with the business idea of efficiency... Economy in administration can be a matter of internal growth as well as a product of external force; possibly the best result will come from the sympathetic coordination of both agencies [efficiency and economy]. (USGS 1921 pp. 4–6)

During the 1920s topographic mapping at the USGS would begin to be reorganized in a way that reduced their dependence on expensive field survey by using new photogrammetric techniques. This shift would not be possible without help from the Air Services' newly acquired capability in the area of aerial photography and mapping.

A central figure who worked to establish and nourish military – civilian cooperation in aerial surveying and mapping during these first few critical years after the war was George W. Goddard. As a young lieutenant in the fledgling Army Air Service, he worked in cooperation with several civilian agencies during the 1920s. In 1921 at the age of 31 he was appointed as the Air Service representative on the Federal Board of Surveys and Maps. This seems to have brought him into contact with civilian agency mappers. In an early project he coordinated with local government agencies and USGS to complete a widely publicized aerial survey of Los Angeles – an area that was seeing rapid growth (Los Angeles Times 1923). This was a precursor to numerous military and civilian cooperative airphoto projects through the 1920s. These included projects in Michigan that led to the first topographic map being compiled wholly from aerial photography, mapping the Atlanta region in cooperation with USGS and local government, mapping Washington DC (Washington Post 1926), and a major project to photograph and map much of southern Alaska that involved the USGS, Coast and Geodetic Survey, and the Navy Department (Chicago Daily Tribune 1925). This occurred as Congress, spurred by strong lobbying efforts from engineering and land development groups, finally approved the Temple Act in 1925 that authorized the topographic mapping of the U.S. over the next 20 years. This act contained an explicit endorsement from the Secretary of War, stating that the Army had adopted the topographic quadrangle as their standard tactical map. "At the time the armistice was declared the United States had some 4,000,000 men under arms; was spending money for war purposes at the rate of about $24,000,000 a day. If, therefore, the possession of suitable maps saves men and time, even a few days, it is evident that the entire expense of mapping the United States is justified as a measure of preparedness, without reference to the many other activities which would be benefited thereby." (U.S. Senate 1925). The total estimated cost of the program was to be $37.2 million through fiscal year 1944 (over $440 million in 2007 dollars).

## Conclusion

In the decade after WWI, civilian mapping agencies were aided in their shift to a completely new approach to detailed mapping by the Air Services of the Department of War. The organizational changes in the way these maps were produced reflected changes in American industry more generally – eventually decentering the field topographer as mapmaker, shifting him to a supporting role; that of establishing horizontal and vertical control for more detailed maps. The photogrammetrist became the mapper, working in an urban office using multiplex stereoplotters through the calendar year, greatly reducing the annual routine of field and office work for surveyors. Cooperation with the newly created Tennessee Valley Authority by USGS during the 1930s accelerated this process. By the late 1930s a reporter would write "Historians of the future will probably point to the present age as the period when maps ceased to be made by skilled hands and were made by precision machinery instead." (Hamilton 1938) Cadres of technical specialists were created in civilian agencies that would be available for the next great national crisis, although it was to be seen that this was insufficient to meet the demand for global mapping during WWII.

In the wake of World War I the United States committed substantial resources to the completion of our national topographic map with the implicit understanding (and hope) that, through new technologies of aerial photogrammetry, the map would be completed in timely fashion and the final product would be highly precise, detailed, and cost effective. As the century progressed this dependence on military technology and more efficient and productive organizational structures by civilian mappers was to become even more deeply embedded, producing a national map that was instrumentally martial, revealing a fundamental characteristic of our national identity. Maps of nations reveal more than the locations of cities, rivers, mountains, factories, farms, prisons, and schools – they also reflect something about the nature of the societies that create them.

## Notes

1. The Kodak advertisement is from the Duke University library collection "Emergence of Advertising in America" at: http://library.duke.edu/digitalcollections/eaa/
   There is a copyright page with proper attribution at: http://library.duke.edu/digitalcollections/eaa/copyright.html
   Proper attribution should include: Advertising Ephemera Collection – Database #K0288
   Emergence of Advertising On-Line Project
   John W. Hartman Center for Sales, Advertising & Marketing History
   Duke University Rare Book, Manuscript, and Special Collections Library http://library.duke.edu/digitalcollections/eaa/
   The page for this image is at: http://library.duke.edu/digitalcollections/eaa.K0288/pg.1/
2. The photographs labelled "Goddard 1" and "Goddard 2" are from the Manuscript Reading Room of the Library of Congress. The photographs are all from the same box – the proper citation is as follows:
   Box 1, George W. Goddard Papers, Manuscript Division, Library of Congress, Washington, D.C.
   A finding aid to Goddard's papers in the LOC can be found at: http://lcweb2.loc.gov/service/mss/eadxmlmss/eadpdfmss/2008/ms008053.pdf
   These photographs are copies of unpublished original materials in the Goddard files. The status of copyright in the unpublished materials of George W. Goddard is governed by the Copyright Law of the United States (Title 17, U.S.C.).

## Biographical Note

**Patrick McHaffie**, served with the Kentucky Geological Survey in the 1980s before joining DePaul University in 1996. Here he chaired the Department of Geography from 2001 to 2007. His research interests and publications include the social history of cartography and GIS, the cartographic labor process, cartographic ethics, the geography of education spending, Appalachian social geography, and the social construction of the global. His current work involves a long term study of change in the cartographic labor process during the 20th century as well as work on automation as policy and ideology in the United States during the cold war. Over the past decade his work has been supported by grants from the National Science Foundation, History of Cartography Project (University of Wisconsin), and the University Research and Quality of Instruction Councils (DePaul University).

## References

USGS (1917) Memorandum for the Director USGS. National Archives and Records Administration, Record Group 57, College Park

USGS (1919) Memorandum for Colonel Marshall. USGS National Archives and Records Administration, Record Group 57, College Park

Atlanta Constitution (1920) Big plane arrives to make topographic map of Bibb County. Atlanta Constitution, November 22, 3

Atlanta Constitution (1921) Air photography work to center about Atlanta. Atlanta Constitution, December 25, 8

Bagley J (1917) The use of the panoramic camera in topographic surveying with notes on the application of aerial photogrammetry to aerial surveys, USGS Bulletin 657. Government Printing Office, Washington, DC

Bagley J (1922) Concerning aerial photographic mapping: a review. Geogr Rev 12(4):628–635

Beaman WH (1890) Geological survey work. The Los Angeles Times, September 17, 6

Boston Daily Globe (1910) War Kites. Boston Daily Globe, May 8, SM7

Boston Daily Globe (1918) Signal Corps calls for 1000 photographers. Boston Daily Globe, March 3, 24

Boston Daily Globe (1919) The stereoscope – the eye behind the lines. Boston Daily Globe, May 4, 48

Chicago Daily Tribune (1906) Airship as agent of War. Chicago Daily Tribune, November 19, 3

Chicago Daily Tribune (1915) Edison to Lead Experts Board. Chicago Daily Tribune, September 10, 5

Chicago Daily Tribune (1916) Air Scouts Hunting Villa. Chicago Daily Tribune, March 16, 1

Chicago Daily Tribune (1918) Evanston photographer to be enrolled in naval reserves. Chicago Daily Tribune, May 22, 6

Chicago Daily Tribune (1925) U.S. Patrol in Alaska opens up Virgin Land. Chicago Daily Tribune, October 25, 14

Evans RT, Frye HM (eds) (2009) History of the topographic branch (division), USGS Circular 1341. USGS, Reston

Gannett H (1906) Manual of topographic methods, USGS Bulletin 307 Series F. Government Printing Office, Washington, DC

Hamilton A (1938) Mapping – Three miles up! Los Angeles Times, January 16, H4

Hinks A (1922) Stereographic survey: the stereoautograph. Geogr J 59(4):273–284

House of Representatives (1912) Appropriation for topographic maps: letter from the Secretary of the Treasury. 62nd Congress 2d Session. Document No. 773. House of Representatives, Washington, DC

House of Representatives (1916) Topographic maps, war department: letter from the Acting Secretary of the Treasury. 64th Congress 1st Session. Document No. 1116. House of Representatives, Washington DC

Jack E et al (1922) Stereographic survey: the autocartograph. Geogr J 59(4):289–292

Los Angeles Times (1918) Develop new type soldiers. Los Angeles Times, June 30, IV11

Los Angeles Times (1921) Accurate surveying now done from airplane. Los Angeles Times, April 17, p. V5

Los Angeles Times (1922) New maps of county requested. Los Angeles Times, October 23, II7

Los Angeles Times (1923) Aviators map county in government survey. Los Angeles Times, March 18, II1

McLeod M (1922) Stereographic survey: the autocartograph. Geogr J 59(4):284–289

New York Times (1902) Century of work for map makers. New York Times, October 5, 29

New York Times (1911) Army fliers to use camera and pencil. New York Times, November 20, 10

New York Times (1919) Photo air map of U.S. New York Times, May 29, 10

Newcombe S (1920) The practical limits of aeroplane photography for mapping. Geogr J 56 (3):201–206

Noble D (1977) America by design: science, technology, and the rise of corporate capitalism. Knopf, New York
Shrader C (2006) History of operations research in the United States Army, Volume I: 1942-62. U.S. Army, Washington, DC
Thurschwell P (2001) Literature, technology and magical thinking. Cambridge University Press, Cambridge
Tufts W (1916-19) Personal papers: materials from the U.S. Geological Survey. National Archives and Records Administration, Record Group 57, College Park.
United States Senate (1917) Military surveys and maps: letter from the Secretary of the Treasury. 65th Congress 1st Session, Document No. 30
United States Senate (1925) Topographical survey of the United States. 68th Congress 2nd Session, Report No. 1011
USGS (1897) Letter to Mr. R.U. Goode from the Director of USGS to Pacific Section of Topography. National Archives and Records Administration, Record Group 57, College Park
USGS (1913) Thirty fourth annual report of the Director of the United States Geological Survey. Government Printing Office, Washington, DC
USGS (1915) Letter to Mr. Frank Sutton from the Director of USGS to Atlantic Section of Topography. National Archives and Records Administration, Record Group 57, College Park
USGS (1921) Forty-second annual report of the Director of the United States Geological Survey. Government Printing Office, Washington, DC
Veterans History Project (1918–19) Letters of Albert Kleinecke. Library of Congress, American Folklife center, Washington, DC
Washington Post (1889) A chance for idle topographers. Washington Post, July 11, 5
Washington Post (1911) Biggest Map Factory. Washington Post, September 10, 4
Washington Post (1921) Army camera men speed up airplane map-making system. Washington Post, December 23, 3
Washington Post (1922) Work of Interior Department put on highly efficient basis. Washington Post, October 8, 6
Washington Post (1926) Army fliers to map city and area for plan group. Washington Post, December 16, p. 24

# Chapter 7
# The Digital Transition in Cartography: USGS Data Innovations, 1970s

**E. Lynn Usery**

**Abstract** The 1970s were a period of data innovation in the evolution of geographic information system technology and geographic information science. Each new form of data innovated new data structures, which in a way is a new model of geographic information, providing a base for new applications of topographic map data. U.S. Geological Survey scientists were actively engaged in this data model development with at least four different objectives: the conversion of existing cartographic data from topographic maps into a dataset built from a data model that incorporated topological structure in the form of a Digital Line Graph; the building of a nationwide database of land cover, also topologically structured, but available in raster format, in the Land Use Data Analysis Program; the automation of producing graphic products including topographic maps with the Digital Cartographic Software System and the Graphic Map Production System; and terrain representation as a Digital Elevation Model, a byproduct of the orthophoto generation process and new digital production techniques. These innovations in data-model development were followed by sustained programs of nationwide data production from these efforts providing digital geographic data that catalyzed the GIScience industry in the 1980s and 1990s. The U.S. Geological Survey efforts in this area continue with the evolution of these 1970s data models to the seamless nationwide databases of today of *The National Map* and the new graphic product called US Topo.

---

E.L. Usery (✉)
U.S. Geological Survey, Rolla, USA
e-mail: usery@usgs.gov

E. Liebenberg and I.J. Demhardt (eds.), *History of Cartography*,
Lecture Notes in Geoinformation and Cartography 6,
DOI 10.1007/978-3-642-19088-9_7, 

## Introduction

Data models were one of the original research areas in automated cartography and geographic information systems (GIS). Goodchild (1988) stated that each new database containing a digital representation of spatial data is essentially a model of geographic reality, and as such provides new and different types of applications with the data. The U.S. Geological Survey (USGS) began data model research and development in the 1970s resulting in models that became standards (USGS 1983, 1987) for the GIS industry. Much of the research work of the USGS during this period was never published in the major cartography and GIS journals, but was documented through fugitive literature such as internal USGS reports and proceedings of major conferences, including the AutoCarto series (CaGIS 2010), and the semi-annual meetings of the American Society for Photogrammetry and Remote Sensing (ASPRS) and the American Congress on Surveying and Mapping (ACSM). It is the purpose of this paper to document the USGS research and development efforts during the 1970s. The remainder of the paper provides the historical legacy, environment, and context of the 1970s developments of the USGS; details the four major data development and automation efforts within the USGS during this period; provides discussion and analysis of those developments in the broader context of the evolution of the GIS industry and the development of the GIScience discipline; and describes the evolutionary developments from these original data efforts to the nationwide datasets of *The National Map* and the introduction of the US Topo product in 2009. A final section draws some conclusions from the documentation of the research and the USGS work.

Any use of trade, product, or firm names is for descriptive purposes only and does not imply endorsement by the U.S. Government

## The Topographic Map Factory and Technological Innovation

In the 1970s the USGS operated four regional mapping centers in Reston, Virginia; Rolla, Missouri; Denver, Colorado; and Menlo Park, California. Each was configured as a map factory with complete field, photogrammetric, cartographic, and photographic reproduction capabilities. The basic mapping process had been honed for an assembly line production process including the basic steps in Fig. 7.1. Each center was responsible for topographic mapping for a specific region of the country. For example, the Mid-Continent Mapping Center in Rolla, Missouri, conducted all topographic mapping in a 14-state region in the central United States. During the 1970s the process of mapping involved field reconnaissance using photos to establish survey control and classify all topographic information, such as buildings, streams, and roads for the topographic map. After the field work was completed, the photo images were transferred from the original negatives with contact printing to film diapositives. The diapositives were then used in the

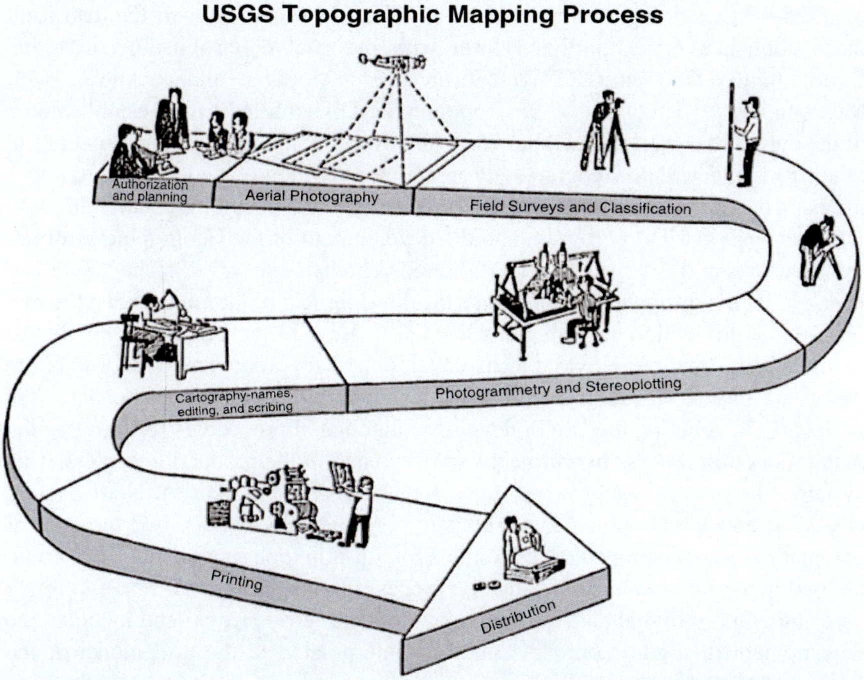


**Fig. 7.1** The stages of the mapping process as implemented in an assembly-line mode

photogrammetric process of stereoplotting to develop a "compilation manuscript" of the map. The compilation manuscript, a piece of scribecoat on which are scribed or drawn in ink all features, including contours from the stereoplotting, to be shown on the map that are acquired from the field survey and the photogrammetric process. The compilation manuscript was then passed to a cartographic operation that included precision scribing of all lines, particularly contour lines, adding all boundary, name, and other information not available from field survey and photography. The map was color separated to red, black, blue, green, and brown plates, and edited extensively in cartography to produce a final set of color separation negatives that were used for photo-lithographic reproduction of the topographic map.

In the midst of this well-developed assembly-line production process, which began with the adoption of photography and scribing in cartographic production in the 1930s to the 1950s and peaked in the 1970s timeframe, computer technology made its debut. The USGS had adopted the computer in the 1960s to drive the AutoPlot, a device that automatically scribed the neatline, the lines of latitude and longitude that bound a USGS quadrangle, to produce the base scribesheet for the topographic map (McHaffie 2002). In the 1970s the USGS began developments of several pilot projects with computers and digital cartographic data. At that time, about 70% of the 48 conterminous United States was mapped at the 1:24,000 scale on a 7.5-min quadrangle format, so the initial effort was to convert the data from those published maps to digital form with the goal of establishing a Digital Cartographic Database (DCDB) (Edson 1975; McEwen and Jacknow 1979; McEwen 1981). This pilot of the Topographic Division led to the establishment of the digital line graph (DLG) data program of the USGS. Almost simultaneous to this effort, the USGS Geography Program, led by James Anderson, initiated a project to acquire land-cover data for the United States resulting in the Land Use Data Analysis (LUDA) Program and the development of the Geographic Information Retrieval and Analysis System (GIRAS) (Mitchell *et al.* 1977). The USGS had been producing orthophotographs since the development of the concept and the first operational production instrument by Russell K. Bean of the USGS in 1962. It was in the 1970s that the USGS began developing a byproduct of the orthophoto process, a digital elevation model (DEM), and archiving these data for later use in the DCDB. Finally, the USGS developed automated procedures for topographic map production process in a project that originated in the production centers. This system, which was actually two systems, the Digital Cartographic Software System (DCASS) and the Graphic Map Production System (GRAMPS), had the goal of automatically producing color separations for 7.5-min topographic maps that could be used in the lithographic reproduction process (Troup and Powell 1979). It relied on acquisition of digital data during the photogrammetric process and included the development of digital contours that were interpolated to the grid model of the DEM and placed in the DCDB.

At the dawn of the GIS era, the USGS was invested heavily in traditional cartographic production processes with field surveying conducted with theodolites, levels, and electronic distance measurement (EDM) units; photogrammetric procedures on scores of analog stereoplotters including Kelshes, Kern PG-2s, and Wild B8s; cartographic procedures using colored pencils and manual editing of scribecoat over scores of light tables; and map reproduction on a five-color press established in the USGS National Center in Reston, Virginia. It would take several decades to replace this extensive production plant with new computer-based technology for producing topographic maps. The four research and development efforts mentioned above, which paved the way for the digital geographic databases and cartographic products of today (*e.g.*, *The National Map* and US Topo) and provided public domain data in standard formats that helped catalyze the GIS industry are discussed below.

## USGS Research and Development Projects in the 1970s

The USGS research and development efforts in the 1970s innovated data models for vector-based line data that became known as DLG. The LUDA program also developed vector-based line data for the boundaries of polygons of land cover. These data became known by the name of the system used to generate them, GIRAS. The LUDA program also produced a raster version of the data known as

Composite Theme Grid (CTG). The DEM raster matrix became a standard data model developed by the USGS that was widely accepted. Data for DEMs were generated from three separate projects, the Digital Profile Recording and Output System (DPROS), a manual profiling system for orthophoto production; the Gestalt PhotoMapper (GPM), an automated analytical stereoplotter that generated orthophotos and DEMs; and DCASS, a system that generated digital contours and color separations for the map production process.

## *Digital Line Graphs*

The DLG program of the USGS Topographic Division began in the early 1970s, becoming an official program with the formation of the Digital Applications Team (DAT) in 1977 (Beck and Plasker 1979; USGS DAT 1977a). The program had the objective of developing digital line graph data files to represent the linework on USGS 1:24,000-scale, 7.5-min topographic map quadrangles. The DLG data model was developed by Atef A. Elassal of the USGS who also developed the Unified Cartographic Line Graph Encoding Software (UCLGES) for processing the raw digitized data to create the DLG format. The design of the DLG data format included complete topological structure.

The DLG data model uses points, lines, and polygons to represent real-world geographic features. For example, roads and streams are represented as lines. Wells, springs, and buildings are represented as points and lakes, reservoirs, and other features with areal extent are represented as polygons. The digital line graph specification included three levels of data depending on the extent of error removal, processing, and topological structure:

DLG-1: Line map information that was collected and coded to prescribed standards and edited to remove data-acquisition blunders;
DLG-2: Line map information that was interactively edited to add additional attribute codes and to remove visible errors and inconsistencies; and
DLG-3: DLG-2 data that were spatially structured to define all topological relations (USGS DAT 1977a).

Interestingly, whereas the USGS defined three levels of DLG data, only the DLG-3 level data were acquired, processed, and distributed. The original DLG Standard format was a binary IBM mainframe, EBCDIC-based representation containing minimal topology. The USGS developed the DLG-Optional format as an ASCII-based, text format that could be ingested in a variety of software and hardware systems (USGS 1987).

With the initiation of the DLG-collection program and the DAT, the USGS developed a digital cartographic data base (DCDB) as a unified set of smaller generic data bases managed by the System 2000 data base management system (McEwen and Jacknow 1979; McEwen 1981). System 2000 was selected because it already was operational on the USGS IBM 370/155 computers in Reston, Virginia, and the hierarchical structure was appropriate and effective. Data were organized by layers, each layer including a specific feature category. For example, the DLG hydrography layer included all hydrographic features. A Public Land Surveys layer included the lines and points representing the Public Land Surveys System (PLSS) (Edson and Lee 1977).

Whereas the primary DLG database was established to archive data from the 1:24,000-scale, 7.5-min topographic quadrangles, a second database for National Atlas data was developed at 1:2,000,000-scale (Stephens *et al.* 1979; Stephens 1980). This database was built with the same DLG data format and included the many layers of data from *The National Atlas of the United States* (USGS 1970).

In the 1970s, creating digital line representations of cartographic features required manual line-following on a digitizing table. Early technology used by the USGS included the Gradicon digitizer, manufactured by Instronics, Ltd., Stittsville, Ontario, Canada, which sold for $25,000.00. The Gradicon was a cross-slide design including a movable bar for the x-axis and coordinates and a perpendicular movable bar for the y-axis and coordinates. The Gradicons were quickly replaced with the Altek digitizing tables that included an embedded grid of electric wires in glass. The glass could be backlit and a curser with a magnetic ring placed on the glass sensed the embedded electric wires and determined coordinate positions to an accuracy of 0.001 in. Initially, all digitizing was conducted in a blind digitizing mode, *i.e.*, as data were digitized they automatically were written to magnetic tape, and the operator could not see a plot or display of the data. Blind digitizing from the 1:24,000-scale and the 1:2,000,000-scale maps using Altek AC 40 digitizer units became the standard production process for DLG-3 data. It was not until 1980 with the implementation of the Digital Data Edit System from M&S Computing, Inc. (which later became Intergraph) that interactive digitizing and editing became possible.

The digitized data were processed with the Unified Cartographic Line Graph Encoding System (UCLGES), an IBM mainframe-based program that was used to check the topology, alignments, and attributes of various geographic features for correctness and accuracy. Data were validated by plotting the digital coordinates from the magnetic tape. The plots were generated on a Gerber 4477 plotter with a 12-foot table (Webster 1980). The Gerber 4477 could be used for plotting ink on paper or scribecoat, scribing the scribecoat, or through the use of a photohead, the plotter could expose film negatives for use directly in the cartographic reproduction process. This plotter became the workhorse for the USGS digital cartography programs and supported not only DLG production, but was used for LUDA, DCASS/GRAMPS, and other digital plots.

## *Land Use Data Analysis Program*

The availability of the original Landsat multispectral scanner data beginning in 1972 with the National Aeronautics and Space Administration high altitude aerial photographs and high altitude photography with the National High Altitude Photography (NHAP) program becoming operational in 1978, led to the development of a land-cover database for the United States by the USGS. The land-cover data were acquired in an hierarchical classification system that became known as the Anderson classification, or the USGS Land Cover Classification System (Anderson *et al.* 1976). The land-cover data were extracted by manual interpretation of photos from a film-positive base, usually enlarged to a scale of approximately 1:125,000 at a resolution or minimum mapping unit of 40 acres (10 acres in urban areas). The result was an inked manuscript on a stable-base that was digitized using manual line-following on a Cart/8 system developed by Ray Boyle. However, most of the work was contracted for raster scanning, predominately, using the automated laser line-following scanners operated by IOMetrics in Mountain View, CA (Fegeas 1978). The data-collection and processing effort, which became known as the Land Use Data Analysis (LUDA) program, generated land-cover maps for the conterminous 48 states of the United States and Hawaii at scales of 1:250,000 and 1:100,000. In addition to land cover, political units, hydrologic units, census county subdivisions, Federal land ownership, and State land ownership (optional) were collected and included in the databases. The processing was achieved with a USGS developed system, GIRAS (Mitchell *et al.* 1977; Hallam 1979). The GIRAS processing system created vector data as topologically-structured land-cover polygons with complete specification of the relations of the nodes, lines, and areas. The GIRAS data also were converted to a raster format, CTG, which included all six data layers from the GIRAS data. As with the DLG, the USGS placed these data (GIRAS and CTG), in the public domain and distributed them through the National Cartographic Information Centers that had been established in 1974 as public outlets for USGS publications, images, and digital data.

## *Orthophotographs and DEMs*

The USGS has been producing orthophotographs since the 1960s, with the primary product being the geometrically-corrected and printed orthophotograph or orthophotoquad, as they became known since the orthophotos were created centered on 7.5 min quadrangles and represented the standard 7.5-min quadrangle area. From the mid-1960s, increasing user acceptance of and demand for orthophoto products spurred efforts to develop a means of using digital input data to control automated operation of the T-61 and T-64 Orthophotoscopes developed by the USGS. The use of digital profile data to control the z-motion of the exposing projector of the orthophotoscope was seen as a means to increase production and improve image quality. This effort culminated in the development of the DPROS, which recorded digital profiles on magnetic tape in a format that could be used to produce DEMs as well as orthophotographs (Lemeshewsky and Shope 1977).

As a part of the process of creating the orthophoto, an elevation model is extracted from the photo stereopair and used to remove relief distortion in the photographs. In the 1970s, this elevation model became a standard product and was entered into the DCDB. The data format of the elevation model developed from the process that was used to create the orthophoto. The earliest methods required manual profiling, a technique in which a photogrammetric operator successively scans profiles or strips of the orthophoto, in stereo, keeping a floating mark on the ground in the stereoimage. The position of the mark on the ground determines the amount of displacement in the image caused by the relief of the position being examined. Thus, each profile that was scanned created a digital profile of elevations. When all profiles for an image were collected, they were resampled to a regular grid to form the DEM.

Another system, the GPM, generated orthophotos and DEMs using an automatic correlation process. The system basically used an automated analytical stereoplotter to scan 9 × 8 mm patches of roughly equivalent positions in the two photos of a stereopair. A set of 2,444 points were identified in each of the two patches and a correlation coefficient was computed based on these 2,444 points in the two patches. The highest correlation coefficient identified corresponding positions in the left and right photos of the stereopair. Using those positions the x-parallax could be computed between the two corresponding positions and based on the photo and equipment parameters, the elevation difference between the two points determined. Computing elevations for all patches in a stereopair of NHAP photos and resampling them to a regular grid, a DEM for a 7.5-min quadrangle was generated.

The criteria for DEM processing were (1) Universal Transverse Mercator (UTM) coordinate system, (2) horizontal point spacing in easting and northing at 30 m, (3) inclusion of only the points within a 7.5-min quadrangle that are along 30-m grid lines (500,010; 500,040; 500,070; etc.), and (4) an estimated vertical accuracy of 7 m root mean square error (Allder *et al.* 1982). Each resulting file contains between 138,000 points (at a latitude of 50°) and 195,000 points (at a latitude of 25°) for a 7.5-min quadrangle. The 7-m estimated accuracy criterion resulted in rejecting many models of GPM data; however, a separate part of the DCDB was reserved for future inclusion of files with an estimated accuracy between 7 and 15 m. When the accuracy exceeds 15 m the model is rejected and rescheduled for data acquisition.

A third method of producing DEMs from digitally-encoded vector contour data was developed using the DCASS software (Boyko 1982). Analog stereoplotters, interfaced with three-axis digital recording modules, were used to collect vector contour data while the instruments were being used for 1:24,000-scale photogrammetric stereocompilation. During the acquisition phase, the contours were assigned elevation values (attributes). The vector contour data were then processed into scan lines and the elevation matrix formed using bilinear interpolation between the intersections of scan lines with the contour vectors. DCASS initially was developed to perform automated map compilation with a primary goal to produce color separates for map reproduction. The availability of the digital contours allowed a DEM to be generated as a byproduct, costing only operator setup and machine time.

As with the DLGs, the DEMs were classified into three levels according to the quality of the model:

Level 1– Raw elevation data sets in a standardized format. Only gross errors in data capture have been eliminated;
Level 2– Elevation data sets that have been smoothed for consistency and edited to remove random errors; and
Level 3– Elevation data sets that have been edited and modified to ensure positional consistency with planimetric data categories such as hydrography and transportation (USGS DAT 1977b).

Virtually all of the DEM's produced in the 1970s are considered to be level 1 (McEwen and Jacknow 1979; McEwen 1981; Allder *et al.* 1982).

## *Automating Topographic Map Production*

The automation of the assembly-line topographic mapping production process, shown in Fig. 7.1, also began in the 1970s with automating stereocompilation (Troup and Powell 1979). Analog stereoplotters, Kern PG-2 and Wild B-8, were retrofitted with 3-axis Altek AC-189 digitizers with magnetic tape recorders, connected to a Data General Nova Eclipse minicomputer and a Votrax voice entry system (Borgerding *et al.* 1982; Powell *et al.* 1982). As the stereoplotter operator moved the platen on the instrument, a series of x and y coordinates reflecting the platen position were recorded on the magnetic tape. The voice entry system allowed the operator to say an attribute name at the beginning of each line, and the attribute code was written to the tape followed by the string of x and y coordinates. For example, if the operator were collecting roads, he/she could say "class one road" and the attribute code for a class one road would be written to tape. The operator then began tracing the road in the stereo instrument keeping the floating mark on the ground and following the track of the road. Thus, the xy coordinate ground trace of the road would be recorded on the tape in precise coordinates. Later, the tape was processed with geometric correction, error detection, and symbolization programs and plotted on a Gerber 4477 plotter. For contours, the stereoplotter operator locked the z-value of the stereoplotter at a specific contour value, for example, 500 ft. The operator then moved the platen to keep the floating mark on the ground, and thus traced the 500-foot contour. Through this process, all data layers for the topographic map available from the photos were recorded onto magnetic tape, essentially producing a compilation manuscript in digital form.

Following the traditional production process, the digital compilation manuscript on magnetic tape was passed to the Cartography Branch where map editing, name placement, boundary placement, and other cartographic finishing operations were applied in digital form to create final publication separates. The cartographic editing and manipulation software, GRAMPS, was developed in 1980 on Intergraph dual monitor interactive graphics workstations. Resulting from the entire DCASS/GRAMPS process was a set of color separations in digital form that could drive the Gerber photohead plotter and create the necessary film negatives for the five-color printing press. As a byproduct, the digital contour layer was processed to generate a DEM as described above.

The functions of DCASS, that is, stereocompilation of digital topographic map data, were soon replaced first with analytical stereoplotters, and then in the late 1980s with softcopy photogrammetry, which included the development of digital orthophotos. The GRAMPS tasks of interactive cartographic construction and editing were soon a part of commercial GIS functionality and improved to the point that today (2010), automated cartography is an essential part of a GIS.

## USGS Data Developments in the 1970s and GIS Evolution

The digital cartography data and system developments of the USGS in 1970s led to production capabilities and data archiving into the NDCDB in the 1980s. For example, from 1980 to 1984, the USGS added 8,000 DEMs and 16,000 DLGs to the NDCDB (Frye 1984). Based on the data innovations, developments, and production in the 1970s, the USGS issued data standards for DLGs and DEMs in the early 1980s (USGS 1983). The data standards evolved to Data Users Guides (USGS 1987, 1990). The effects of the USGS data innovations in the 1970s became evident in the 1980s with the rapid adoption of GIS technology with the advent of minicomputers and microcomputers. The early diffusion of GIS technology from research labs to large government organizations, to the rapid spread of microcomputer technology in the 1980s and a corresponding spread of GIS implementations and applications, occurred before the wide availability of geospatial data. Data acquisition costs were prohibitive for many GIS implementations. The base of digital geographic data made available by the USGS and others, such as the Census Bureau, and the standards developed for the data provided a starting point for GIS users. The USGS continued with further innovations using the 1:100,000-scale topographic maps designed and compiled in the 1970s with the idea that they eventually would be digitized. These maps, designed with continuous-line and other symbols and colors compatible with raster scanning and automatic recognition were scanned on SciTex Response 250 scanners. The SciTex systems, designed for the textile industry, were adapted to scanning map separates of the 100,000-scale topographic maps. The scanned data were edited, thinned and vectorized on a Response 250 Color Design Console. The edited data were then topologically-structured and converted to DLG-3 format with a USGS-developed software package called PROSYS, which was the UCLGES software moved to and optimized for a super-minicomputer. The vectorized, topologically-structured and attributed data were passed to the Census Bureau (Theis 1982; Fegeas and Pearsall 1984). The Census Bureau added attributes and built the 1:100,000-scale digital data into the Topologically Integrated Geographic Encoding and Retrieval

(TIGER) system that became the base for the 1990, 2000, and now with adjustment, 2010 censuses (Anderson and Marx 1985).

The USGS innovations and data archives were a part of a much broader context for the development of GIS and GIScience. Coppock and Rhind (1991) describe the early developments of computer-based GIS in the 1960s, mostly originating in North America. They elaborate the roles of key organizations, including the U.S. Bureau of the Census, the U.S. Geological Survey, the Harvard Laboratory for Computer Graphics, and the Experimental Cartography Unit. Fueling the research and developments in data, software, and hardware for GIS and geospatial applications were military developments such as the Global Positioning System (GPS) and commercial developments from companies such as Environmental Systems Research Institute and Intergraph. Suppliers of geospatial data were few and the market for data did not support the high price of collection. Geospatial data provided by the USGS as DLGs, DEMs, GIRAS, and CTG, became the base of many GIS applications. Many, if not all, of the major and minor software developers for GIS and image processing included import capability for some, if not all, of the geospatial data formats of the USGS. The USGS also supplied remote sensing data in the form of Landsat images and digital tapes that supported many land-cover studies as a part of building a GIS. The USGS data had been developed to serve the Nation, but the appetite of GIS for data continued until the 1990s, when geospatial data became widely available and at continually higher resolution and levels of detail.

## Evolution from USGS Digital Cartographic Data to *The National Map*

The DLG, DEM, and LUDA data structures developed in the 1970s evolved to the nationwide coverages the USGS now provides as the data layers of *The National Map* (Groat 2003; Kelmelis *et al.* 2003). The DLG data for hydrography, the blue lines from the topographic maps, were assembled into basins and sub basins from existing DLG sources, and the remaining data were collected to form complete nationwide coverage at the 1:100,000 and 1:24,000 scales. These data became the National Hydrography Dataset. Similarly, the DEM data were assembled and supplemented to achieve nationwide coverage and become the National Elevation Dataset with 30 m and 10 m resolutions available. The LUDA data were completed in the early 1980s, to be followed by other USGS programs for nationwide land cover with 30 m resolution. The National Land Cover Dataset (NLCD) is available at 30 m resolution and versions created from 1991 to 1992 and 2000 to2001 Landsat Thematic Mapper scenes are available. These three National datasets form three of the eight layers of *The National Map*, and have their origins in the 1970s data developments of the USGS.

The USGS data developments led to nationwide databases of geospatial information. One measure of the effect of the USGS developments of DLGs, DEMs, and LUDA, is the extent to which these data were used. Varanka (2010) has analyzed the use of USGS data in published scientific works finding a significant dependence on USGS datasets from 1980 to 2010. Such use, new research and development of geospatial data concepts, and the data archives of the USGS will continue to be a major force in the GIS industry.

## Conclusions

The USGS developed innovations in geospatial data structures and models that became the base for nationwide data acquisition programs. The development of DLG, DEM, and LUDA data and production innovations with the advances in computer and network technology led to the nationwide databases available in *The National Map*. Those developments and the massive data-collection and distribution efforts of the USGS affected the broader GIS industry by supplying free, public domain data in a data-scarce environment. The initial development of geospatial datasets by the USGS in the 1970s had significant effects on the GIS industry and helped create a geospatial environment in which nationwide databases, public domain access, and innovation can flourish.

## Biographical Note

**E. Lynn Usery** is a Research Geographer and Director of the Center of Excellence for Geospatial Information Science with the U.S. Geological Survey (USGS). Previously, he was a professor in the Department of Geography at the University of Georgia and at the University of Wisconsin-Madison where he taught cartography and geographic information systems (GIS) and helped establish GIS certificate programs. Before his academic work, Dr. Usery was a cartographer and geographer for the USGS from 1977 to 1988 focusing on developing automated cartographic production systems and GIS, and conducting GIS analysis projects. In addition, he is a former President of the University Consortium for Geographic Information Science (UCGIS) and the Cartography and Geographic Information Society (CaGIS). He was editor of *Cartography and Geographic Information Science* and currently is the Chair of the U.S. National Committee to the International Cartographic Association. He is a Fellow of CaGIS and the American Congress on Surveying and Mapping, and the University Consortium for Geographic Information Science. He currently is working on ontologies, data models, data integration, and grid computing for *The National Map* and is involved in research in geographic information science, with publications on theoretical aspects of geographic representation, human cognition of geographic phenomena, automatic feature extraction from images, map projections, and visualization, and applications of geographic information science to precision farming, watershed modeling, and water quality. Lynn received his BS in geography from the University of Alabama and MA and Ph.D. degrees in geography from the University of Georgia.

## References


Allder WR, Caruso VM, Pearsall RA, Troup MI (1982) An overview of digital elevation model production at the U.S. Geological Survey. In: Proceedings, AutoCarto V. American Congress on Surveying and Mapping, Crystal City, pp 23–32

Anderson KE, Marx RW (1985) A prospective for a national land data system: ten years later. In: Proceedings, AutoCarto VII. American Congress on Surveying and Mapping, Washington, D.C., pp 1–10

Anderson JR, Hardy EE, Roach JT, Witmer RE (1976) A land use and land cover classification system for use with remote sensor data. U.S. Geological Survey Professional Paper 964, U.S. Government Printing Office, Washington, D.C., p 28

Beck FJ, Plasker JR (1979) USGS digital cartographic pilot projects. In: Proceedings, American Congress on Surveying and Mapping, 39th Annual Meeting, American Congress on Surveying and Mapping, Washington, D.C., pp 300–305

Borgerding LH, Lortz FE, Powell JK (1982) Computer-assisted map compilation, editing, and finishing. In: Proceedings, AutoCarto V. American Congress on Surveying and Mapping, Crystal City, pp 141–146

Boyko KJ (1982) Production of digital elevation models by stereomodel digitizing. In: Proceedings, American Society for Photogrammetry, 48th Annual Meeting, American Society for Photogrammetry, Denver, pp 38–47

CaGIS (2010) Cartography and Geographic Information Society, AutoCarto Proceedings. http://www.cartogis.org/autocartoarchive/autocarto-proceedings. Accessed 16 Aug 2010

Coppock JT, Rhind DW (1991) The history of GIS. In: Maguire DJ, Goodchild MF, Rhind DW (eds) Geographical Information Systems: Principles and Applications, Vol 1. Longmans Publishers, London, pp 21–43

Edson D (1975) Digital cartographic database preliminary description. In Proceedings, AutoCarto II, American Congress on Surveying and Mapping, Reston, pp 523–538

Edson DT, Lee GYG (1977) Land line survey data within a digital cartographic database. In: Proceedings, American Congress on Surveying and Mapping, 44th Annual Meeting, American Congress on Surveying and Mapping, Washington, D.C., pp 493–507

Fegeas RG (1978) Data editing. In: Proceedings, AutoCarto III, San Francisco, pp 457–471

Fegeas RG, Pearsall RA (1984) Topological structuring of raster-scanned line map data. In: Proceedings, American Congress on Surveying and Mapping, 44th Annual Meeting, American Congress on Surveying and Mapping, Washington, D.C., pp 264–272

Frye CJ Jr (1984) U.S. Geological Survey Report. In: Olson JM (ed) U.S. National Report to the ICA, 1984, The American Cartographer, Supplement to Volume 11. pp 10–15

Goodchild MF (1988) A spatial analytic perspective on geographical information systems. Int J Geograph Inf Syst 1:327–334

Groat CG (2003) *The National Map* – a continuing critical need for the nation. Photogrammetric Eng Remote Sensing 69:1087–1090

Hallam CA (1979) The USGS geographic information analysis and retrieval system: overview. In: Proceedings, American Congress on Surveying and Mapping, 39th Annual Meeting, Washington, D.C., pp 229–246

Kelmelis JA, DeMulder ML, Ogrosky CE, van Driel NJ, Ryan BJ (2003) *The National Map* – from geography to mapping and back again. Photogram Eng Remote Sensing 69:1109–1118

Lemeshewsky GP, Shope AR (1977) Digital profile recording and output system. Paper presented at the American Society for Photogrammetry Annual Meeting, Washington, D.C.
McEwen RB (1981) Information potential of a digital cartographic database. In: Proceedings, American Society for Photogrammetry, 47th Annual Meeting, Washington, D.C., pp 150–153
McEwen RB JacknowHR (1979) USGS digital cartographic database. In: Proceedings, AutoCarto TV, American Congress on Surveying and Mapping, Reston, pp 225–235
McHaffie PH (2002) Towards the automated map factory: early automation at the U.S. Geological Survey. Cartogr Geogr Inf Sci 29:193–206
Mitchell WB, Guptill SC, Anderson KE, Fegeas RG, Hallam CA (1977) GIRAS—a Geographic Information Retrieval and Analysis System for handling land use and land cover data. U.S. Geological Survey Professional Paper 1059, p 16
Powell JK, Osick GL, Miller TM (1982) Practical experience in integrating conventional photogrammetric compilation and digital mapping techniques. In: Proceedings, AutoCarto V, American Congress on Surveying and Mapping, Crystal City, pp 25–41
Stephens Mi (1980) The USGS 1:2,000,000-scale digital database. In: Proceedings, American Congress on Surveying and Mapping, 40th Annual Meeting, St. Louis, pp 436–443
Stephens MJ, Domaratz MA, Schmidt WE (1979) The development of a national small-scale digital cartographic database. In: Proceedings, AutoCarto IV, American Congress on Surveying and Mapping, Reston, pp 345–352
Theis RF (1982) Raster data acquisition and processing. In: Proceedings, AutoCarto V, American Congress on Surveying and Mapping, Crystal City, pp 667–676
Troup MI, Powell JK (1979) Interactive compilation of topographic color separates. In: Proceedings, American Society for Photogrammetry, 45th Annual Meeting, Washington, D.C., pp 112–115
USGS (1970) The national atlas of the United State. U.S. Geological Survey, Reston
USGS DAT (1977a) Digital Applications Team, computer files and attribute codes for digital line graphs. USGS Topographic Division, Reston
USGS DAT (1977b) Digital applications team, computer files and attribute codes for digital elevation models. USGS Topographic Division, Reston
USGS (1983) USGS circular 895-A Overview and USGS Activities; USGS Circular 895-B Digital Elevation Models (DEM); USGS Circular 895-C Digital Line Graphs (DLG) from 1:24,000-Scale Maps; USGS Circular 895-D Digital Line Graphs (DLG) from 1:2,000,000-Scale Maps; USGS Circular 895-E Land Use and Land Cover Digital Data; USGS Circular 895-F Geographic Names Information System USGS Circular 895-G Digital Line Graph (DLG) Attribute Coding Standards, U.S. Geological Survey (USGS), Reston
USGS (1987) US GeoData Data Users guide: (1) Digital line graphs from 1:24,000-scale maps; (2) Digital line graphs from 1:100,000-scale maps;(3) Digital line graphs from 1:2,000,000-scale maps; (4) Land use and land cover from 1:2,000,000-scale maps; (5) Digital elevation models; (6) Geographic names information system; (7) Alaska interim land cover mapping program, U.S. Geological Survey (USGS), Reston
USGS (1990) Land use and land cover digital data from 1:250,000- and 1:100,000-scale maps, data users guide 4. http://www.vterrain.org/Culture/LULC/Data_Users_Guide_4.html. Accessed 16 Aug, 2010
Varanka D (2010) The use of USGS digital geospatial data products for science research. In: Proceedings, 3rd International Symposium on the History of Cartography, University of Texas at Arlington, Arlington
Webster R (1980) The development of photoplotting in automated cartography. In: Proceedings, American Congress on Surveying and Mapping, 40th Annual Meeting, St. Louis pp 316–322

# Chapter 8
# The Use of U.S. Geological Survey Digital Geospatial Data Products for Science Research

**Dalia Varanka, Carol Deering, and Holly Caro**

**Abstract** The development of geographic information system (GIS) transformed the practice of geographic science research. The availability of low-cost, reliable data by the U.S. Geological Survey (USGS) supported the advance of GIS in the early stages of the transition to digital technology. To estimate the extent of the scientific use of USGS digital geospatial data products, a search of science literature databases yielded numbers of articles citing USGS products. Though this method requires careful consideration to avoid false positives, these citation numbers of three types of products (vector, land-use/land-cover, and elevation data) were graphed, and the frequency trends were examined. Trends indicated that the use of several, but not all, products increased with time. The use of some products declined and reasons for these declines are offered. To better understand how these data affected the design and outcomes of research projects, the study begins to build a context for the data by discussing digital cartographic research preceding the production of mass-produced products. The data distribution methods used various media for different system types and were supported by instructional material. The findings are an initial assessment of the affect of USGS products on GIS-enabled science research. A brief examination of the specific papers indicates that USGS data were used for science and GIS conceptual research, advanced education, and problem analysis and solution applications.

## Introduction

Geographic information systems (GIS) played a central role in transforming geographic research. How this happened and the implications of GIS on geographic science research is an important and broad topic for historical investigation. This

D. Varanka (✉) • C. Deering • H. Caro
U.S. Geological Survey, Reston, VA, USA
e-mail: dvaranka@usgs.gov

E. Liebenberg and I.J. Demhardt (eds.), *History of Cartography*, Lecture Notes in Geoinformation and Cartography 6,
DOI 10.1007/978-3-642-19088-9_8, 

paper begins to examine one area of GIS development, the role of U.S. Geological Survey (USGS) digital geospatial data products within the understanding of its effect on the broader field of geographic information science in the United States. Data are designed by institutions, and agencies like the USGS have communicated analytical practices, standards, and term definitions (Robinson 2008), but the meaning and effectiveness of data are found in their application and use.

As a result of the Office of Management and Budget Circular No. A-16 revision of 1967, the USGS was placed in a lead position for coordinating Federal geospatial policy and resources to be shared among agencies (Office of Management and Budget 1990; Shuler 2007). Concurrently, U.S. government policy regarding the creation and distribution of geographic data was intended to support the initial investment costs of new technology, allowing non-public entities to add value to resources and more widely distribute them through sales (Lopez 1998). Within this broader policy context, the USGS produced digital data products for the emerging GIS technological community. The mapping program also was intended to provide low-cost reliable data for the public.

The characteristics of these products developed in multiple contexts; early technological research driven by USGS objectives, but also by government and commercial GIS institutions through contracts with the USGS. They were used by scientists and educators to perform tasks and convey ideas. The products were technological practices that were developed for social (including economic), scientific, and political objectives and whose representational schemas are used and changed with time for various approaches involving geo-spatial practices (Kitchen and Dodge 2007).These data schemas and related technology were shaped by the conditions of their development and use, and whose characteristics and accessibility had implications for interested users. Similar objectives for such research were set by the GIS History project (Mark et al. 1996).

Few of the publications counted in this study were analyzed for the way that USGS products were used in the reported project. A quick examination of a random selection of papers indicates that key roles of the products were to meet needs for science measurement and visualization. Sometimes topographic maps were used for conceptual rather than empirical value. For example, a nested system of grids using the 7.5-minute quadrangle was used to create, store, and analyze data at the selected appropriate levels of scale within the Oak Ridge Regional Modeling Information System (Tomlinson et al. 1976). Though diverse maps were used in this project, the topographic quadrangle served as the framework for data integration.

The relative value of USGS topographic data products to the growth of the industry and technology is variably regarded (Coppock and Rhind 1991). The historiography of digital mapping and data technology has been regarded at times as creating the rationalization for its own existence. To attempt to evaluate this effect, this paper examines the evidence of the development, the resulting products, and access. The applications are examined through the citation of USGS products by science research literature from roughly 1975–1995. The sections that follow address early research on GIS and digital mapping and the technology that was produced by the USGS for these purposes, public access to the data, and the use of digital USGS data products in research, education, and application.

In this study, whether or not use increased is assessed, and reasons for the trend are suggested. This paper assumes a generally chronological sequence of events, despite expectations that the character of certain events will require a later re-examination of earlier affects. Another limitation of this historical narrative is that it is written from the internal perspective of USGS science culture and so may be vulnerable to certain self-rationalization and may lack a critical external viewpoint. Despite these weaknesses, this paper begins to organize preliminary historical information to contribute to a broader dialogue of the affect of digital geographic information on geographic science knowledge.

## Early Research and Development

The USGS began the transition from topographic maps to digital data by involvement with the broader GIS research community. USGS sponsored and participated in conferences such as AutoCarto, and sponsored workshops on GIS and automated mapping (Dutton 1978; Chrisman 2006). Research presented in international conference proceedings involved using USGS products. For example, an early shaded relief image derived from digital elevation data developed by the USGS was highlighted as display output for computerized cartographic models (Meyers and Voelker 1972). Topographic maps were used in research on manual and scanned graphic digitizing (Calkins 1978). Before computerized cartography, spatial data existed mostly in map form. Digitization was a technical challenge because computers required error-free, logical, digital data derived from analogue, graphical data (Tomlinson et al. 1976; Robinove 1986). Eventually, complex mapping systems were designed (Cowen and White 1989). Findings from these studies by the USGS and other researchers, helped shaped the form of products for public use.

States commonly played a role in product design and generation. National coverage of topographic maps formed the base map for an early production system, the Land Use and Natural Resources Inventory of New York State, and later for other systems employing manual digitizing (Tomlinson et al. 1976). The Federal Mapping Task Force, convened in 1972, was headed by Office of Management and Budget to play a coordination role. There was no agency with the responsibility to coordinate mapping among the federal agencies (Southard 1974). The data met established standards that were developed with the advice and support of professionals and the general public (Chrisman and Moellering 1983).

The USGS digital cartographic data products reviewed in this paper fall within three primary programs. The vector or line/graph program produced Digital Line Graph (DLG) data and a much later product called the National Hydrography Database (NHD).The Land Use Data Analysis (LUDA) program, based on the Geographic Information Retrieval and Analysis System (GIRAS), was remote sensing data converted to vector data. Data from the LUDA program was referred to as GIRAS for the vector data and Composite Theme Grid (CTG) for the raster data. Derivative data from the LUDA programs were the National Land Cover Dataset (NLCD). Digital Elevation Models (DEM), and later the National Elevation Database (NED), emerged from the production of Digital Orthophotoquad (DOQ) photography. Other innovations contributing to the development of GIS include the Aerial Profiling of Terrain System (APTS) or conversion tools such as the Spatial Data Transfer System (SDTS) and the General Cartographic Transformation Package (GCTP) (Usery and others 2009; Usery 2010; Cyran 1983). Most of these programs were organized and funded with contributions from partners.

## Data Accessibility to the General Public

The USGS aimed to assist the development of local and private applications from the digital data, but wide-spread use of the data required that institutional managers be convinced of the benefits of implementing such systems (U.S. Geological Survey 1986a). Free or low-cost data, such as the USGS provided, facilitated capital investment in GIS. Product innovations by the USGS were diffused to users and supported using various means. Initially, nine-track magnetic tapes were used to distribute the data. Later, free data downloads were available through Internet-based File Transfer Protocol (FTP) access in the 1980s. The data had users' manuals so that users understood the characteristics of the products (U.S. Geological Survey 1986b, 1987a, b, c). Eventually, distributed data networks led to the development of a national spatial data infrastructure (NSDI) coordinated by the USGS with spatial data infrastructure portals managed by local partners.

The use of data required the development of software training and worker education. Training and user manuals for GIS software included format descriptions and import options for digital USGS data (ESRI 1990). GIS classes and text books included discussions of USGS data (Domaratz et al. 1984; Antenucci et al. 1991; Star and Estes 1990).

The use of USGS digital cartographic data was supported through the mutual development of commercial products that built upon USGS results and were, in turn, implemented by the USGS through contracts and acquisition. United States policy in the 1980s required certain technical production tasks to be contracted to private industries. By contracting this work, the USGS helped build these industries in the 1980s. As agencies became interested in GIS support for local governance, these markets required larger scale data. Private companies such as GDT, founded by Don Cooke in 1980, modified public data on their own computers to serve geographically local and regional clients, or clients with specific business interests, such as banks. Etak Inc. conformed to the U.S. National Map Accuracy Standard for 1:24,000 and 1:100,000 scale maps (Tele Atlas North America 1999), as resellers of value-added data of 1990 U.S. Census Bureau Topologically Integrated Geographic Encoding and Referencing (TIGER) transportation files that were produced by the USGS. Information regarding the use of USGS data in commercial products was proprietary, however, and so was not easily available. The USGS purchased the software products built around GIS data, having at least three GIS systems at Earth Resources Observation and Science (EROS) Data Center by 1984. These were: Automated Geographic Information System/Generalized Balanced Ternary Record Access Manager (AGIS/GRAM) by Interactive Systems Corp. (ISC); ARC/INFO by Environmental Systems Research Institute (ESRI); and MOSS (Map Overlay and Statistical System) by the U.S. Fish and Wildlife Service, Bureau of Land Management, and Autometrics, Inc. In the 1990s, the USGS built private products through Cooperative Research and Development Agreements (CRADA). CRADAs allow licensing of exclusive development to a single reseller. Two examples are the Terra Server with Microsoft and Revision Production and Graphics (RevPG) with ESRI.

## Digital Data Products in Science Literature

An estimate of the extent of USGS digital cartographic data product use was assessed by graphing the number of science publications that cite the products from 1981 to 2008. Limitations of this data-collection method do not guarantee that every returned result for an acronym actually is the acronym in context of the full term of the search. All terms were searched in conjunction with "USGS" to try to reduce this margin of error. Results include the expanded names and acronyms because the full name and the acronym likely appeared in the publication.

Citations took various forms. Some of the references do not cite a data product, but make reference to a User's Guide or Instructions or Standards or Metadata. The assumption is that if an author is citing a guide or instructions or standards or metadata, he or she is using the data. Sometimes authors will mention data products in the title, abstract, full text, or keywords, but not cite those products in the reference list. Some authors may reference the product as online resources and may not include the title of the dataset, or the official USGS title.

A search of Google Scholar (Google) to retrieve general citations counts by year for ten products indicated a range of results. Citations of USGS products in the scientific literature about digital mapping and GIS began about the early 1980s and increased in the mid-1990s (Table 8.1), though the total number of citations increased over all (Fig. 8.1). Product headings in Fig. 8.1 are organized in roughly chronological order of production, appearing in column headings from left to right, beginning with DOQs (analogue DOQs were produced in the 1960s and digital DOQs appeared in the 1990s) to NED, a twenty-first-century product. Elevation data products, the DEM and NED, were cited regularly by 1980, and cited most often.

Citations dated after 2008 were excluded. The decline in a curve in the most recent years may reflect bibliographic indexing factors in addition to possible decreases of citations. A lag can occur in getting products published and/or available to search engines, and results could be low, even when adjusted, for this same

**Table 8.1** Citations of ten primary USGS products, by year

| | DOQ | DLG | DEM | GIRAS | CTG | GNIS | NLCD | DRG | NHD | NED |
|---|---|---|---|---|---|---|---|---|---|---|
| 1981 | 0 | 0 | 39 | 1 | 0 | 0 | 0 | 0 | 0 | 17 |
| 1982 | 0 | 3 | 25 | 5 | 0 | 1 | 0 | 0 | 0 | 11 |
| 1983 | 0 | 1 | 25 | 2 | 0 | 0 | 0 | 0 | 0 | 9 |
| 1984 | 0 | 1 | 36 | 4 | 0 | 1 | 0 | 1 | 0 | 16 |
| 1985 | 1 | 3 | 36 | 1 | 0 | 1 | 0 | 2 | 0 | 12 |
| 1986 | 0 | 6 | 39 | 4 | 2 | 3 | 0 | 0 | 0 | 15 |
| 1987 | 0 | 11 | 47 | 1 | 0 | 2 | 0 | 1 | 1 | 14 |
| 1988 | 0 | 10 | 40 | 0 | 0 | 0 | 0 | 1 | 1 | 6 |
| 1989 | 1 | 12 | 50 | 3 | 1 | 3 | 0 | 1 | 0 | 6 |
| 1990 | 1 | 11 | 68 | 2 | 0 | 3 | 0 | 1 | 2 | 10 |
| 1991 | 0 | 22 | 74 | 0 | 1 | 3 | 0 | 2 | 0 | 22 |
| 1992 | 2 | 18 | 69 | 4 | 2 | 4 | 0 | 1 | 0 | 27 |
| 1993 | 1 | 41 | 105 | 1 | 0 | 6 | 0 | 1 | 1 | 36 |
| 1994 | 0 | 37 | 128 | 1 | 2 | 5 | 0 | 3 | 0 | 34 |
| 1995 | 9 | 41 | 144 | 1 | 0 | 3 | 0 | 2 | 3 | 21 |
| 1996 | 7 | 51 | 193 | 7 | 3 | 3 | 1 | 6 | 1 | 27 |
| 1997 | 17 | 37 | 225 | 12 | 4 | 10 | 0 | 8 | 2 | 74 |
| 1998 | 10 | 52 | 289 | 13 | 3 | 9 | 0 | 14 | 5 | 131 |
| 1999 | 27 | 71 | 415 | 12 | 2 | 23 | 0 | 30 | 11 | 123 |
| 2000 | 31 | 67 | 508 | 14 | 39 | 47 | 2 | 37 | 18 | 166 |
| 2001 | 42 | 65 | 605 | 8 | 9 | 28 | 19 | 47 | 16 | 193 |
| 2002 | 49 | 79 | 784 | 14 | 7 | 30 | 57 | 77 | 34 | 306 |
| 2003 | 52 | 82 | 863 | 8 | 12 | 46 | 75 | 57 | 28 | 165 |
| 2004 | 62 | 70 | 977 | 14 | 16 | 41 | 104 | 73 | 46 | 165 |
| 2005 | 58 | 68 | 1,070 | 7 | 13 | 33 | 131 | 63 | 44 | 173 |
| 2006 | 67 | 51 | 1,140 | 19 | 11 | 33 | 143 | 58 | 63 | 191 |
| 2007 | 44 | 48 | 1,160 | 13 | 15 | 26 | 131 | 47 | 87 | 218 |
| 2008 | 26 | 28 | 1,080 | 7 | 10 | 29 | 159 | 79 | 103 | 186 |

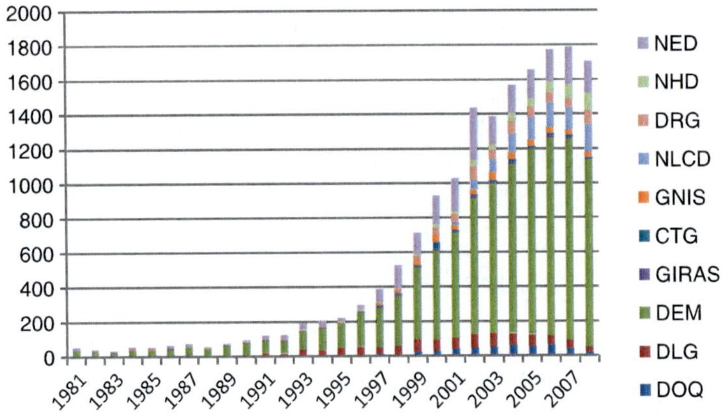


**Fig. 8.1** Trends in USGS product citations, by year

reason. Additional reasons for declines in citation in recent years are that once third-party vendors appropriate the USGS technology, a decline could be expected to appear in the USGS customer base. Though the term "Digital Elevation Model" referred to the original USGS product, its use entered the GIS vocabulary as a generic product (Chrisman 1997, 41). Also, a new design for the national topographic mapping was introduced in 2001 called *The National Map*. Though twentieth-century products continued to be used, their citation may appear under the name of this new program.

The citation data in the following sections is drawn from several research databases in addition to Google Scholar. These include Science Direct (Elsevier B V a), ISI Web of Knowledge (Thomson Reuters), and SCOPUS (Elsevier B V b).

## *DLG and NHD*

DLG data were digital topographic data in a vector data model. Various versions of digital data were attempted; the data that were widely distributed were called DLG-3. For this version, the databases offered everything that could be collected and that was already on the maps. Revision procedures called for features to be collected from air photos. This meant that greater reference needed to be made to feature definitions and instructions, supporting the development of standards that facilitated the use and interoperability of these data with other systems.

The term "Digital Line Graphs" had 467 references; of those, 71 were USGS authored/published. Of the remaining 396, 241 were confirmed as having the relevant search words in the title, abstract, keywords, references, and/or full-text; 91 could be confirmed; 56 were not confirmed because to do so would have required searching full-text via Inter Library Loan. Eight were Web pages pulled from the GeoRef (American Geological Institute) database. After confirmation of some search results using Google Scholar, the final confirmed body of work was a bibliography of 305 references, all confirmed to include "Digital Line Graph" or DLG in the bibliographic information or in the full-text. Though the USGS publishes science findings of its own research projects, USGS publications were not counted, since the objective of this study was to trace the diffusion of information to the public. Internal applications of USGS digital cartographic data product constitutes a corpus of work in addition to those discussed in this paper.

The citations of DLGs increased slowly in the 1980s, but quickly between 1991 and 2005.This may be generally because of the rise of GIS on microcomputers in the 1990s (Fig. 8.2). DLGs continued to be used into the twenty-first century (Clapham 2005), even after the data currency was no longer maintained. A feature-based data model version of DLG called DLG-F was designed and implemented for the NHD database. Citations of those data notably rise after 2003 (Fig. 8.3).

Random samples from the citations indicate physical science applications for the data included climate studies (Carbone and James 1993), aiding production estimates for the U.S. Department of Agriculture (Cheng et al. 1991), and land-use

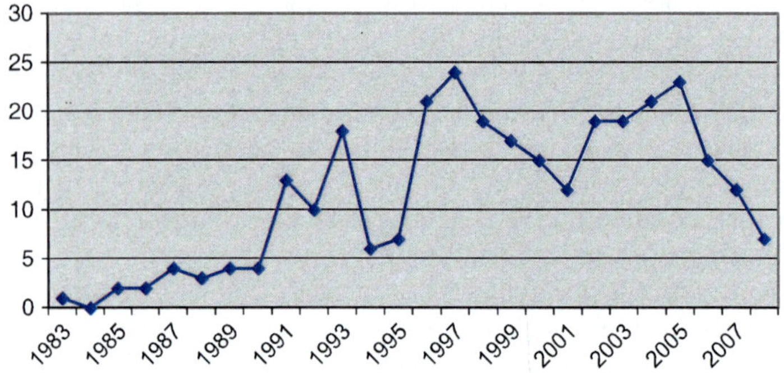


**Fig. 8.2** Citations of DLG data in science literature from 1983 to 2007

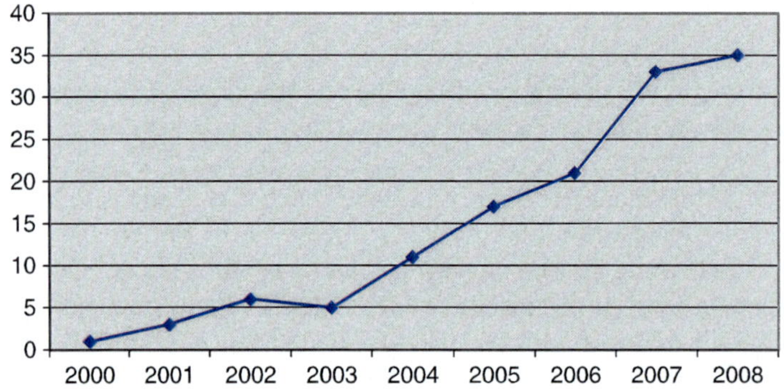


**Fig. 8.3** Citations of NHD data in science literature 2000–2008

planning (Coleman 1992; Cowen and Shirley 1991). Educational uses of the data involved the development of earth science education at the university level (Abolins 1997). As digital data, the use of DLGs in research aided the improvement of geospatial methods and techniques. DLGs served as subject data for studies of advancing generalization methods (Nickerson and Freeman 1986). When used together with remote sensing techniques, a strong area of DLG applications were in data conflation or data merging (Bernstein and Hanson 1985; Civco and Hurd 1991), where they may have also aided visualization and interpretation (Capraro et al. 1997).

## *DEM, DOQ, NED*

DEMs were the most widely used digital product produced by the USGS in science research. Its use steadily increased in the period between 1979 and 2008 (Fig. 8.4).

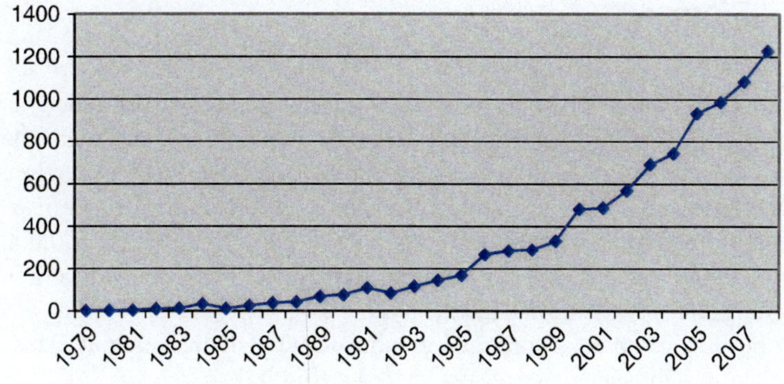


**Fig. 8.4** Citations of DEM in science literature 1979–2007

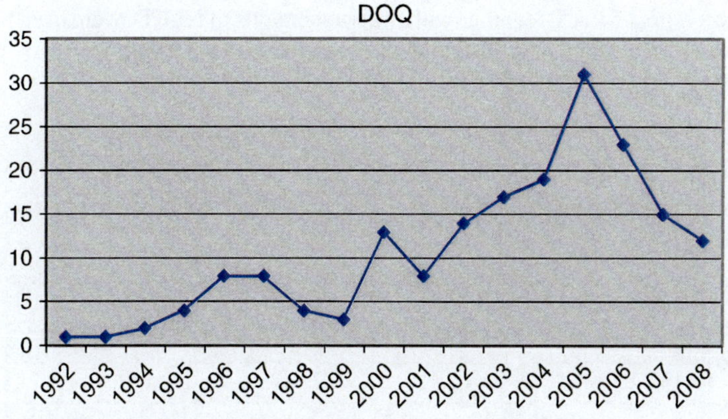


**Fig. 8.5** Citations of DOQ in science literature 1992–2008

Citations of digital DOQs sharply increase after 1990, but decline after 2005 (Fig. 8.5). One reason for this decline may be that Digital Elevation Models originally introduced with a USGS name assimilated into the discipline so that DOQ became the commonly-used term in general for the particular technology. This would not be reflected in the counts because the search was for USGS affiliation. Other reasons for this decline may be attributed to the rise of high-resolution commercial imagery, such as those produced by the IKONOS or Quickbird satellites, launched in 2000, with 1 m resolution, and a rapid revisit rate, or that the U.S. Department of Agriculture's National Agriculture Imagery Program replaced the USGS coverage through the National Digital Orthophoto Program. Similarly, citations of NED data level off, indicating a leveling curve, perhaps because of increases in the use of Lidar data (Fig. 8.6).

## *LUDA and NLCD*

The Land Use Data and Analysis (LUDA) program was first funded in 1974 (Mitchell 1974). The original objectives were to map land use and four other data categories for the entire United States (Federal and State-land ownership, river basins and subbasins, counties, and census county subdivisions) for 5 or 6 years. The primary digital maps would consist of 635 sheets at a scale of 1:250,000, with sheets at 1:50,000 and 1:25,000 for areas of more rapid change and concern in land-use planning and resource management. The data were digitized from analogue maps to build a database to produce graphical and statistical output. The use of GIRAS data in published science was sporadic (Fig. 8.7).

In contrast to GIRAS/LUDA land-cover data, the use of NLCD data increased steadily since its initial release in 1992 and a revised version in 2000 (Fig. 8.8). GIRAS/LUDA was raster data from aerial photography that was vectorized manually, in contrast to NLCD, which was obtained from satellite imagery and vectorized in remote sensing software. GIRAS, together with the two versions of NLCD, were used together for land-use change studies, a discipline that requires historical data for trend analysis.

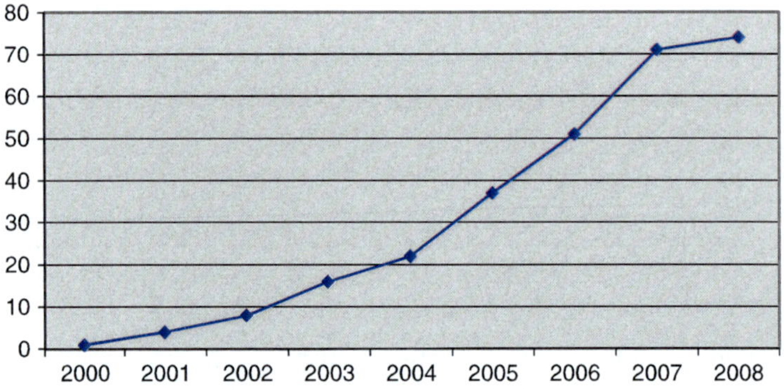


**Fig. 8.6** Citations of NED in science publications

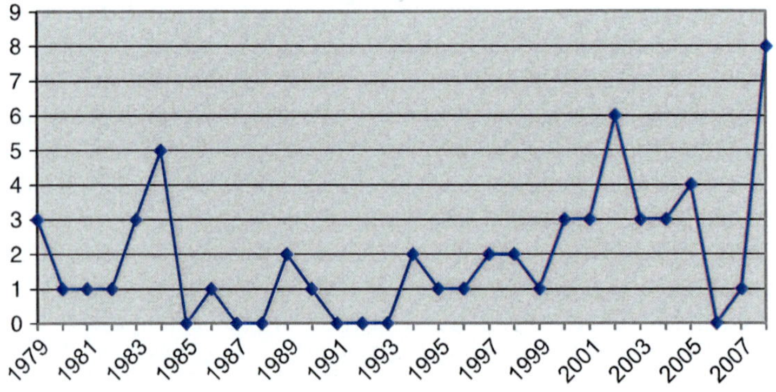


**Fig. 8.7** Citations of GIRAS data in science literature

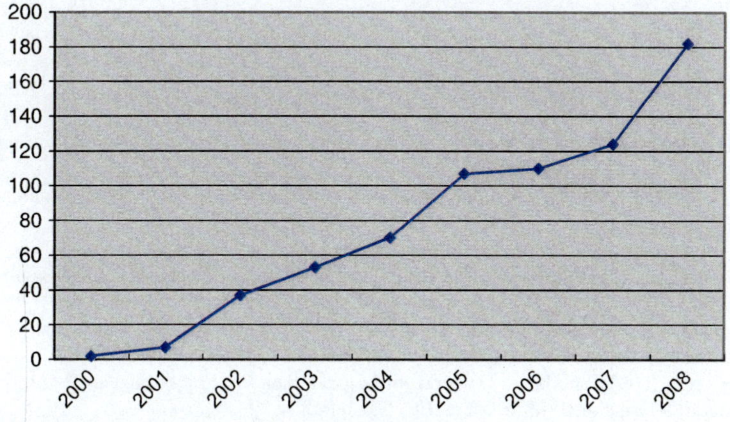


**Fig. 8.8** Citations of NLCD in science literature 2000–2008

## Summary

The USGS introduced several digital geospatial data products during the rise of GIS in the United States. Graphs of the numbers of product citations in science literature from on-line databases indicated that the use of most USGS products increased with time. Citations of historical data used for temporal change analysis varied, as did the use of technology that was common for a time, but may have become outdated because of data currency or competing technological innovation. The use of these products may be attributable to their low cost, which was important in the context of initial capital investments required to implement new GIS projects, their systematic coverage, and controlled quality. "In value of annual sales, the GIS data industry is much more significant than the software industry" (Longley et al. 2001, 25). An advantage of offering national coverage was that different geographical areas of the country could equitably access some data regardless of local wealth or profit motive. Though the factors affecting the effect of a national topographic mapping program to the United States public are complex, preliminary evidence indicates that the public considered USGS products to support science research.

## Biographical Note

**Dalia Varanka**, is a Research Geographer with the U.S. Geological Survey. She received her B.A. degree from the University of Wisconsin-Green Bay in Regional and Urban Analysis (1978), her M.A. degree in Geography at the University of Illinois at Chicago (1987), and her Ph.D. degree in Geography from the University of Wisconsin-Milwaukee (1994). Dr. Varanka began her Federal career with the Bureau of Land Management in 1993 and has been with the U.S. Geological Survey since 1997.

## References

Abolins MJ (1997) Using free digital data to introduce volcanic hazards. J Geosci Educ 45 (3):211–215

American Geological Institute. GeoRef information services. American Geological Institute. http://www.agiweb.org/georef/

Antenucci JC et al (1991) Geographic information systems, a guide to the technology. Van Nostrand, New York

Bernstein R, Hanson WA (1985).Advances in landsat image processing and mapping. Vol. 550 of Sensor design using computer tools II. SPIE-The International Society for Optical Engineering, Arlington, 1985

Calkins HW (1978). Creating large digital data files from mapped data. In proceedings, UNESCO Conference on Computer Mapping of Natural Resources, Mexico City, 1978

Capraro G et al (1997) Intelligent visualization of radar data. In: Proceedings of the 1997 Radar Edinburgh International Conference, IEE, Edinburgh

Carbone G, James LA (1993) Applying GIS technology to climate data in South Carolina. Geo Info Syst 3(9):52–59

Cheng TD et al (1991) Compiling and editing agricultural strata boundaries with remotely sensed imagery and map attribute data using graphics workstations. In: 1991 International Geoscience and Remote Sensing Symposium – IGARSS'91, IEEE, Espoo, 1991

Chrisman N (1997) Exploring geographic information systems. Wiley, New York

Chrisman N (2006) Charting the unknown, how computer mapping at Harvard became GIS. ESRI Press, Redlands

Chrisman NR, Moellering H (1983) Issues involved in the development of national digital cartographic data standards. In: Proceedings 1983 Harvard Computer Graphics Conference, Harvard University Laboratory for Computer Graphics and Spatial Analysis, vol 1. Cambridge, 1983, pp 1–10

Civco DL, Hurd JD (1991) Multitemporal, multisource land cover mapping for the state of Connecticut. In: 1991 Proceeding ACSM-ASPRS Fall Convention, Atlanta, 1991

Clapham WB Jr (2005) Quantitative classification as a tool to show change in an urbanizing watershed. Int J Remote Sens 26:4923–4939

Coleman TL (1992) Three-dimensional modeling of an image-based GIS to aid land use planning. Geocarto Int 7(4):47–53

Coppock JT, Rhind DW (1991) The history of GIS. In: Maguire DJ, Goodchild MF, Rhind DW (eds) Geographical information systems: principles and applications, 1st edn. Longmans Publishers, London, pp 21–43

Cowen DJ, Shirley WL (1991) Integrated planning information systems. In: David J, Maguire Michael, Goodchild F, David WR (eds) Geographical information systems: principles and applications, 2nd edn. Longman Scientific & Technical, Essex, pp 297–310

Cowen DJ, White TR (1989) A versatile mapping system for the USGS 1:100 000 DLGs. In: Autocarto 9. Proceedings Symposium, Baltimore, 1989, pp 705–714

Cyran EJ (1983) Aerial profiling of terrain system. J Surv Eng 109(2):136–150

Domaratz MA et al (1984) USGS digital cartographic data standards, Digital Line Graphs from 1:2,000,000-scale maps. In: Marble DF, Calkins HW, Peuquet DJ (eds) Basic readings in geographic information systems. SPAD Systems, Ltd, New York

Dutton G (ed) (1978) First international advanced study symposium on topological data structures for geographic information systems, volume one – summary of proceedings. Laboratory for Computer Graphics and Spatial Analysis, Cambridge, 1978

Elsevier BV. a Science Direct. Elsevier BV http://www.sciencedirect.com/

Elsevier BVb SCOPUS. Elsevier BV http://www.info.scopus.com/

ESRI (1990) Understanding GIS, the ARC/INFO method. Environmental Systems Research Institute, Inc, Redlands

Google (2009) Google Scholar (beta). Google Inc. http://scholar.google.com/schhp?hl=en&tab=ws

Kitchen R, Dodge M (2007) Rethinking maps. Prog Hum Geogr 31:331–344

Longley PA et al (2001) Geographic information systems and science. Wiley, New York

Lopez X (1998) The dissemination of spatial data, a North American-European comparative study on the impact of government information policy.. Ablex Publishing Corporation, Greenwich

Mark DM et al (1996) The GIS history project. National Center for Geographic Information and Analysis. http://www.ncgia.buffalo.edu/gishist/bar_harbor.html

Meyers Jr CR, Voelker AH (1972) Introduction to computer display applications using spatial data. In geographical data handling, Tomlinson RF (ed) 891–917. International Geographical Union Commission on Geographical Data Sensing and Processing for the UNESCO/IGU Second Symposium on Geographical Information Systems, vol 2. Ottawa, August 1972

Mitchell WB (1974) Geographic information systems panel. Auto-Carto I, International Conference on Automation in Cartography, December 9–12, 1974, Reston. http://mapcontext.com/autocarto/proceedings/auto-carto-1/index.html

Nickerson BG, Freeman H (1986) Development of a rule-based system for automatic map generalization. In: Proceedings of the Second International Symposium on Spatial Data Handling held in Seattle, Washington, 5–10 July 1986. International Geographical Union Commission on Geographical Data Sensing and Processing, Williamsville, 5–10 July 1986

Office of Management and Budget (1990) Circular A-16: coordination of surveying, mapping, and related spatial data activities. OMB, Washington DC. http://www.whitehouse.gov/omb/circulars/a016/a016.html

Robinove CJ (1986) Principles of logic and the use of digital geographic information systems, U.S. geological survey circular 977. Washington DC

Robinson M (2008) A history of spatial data coordination. Federal Geographic Data Committee. http://www.fgdc.gov/ngac/a-history-of-spatial-data-coordination.pdf

Shuler JA (2007) Brief history of spatial information policies in the United States. Cartography Geogr Inf Sci 34(2):137–148

Southard R (1974) Governmental implications of automation panel. In Auto-Carto I, International Conference on Automation in Cartography, December 9–12, 1974, Reston. http://mapcontext.com/autocarto/proceedings/auto-carto-1/index.html

Star J, Estes J (1990) Geographic information systems, an introduction. Prentice Hall, Englewood Cliffs

Tele Atlas North America (1999)"Etak enhances its digital map databases nationwide to extend functionality" Press release, June 07, 1999. Directions Magazine. http://www.directionsmag.com/press.releases/index.php?duty=Show&id=599&trv=1

Thomson Reuters. ISI web of knowledge. Thomsom Reuters. http://www.isiknowledge.com

Tomlinson RF, Calkins HW, Marble DF (1976) Computer handling of geographic data, an examination of selected geographic information systems. The UNESCO Press, Paris

U.S. Geological survey (1986a) Large-scale mapping guidelines, Open File Report 86–005

U.S. Geological Survey (1986b) Land-use and Land-cover digital data from 1:250,000- and 1:100,000-scale maps: United States department of the interior U.S. Geological Survey data users guide 4, Reston

U.S. Geological Survey (1987a) Digital Line Graphs from 1:2,000,000-scale maps: United States Department of the Interior U.S. Geological Survey data users guide 3, Reston

U.S. Geological Survey (1987b) Digital Elevation Models, United States Department of the Interior U.S. Geological Survey data users guide 5, Reston

U.S. Geological Survey (1987c) Geographic names information system: United States Department of the Interior U.S. Geological Survey data users guide 6, Reston, Virginia

Usery EL (2010) The digital transition in cartography: USGS data innovations, 1970s. In: Proceedings, 3rd International Symposium on the History of Cartography, University of Texas at Arlington, Arlington

Usery EL, Varanka D, Finn MP (2009) From the dawn of digital to the national map, 125 years of topographic mapping, Part 2, in ArcNews 31(4):39. http://www.esri.com/news/arcnews/winter0910articles/125-years.html

# Part IV
# New Mexico

# Chapter 9
# Colonizing Chaco Canyon: Mapping Antiquity in the US Southwest

**Berenika Byszewski**

**Abstract** In 1849, following victory in the Mexican War, the United States set out to assert its newfound sovereignty over the Navajo Indians in the uncharted territories of the Southwest. The map and associated journal from the expedition has been regarded by historians as the first sustained window into the region and its people. A major focus of the mapping effort was the discovery and interpretation of ancient ruins. In this paper, I follow the cartographer Lieutenant James H. Simpson's survey party into the Navajo borderlands, tracing their discovery and mapping of the monumental ruins of Chaco Canyon. Through an analysis of Simpson's map and journal, I suggest that the mapping effort marked Chaco as a site of knowledge production and part of a new geography of antiquity. This mapping constructed Chaco as a resource for new national imaginaries and laid the foundation for increasing forms of protection that reproduce the logic of Native removal.

## Introduction

On August 14, 1849, just as he arrived in Santa Fe after mapping a wagon route there from Fort Smith, Arkansas, Lieutenant James H. Simpson of the US Topographical Corps of Engineers was ordered to accompany a punitive expedition into the Navajo borderlands, to "make such a survey of the country as the movements of the troops would permit" (Simpson 1852). This was the first official US mapping of this part of the Southwest after the Mexican Cession, and has been recognized by historians as the first sustained window into the region and its people (e.g. Goetzmann 1966). In this paper, I follow Lieutenant Simpson's survey party, tracing their discovery and mapping of the ancient ruins of Chaco Canyon.

B. Byszewski (✉)
Department of American Studies, University of New Mexico, Albuquerque, NM, USA
e-mail: berenika@unm.edu

E. Liebenberg and I.J. Demhardt (eds.), *History of Cartography*,
Lecture Notes in Geoinformation and Cartography 6,
DOI 10.1007/978-3-642-19088-9_9, 

Through an analysis of Simpson's map and journal, I suggest that the mapping effort served to fix Chaco as a site of knowledge production and part of a new geography of antiquity. This mapping constructed Chaco as a resource for new national imaginaries, fixing the ruins in prehistory and the moment of their discovery, and laid the foundation for increasing forms of protection that continue to reproduce the logic of Native erasure and dispossession.

While previous studies have examined the exploration and mapping of the US Southwest (e.g. Reinhartz and Saxon 2005), few have considered how the mapping of antiquity worked in the development of the nation. Chaco Canyon – a unique complex of monumental stone structures inhabited approximately 1,000 years ago – is perhaps the most extensively mapped archaeological site in the US, and a site constructed as a metaphorical "blank spot on the map." It has been variously mapped and imagined as a prehistoric empire, an historic homestead, a cradle of Southwestern archaeology, a world heritage site, and an archaeo-astronomical complex. Currently managed as a national park, is also an ancestral centerplace for two dozen Native American groups, and the historical homeland of a displaced Navajo community. As such, it is an important site in which to investigate the ways mapping helps produce particular meanings of the past emerging from a politics of the present.

In this summary, I outline some key elements in my analysis of the discovery and mapping of Chaco in the context of the Navajo Expedition. I begin with a consideration of the unstable knowledge of the Navajo borderlands that preceded Simpson's survey as it was translated on maps of the region. Then I focus on the fixing of the Chaco ruins through the technology of the traverse map. In both cases, I consider how the mapping of antiquity served to reproduce forms of Native removal from Chaco.

## Unstable Knowledge of Navajo Borderlands

Rather than a blank spot on the map, the world Simpson "discovered" had been variously mapped by Spanish conquistadors and missionaries, naturalists, and international cartographers. A review of the 'Navajo borderlands' (Brooks 2001) depicted on earlier maps reveals an unstable knowledge of the area and location of Chaco: place–names, topography, settlements and whole regions shifted and morphed, disappeared and reappeared between maps, not unlike the semi-nomadic people they were trying to convey.

These maps also reveal that myths, legends, and ruins became important sites to place on a map, even if their whereabouts were completely unknown. For example, in Nicolas Sanson's 1650 map of North America (Fig. 9.1), Cibola, the fabled cities of gold, floats in the white space outside the edge of empire. In Alexander von Humboldt's well-known map of New Spain (1804, published 1811), he places Atzlán – the lost city of the Aztecs – at roughly 36° latitude and between 111° and 113° longitude. This vague location of the mythical city was incorporated into a

**Fig. 9.1** *Amerique Septentrionale*, Nicholas Sanson, 1650

popular 1849 commercial map of the United States published by J.H. Colton (Fig. 9.2). This is the map Simpson refers to in his journal and likely carried with him on the Navajo Expedition.

The salience of ruins as part of the geographical imaginary is also evident in the cartouche that decorates Colton's map (Fig. 9.3). The cartouche is suffused with the timeless quality of nature and antiquity, where soldiers and Indians gaze out over a placid lake from which emerges a huge rock in the shape of an Egyptian pyramid. This image is at once utopian and nostalgic, a culmination of Manifest Destiny where the natural and cultural worlds blend into each other. Taken together, these maps underscore the ways geographical knowledge floats between maps and across empires, and shows that myths and ruins were already a part of spatial discourse of empire and nation.

Many scholars have considered the hidden indigenous knowledge inherent in such maps (e.g. Pickles 2005; Pratt 1992). In addition to having knowledge of the terrain, multiple Native groups – including the Utes and Navajos, and Pueblos such as Hopi, Acoma, Zuni, Jemez, and others – traced various connections to landscapes and ruins in this region. Expressed through migration stories, ceremonialism, and pilgrimage, these sacred geographies rarely emerge in colonial cartographic practices. The incommensurability of sacred geographies and imperial mapping techniques reproduces a form of Native removal: the exclusion of non-visual attachments to places in the representation of Southwestern landscapes.

**Fig. 9.2** Detail of *Map of the U.S.*, J.H. Colton, 1849, showing the approximate location of Chaco

**Fig. 9.3** Detail depicting the cartouche of the *Map of the U.S.*, J.H. Colton, 1849

## Fixing Chaco Through the Traverse Map

The map and associated journal produced during the Navajo Expedition is not silent about the reasons for entering Navajo country; the map charts a bold red line through new terrain, identifies potential resources, and claims the landscape (Fig. 9.4). In the afterglow of victory in the Mexican War, the US mounted what equated to a "shock-and-awe" campaign, amassing about 400 men, including Mexican and Pueblo Indian militias. The expedition circled the Navajo stronghold

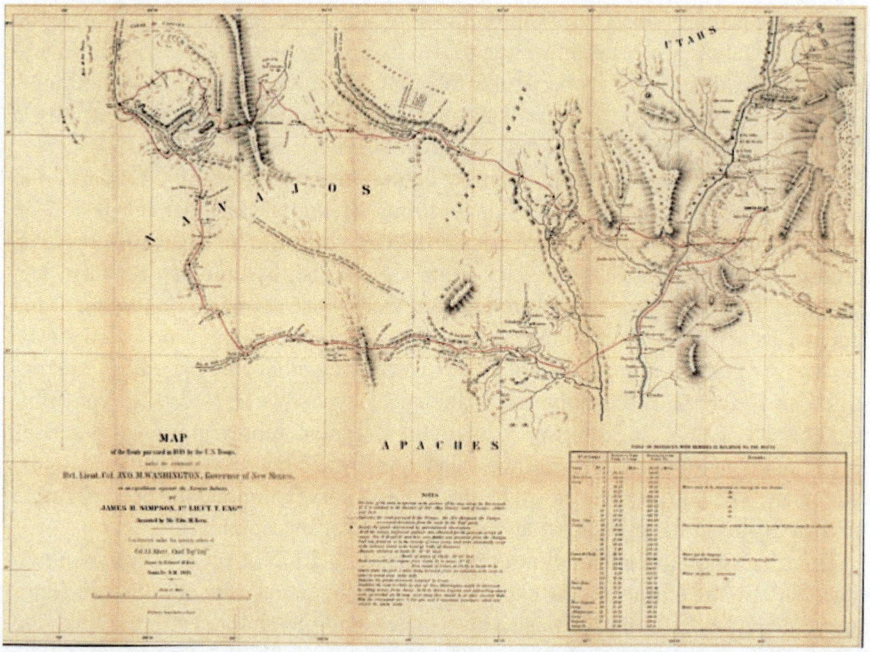


**Fig. 9.4** Map of the Route pursued in 1849 by the U.S. Troops, under the command of Bvt. Lieut. Col. Jno. M. Washington, Governor of New Mexico, in an expedition against the Navajos Indians, Lieutenant Simpson, 1849; Drawn by E. Kearn, 1849

of Cañon de Chelly – the "ultima thule" of the expedition – with the goal of signing a comprehensive treaty with the tribe. One of the stipulations claimed the right of the US to have their "boundaries fixed and marked, so as to prevent any misunderstanding on this point between them and their neighbors" (Simpson 1852; 55). Thus, the map, like the treaty, can be seen as laying the groundwork for future boundary making.

Unlike the maps of Humboldt and Colton, which tried to fill in the blanks of continent and nation, Simpson's map is partial and incomplete, leaving large spaces on the map blank. This representation emerges from the techniques of traverse mapping and is similar to other soldier–engineer surveys of the Southwest.[1] While this partial view revealed the limits of geographical knowledge, it also underscored the increased truth and accuracy of the observational approach. Simpson is explicit about the importance of first-hand knowledge, and is careful to point out when he

[1]See Emory's map (William H. Emory, US GPO: *Military Reconnaissance of the Arkansas Rio Del Norte and Rio Gila By W.H. Emory, Lieut. Top. Engrs....* 1847) for a similar traverse map, and Abert's map (James W. Abert and William G. Peck. *Map of the Territory of New Mexico... 1846–7*. Washington, D.C.: Senate Ex. Doc. 23, 1848) for a contrasting example that fills in the spaces around the survey.

was not the primary observer, both on the map and in the journal. For example, on the map, he labels an uncertain wagon route as being "said to exist, having a general direction like this, but of its particular location and character US knows nothing." Thus, the mapping effort began a Chaco imaginary based on observable truths; a lost city whose treasure was knowledge and not gold.

Although the map did not claim a comprehensive view of landscape, the accompanying journal integrates many forms of knowledge that suggested an exhaustive reconnaissance of the route. In the time prior to specialist surveys, Simpson also functioned as the expedition's geologist, hydrologist, botanist, and ethnographer, while the Kern brothers created illustrations of landscapes and Indians. In her analysis of Napoleon's *Description de l'Egypt*, Anna Godlewska emphasizes that the map, image, and text worked together ideologically to reconstruct an "eternal and immutable" ancient Egypt that replaced Egypt itself (Godlewska 1995). In Simpson's mapping, similar representational practices carved out a window on the Southwest that was framed by a certain settler colonial rationality. The partial view traced by the traverse survey implied the need of future surveys to add to the growing body of knowledge of the Southwest.

The traverse map embodied not only an accumulation of observational knowledge, but also the journey of the exploration, and movement over new terrain. Notes on available fodder, camp sites, alternate routes, and productive lands are linked to spots on the map through the journal, and assume a repeatability of the route by both future military engagements and potential settlers. Along the way, Simpson marked resources important to this settler colonial logic: settlements, minerals, water holes, and *ruins*. In this context, ruins became a particularly potent resource, an arena in which to create and recreate attachments to nation and region.

It is in this context of journey and repeatability that Simpson relates the discovery of Chaco. The movement of the small exploratory party that detached from the regiment is captured on the map as a faint dashed line that makes its way through the canyon, ruin by ruin (Fig. 9.5). The major ruins are individually marked and numbered on the map in the order they were discovered. In addition, this approach marks Chaco as a region, as opposed to a mere point on a map, defined by named topographic features including Cañon de Chaco, Rio de Chaco, and Mesa Fechada. The mapping of the ruins included Romantic sketches of landscapes, plan view drawings composed of an orderly repetition of room shapes and sizes, and architectural drawings of masonry styles. The combination of the observational approach with the journey of discovery frame Chaco as a place to be revisited and rediscovered.

But as a whole, the dominating and teleological progress of the line orders the underlying instability of the expedition, and the map serves to obscure the utter dependence on Native and Mexican guides, translators, scouts, and militia to the movements and survival of the US military. Rather than a single line of movement, the expedition took multiple paths, gathered militias en route, and suffered numerous defections. Scouts moved forward and back, and lookouts often mistook their own troops for Navajo ambushes, firing more than once at their own men. The excursion through Chaco Canyon was similarly fraught with misunderstandings that left the survey party fragmented and lost for an extra night. While at the first ruin, Simpson

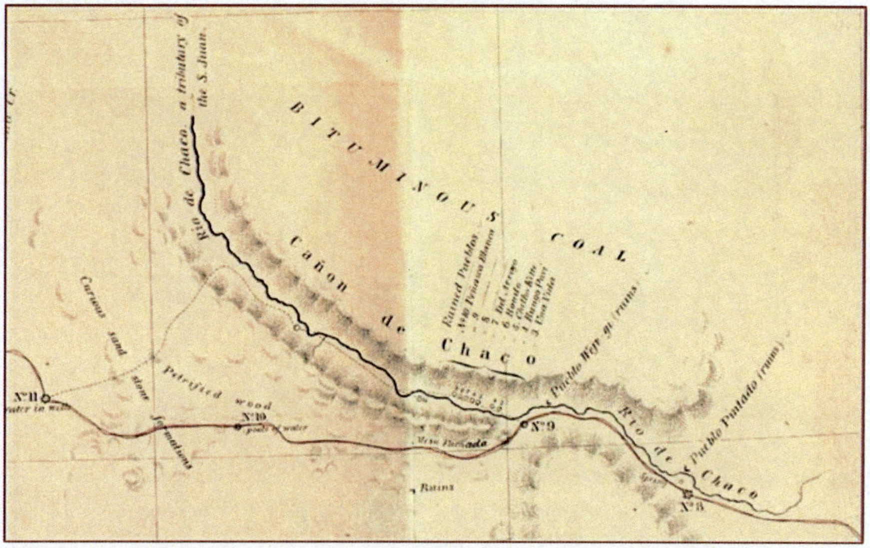


**Fig. 9.5** Detail of Chaco Canyon, *Map of the route pursued in 1849*, E. Kearn, 1849

recounts half a dozen Indian and Spanish names for it from half a dozen people, but settles on Jemez chief Hosta's name, Pueblo Pintado. Simpson uses Hosta as a guide for the remainder of the canyon, while the others, including Navajo Chief Sandoval, return to the regiment (Simpson 1852). In this way, the map serves to erase the complex power relationships between diverse groups, as well as various forms of geographical and sacred knowledge that intersect and compete in places like Chaco.

## Conclusion

Although Simpson frames this fascination with Southwestern antiquity as a side show to the main objective of documenting the expedition, it fits squarely with his larger task of making the natural and human landscapes legible to a new audience. While the social and nomadic landscapes of the Navajo Indians were more difficult to "fix" through the tools and text of the mapping effort, the ruins of Chaco Canyon were marked as an important site to explore questions of civilization, science, and a national past. Unlike Simpson's wagon route maps, which were addressed specifically to California emigrants, the Navajo Expedition journal was aimed at diverse east-coast publics, including soldier, scientist, and "lover of nature." The journal was first published by the US Congress in 1850, as part of the Reports of the Secretary of War, with 3,000 copies printed, 300 of which were reserved for the Topographical Bureau (US Congress 1850). It was later reprinted by Lippincott,

**Fig. 9.6** Northwest view of the ruins of the Pueblo Pintado, Richard Kern, Navajo Expedition Journal (Copyright for this image by The Ewell Sale Stewart Library, Academy of Natural Sciences of Philadelphia)

Grambo and Co. in 1852, and was consumed alongside travel literature, soldier diaries, etiquette manuals, and medical books (Goetzmann 1966).

At Chaco Canyon, Simpson began an origin story of the nation, residing in the antiquity of a civilized past and its discovery. When his Native guides brought him to Pueblo Pintado (Fig. 9.6), Simpson found the ruin to more than meet his expectations, describing the masonry as exhibiting "a combination of science and art which can only be referred to a higher stage of civilization and refinement than is discoverable in the works of Mexicans or Pueblos of the present day," (Simpson 1851; 34). At the bottom of this evolutionary scale of architecture were the "huts" of the semi-nomadic Navajos, who Simpson suggested might be descendents of the civilized Chacoans that subsequently regressed to a semi-nomadic state.

Since the initial mapping, Chaco has been a productive and ever-expanding site for varied ideas of antiquity, civilization, and science, and deemed worthy of increasing forms of national protection and control. It is telling that Chaco Canyon became one of the first National Monuments in the country – established in 1907, 5 years before New Mexico became a state. In the 1930s and 1940s, the remaining Navajos within the national park's 34,000 acres were forced to relocate, while the cultural landscape was remade to reflect the time of Simpson's discovery. Currently located at the powerful nexus of science, state institutions, and government preservation, Chaco can be read as an evolving site of intellectual and physical conquest. Through the mapping effort, alternate attachments to place were eclipsed by the new geography of antiquity which circumscribed boundaries around the knowledge of the past and its protection.

## Biographical Note

**Berenika Byszewski** is a practicing archaeologist and environmental planner in New Mexico. She is currently pursuing her PhD in American Studies at the University of New Mexico. Her research interests include historical geography, critical cartography, and the cultural politics of archaeology in the US Southwest.

## References

Brooks J (2001) Captives and cousins: slavery, kinship, and community in the Southwest Borderlands. University of North Carolina Press, Chapel Hill

Goetzmann W (1966) Exploration and empire: the explorer and the scientist in the winning of the American West. Alfred A Knopf, New York

Godlewska A (1995) Map, text and image: the mentality of enlightened conquerors: a new look at the description de l'Egypte. Trans Inst Br Geogr 20:5–28

Pickles J (2005) A history of spaces: cartographic reason, mapping and the geo-coded world. Routledge, London

Pratt M-L (1992) Imperial eyes, travel writing and transculturation. Routledge, London

Reinhartz D, Saxon GD (eds) (2005) Mapping and empire: soldier engineers on the Southwestern Frontier. University of Texas Press, Austin

Simpson JH (1852) Journal of a military reconnaissance from Santa Fe, New Mexico, to the Navajo country, made with the troops under command of Brevet Lieutenant Colonel John M. Washington, chief of ninth military department, and governor of New Mexico, in 1849. Lippincott, Grambo and Company, Philadelphia

US Congress, War Department (1850) Reports of the secretary of war... with the report of Liet. J. H. Simpson of an expedition in to the Navajo Country. 31st Congress, 1st session, Ex. Doc. 64. Union Office, Washington DC

# Chapter 10
# Bounding a Sacred Space: Mapping the Mt. Taylor Traditional Cultural Property

**Peggy L. Allison**

In the summer of 2006, a group representing Native American tribes from all over the southwestern United States came together with agents of the U.S. Forest Service and created a boundary to delineate a traditional and sacred space. This boundary designation, known as a traditional cultural property or TCP, was created in reaction to the attempted exploitation of the area for uranium mining, an issue that has plagued the region since the mid-1940s. This contested landscape is the peak and mesas collectively known as Mt. Taylor, located in northwestern New Mexico, which sits atop the nation's largest single deposit of high-grade uranium. Aside from its economic value, however, the mountain is also a cultural landscape to which the identity of many tribes is irrevocably linked, through history, tradition, and belief.

But how does one go about creating a boundary around such a sacred geography, which has no physical delineations, only perceived ones? In the case of Mt. Taylor, we are dealing with a landscape that holds many different meanings for many different groups, both indigenous and otherwise. The U.S. Forest Service, the ultimate creator of the boundary, has only been working with TCPs for a relatively short period of time and must deal with the inevitable problems that come with implementing a program still in its infancy. Their expressed goal, however, was to incorporate indigenous knowledge to the highest degree possible (Benedict and Hudson 2008), a goal shared by the closely related research fields of counter-mapping and participatory GIS, or PGIS (Hodgson and Schroeder 2002; Sletto 2009; Chapin 2005). The question then is how successfully the incorporation of this indigenous knowledge was in the creation of the TCP boundary and, in a larger sense, how much success have similar projects experienced in the same realm? Can a complex juxtaposition of indigenous beliefs and United States government bureaucracy result in a successful participatory mapping process?

P.L. Allison (✉)
Department of Geography, University of New Mexico, Albuquerque, NM, USA
e-mail: pallison@unm.edu

E. Liebenberg and I.J. Demhardt (eds.), *History of Cartography*,
Lecture Notes in Geoinformation and Cartography 6,
DOI 10.1007/978-3-642-19088-9_10, 

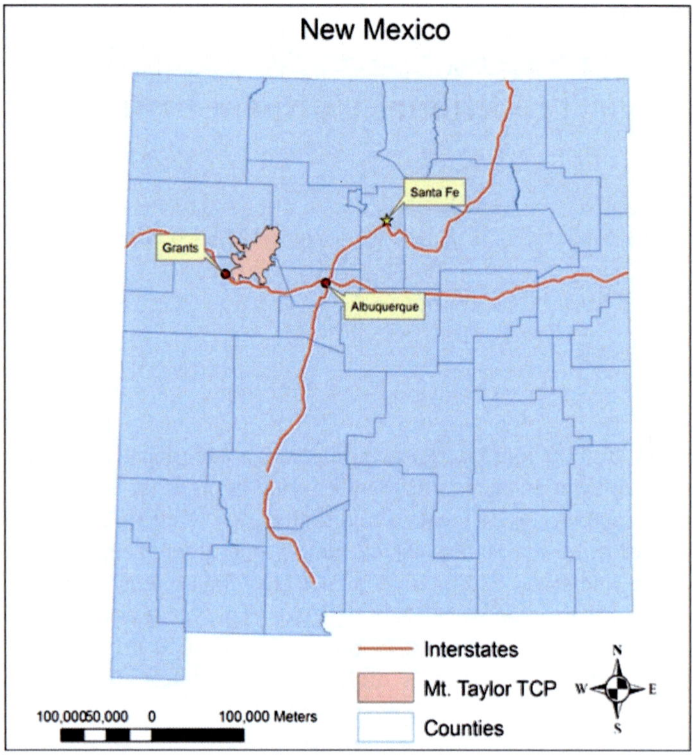


**Fig. 10.1** TCP Location, New Mexico, U.S.A. Copyright by P. Allison

## Mt. Taylor

Mt. Taylor is one of the highest peaks in the State of New Mexico, rising to a height of 3,445 m (11,301 ft) above mean sea level. It is located primarily in Cibola County, just east of the town of Grants and approximately 1 h west by car of Albuquerque. It was formed roughly 80 million years ago and is classified as a stratovolcano, a steep-profiled cone with periodic, explosive eruptions (Kelley 2007). But beyond its basic physical characteristics is a much deeper history of the mountain, the history belonging to the tribes. These tribes are representative of both sedentary and nomadic groups that predated the Spanish colonial settlements in this region. A detailed description of each tribe's beliefs and practices is contained in both the Benedict (2008) and the Benedict and Hudson (2008) report. For the purposes of this paper, only a summary of the major themes is presented. The mountain is believed to be a living entity, inhabited by spirit beings known from the oral histories. It is a place for ceremony and for the gathering of ceremonial and cultural items. It is a place of origin for multiple tribes and an important physical landmark, in many cases delineating one of the four cardinal directions.

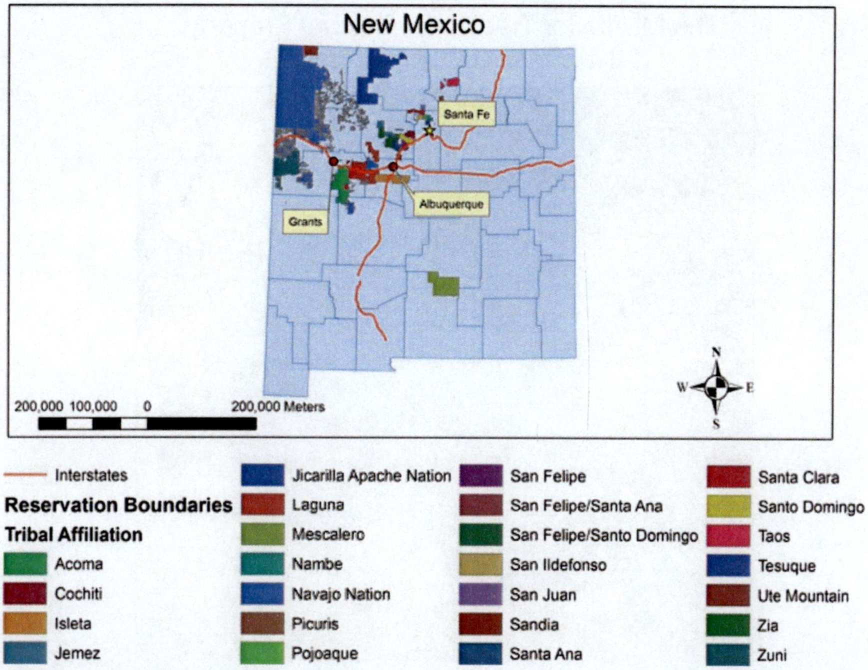


**Fig. 10.2** Copyright by P. Allison

Lands belonging to the Pueblos of Laguna, Acoma, Zia, Zuni, San Felipe, Santo Domingo, Cochiti, Isleta, Sandia, Jemez, Santa Ana, the Jicarilla Apache, and the Navajo Nation are all located within 50 miles of the peak.

All of these characteristics were taken into account when the area was nominated as a traditional cultural property. In order to be eligible for this nomination, a site "must have an association with cultural practices or beliefs of a living community that (a) are rooted in that community's history, and (b) are important in maintaining the continuing cultural identity of the community" (National Register Bulletin 38 1990). If a site meets these criteria, it becomes eligible for inclusion in the National Register of Historic Places (NRHP), the United States' official list of cultural resources worthy of preservation, authorized under the National Historic Preservation Act of 1966 (The National Register of Historic Places 2009). Listing in the National Register does not necessarily equal protection of the property, however. If federal money or a federal permitting process is involved, Section 106 of the National Historic Preservation Act of 1966 is invoked. Section 106 requires the federal agency involved to assess the impact of its actions on historic resources (Advisory Council on Historic Preservation 2009).

The mountain was eventually listed in the NRHP in 2007 and listed in the State Register of Historic Places (SRHP) in 2009. Since 2007, the Nuclear Regulatory Committee (NRC) has received 17 license applications for the building of 26 new nuclear reactors (Smithson 2009), causing a flurry of mining activity in the Four

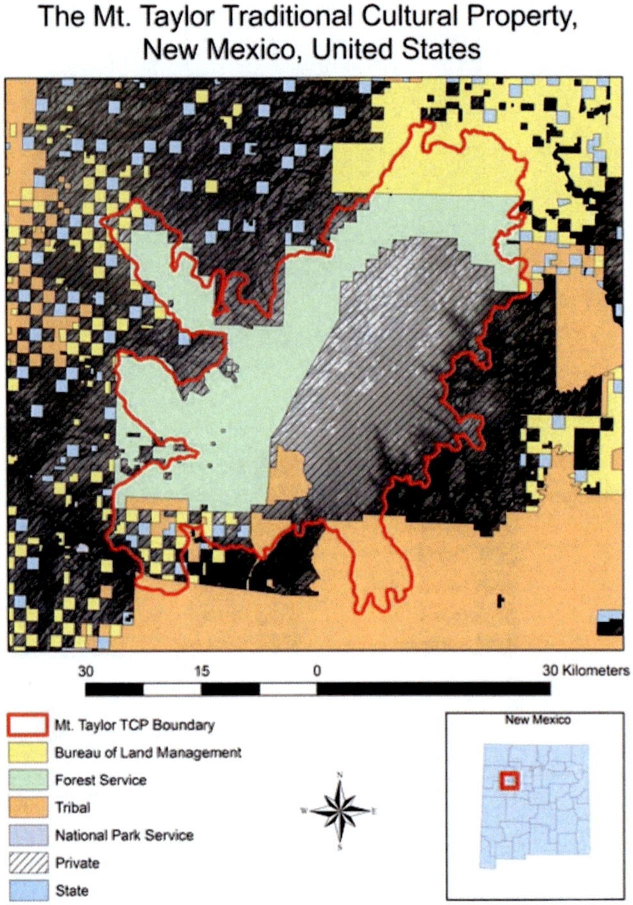


**Fig. 10.3** Copyright by P. Allison

Corners region of the country. Urex Energy Corporation, Western Energy, Hydro Resources, Inc., Strathmore Minerals, Uranium Energy Corporation, and Neutron Energy are just a few of the companies seeking to do exploratory drilling on Mt. Taylor. The new TCP boundary will have a measurable effect on the complexities that mining companies will face if they want to perform any exploratory drilling in the region and may ultimately alter the landscape by discouraging mining in certain areas.

## Mapping Mt. Taylor

Mt. Taylor provides an excellent case study of the some of the pratfalls of attempting to create a boundary around an indigenous space that is primarily spiritual in nature. Similarly to TCPs such as Mount Shasta in Northern California and the Ocmulgee Old Fields of Central Georgia, it represents the spatial delineation of a sacred space that is not only invisible, but varies from tribe to tribe and had to be forced into a single boundary by a third party, the government agency itself.

A total of 16 tribes and 13 chapters of the Navajo Nation were contacted by letter in October of 2007 by the Forest Service to consult on the cultural values and uses of the mountain (Benedict and Hudson 2008). Several of these groups did not respond. Others deferred consultation to Pueblos that are closer in proximity to Mt. Taylor. In the end, a total of eight tribes became involved in the discussion: the Pueblos of Acoma, Isleta, Laguna, Zuni, and Jemez, the Navajo Nation, the Hopi Tribe, and the Jicarilla Apache Nation.

Through a series of public meetings, the importance of Mt. Taylor to each tribe was put forward as support for eligibility as a traditional cultural property. Subsequently, using a combination of oral descriptions, GIS maps, and hand-drawn maps provided by the tribes, the Forest Service created the official boundary of the Mt. Taylor traditional cultural property to include the major contiguous landforms, following the as closely as possible the geographic features of the mountain. While all tribes were in agreement that a larger boundary is more appropriate due to the nature of the space, the chosen boundary has not yet been fully embraced by all of the tribes – Laguna, for example, believes that the spiritual use extends far beyond a physical boundary and is a continuum from the villages to the mountains. In addition, several noncontiguous spaces, while noted, were not included in the boundary, despite statements from representatives of the Navajo Nation of their significance. Several tribes were unwilling to even attempt a boundary, claiming that such a task was impossible (Benedict and Hudson 2008).

In spite of these conflicts, there is by all appearances a much better incorporation of indigenous knowledge and participation in this mapping process than was evident in either the Mount Shasta or the Ocmulgee Old Field cases. While both of these prior cases included consultation and the professed desire to include tribal input, they both resulted in boundaries that were found objectionable and ultimately unsatisfactory by the groups most invested in their protection. Mt. Taylor, in contrast, appears to include all of those areas commonly held by the consulted tribes to be of great importance, with only the few exceptions noted above. It should also be noted that the boundary chosen by the federal government, in this case the Forest Service, is much larger than and incorporates the boundary put forth by the New Mexico State government, due to more input from tribal consultation (Benedict and Hudson 2008). These accomplishments, as measured by the expressed goals of both PGIS and counter-mapping, suggest a positive shift towards indigenous empowerment and inclusive input through mapping.

## Conclusions

Despite the achievements of the Mt. Taylor case, which are hardly insignificant, problems remain with the TCP mapping process as it currently stands. The ability to satisfy all of the indigenous groups involved in consultations is not particularly realistic, and the literature surrounding PGIS and counter-mapping suggests that conflicts will always arise. In addition, although there appears to have been relatively good incorporation of indigenous input in the Mt. Taylor case, evidence suggests this to be the exception rather than the norm. How much of this information is included appears to depend on several factors, not the least of which is the influence of outside parties as well as that of the agency involved.

## Biographical Note

**Peggy L. Allison** is a graduate of the Master's program in the University of New Mexico's Department of Geography. She is also an experienced archaeologist in the American Southwest, having performed multiple inventory projects on or near Mt. Taylor. She currently serves as the GIS Administrator for Dona Ana County, New Mexico.

## References


Advisory Council on Historic Preservation (2009) National Register Evaluation Criteria. ACHP. http://www.achp.gov/nrcriteria.html. Retrieved 11 Apr 2010

Benedict C (2008) Cultural resource survey of 21 drill holes and access roads located on La Jara Mesa, Cibola Country, New Mexico. Cibola National Forest, Albuquerque

Benedict C, Hudson E (2008) Mt. Taylor traditional cultural property determination of eligibility. Cibola National Forest, Albuquerque

Chapin MZ (2005) Mapping indigenous lands. Annu Rev Anthropol 34(1):619–638

Hodgson DL, Schroeder RA (2002) Dilemmas of counter-mapping community resources in Tanzania. Dev Change 33(1):79–100

Kelley SA (2007) Mt. Taylor volcanic field. New Mexico Bureau of Geology and Mineral Resources, Socorro. http://geoinfo.nmt.edu/tour/landmarks/mt_taylor/home.html. Retrieved 1 Dec 2009

National Register Bulletin 38 (1990) Guidelines for evaluating and documenting traditional cultural properties. Patricia L. Parker and Thomas F. King. Interagency Resource Division, National Park Service, Washington D.C.

National Register of Historic Places (2009) The national register of historic places. The National Register of Historic Places. http://www.nationalregisterofhistoricplaces.com/. Retrieved 1 Dec 2009

Sletto B (2009) 'Indigenous people don't have boundaries': reborderings, fire management, and productions of authenticities in indigenous landscapes. Cultur Geogr 16(2):253–277

Smithson S (2009, June 10) Radioactive revival in New Mexico. The Nation, pp 16–20

# Part V
# Brazilian Cartography

# Chapter 11
# The Cartography of the Brazilian Empire

**Paulo Márcio Leal de Menezes and Alan José Salomão Graça**

**Abstract** This paper covers a small section of the historical research on the Cartography and Geodesy of Brazil which is undertaken by GeoCart, the Laboratory of Cartography of the Department of Geography at the Federal University of Rio de Janeiro, Brazil.

The main theme of this paper is the various mapping projects which were developed up to 1822 by the commissions which were appointed to organize the cartography of the Brazilian Empire and of the Rio de Janeiro State, including the Neutral Municipal district of Rio de Janeiro. It was considered necessary to also investigate all of the early initiatives which gave rise to the mapping systems used in the eighteenth century before Brazil became independent from Portugal.

After introducing the early attempts at accurate cartography, the cartographic organization of the Empire of Brazil, albeit without geodetic support, is discussed. The next topic is the 1862 Commission for the General Map of the Brazilian Empire, which was appointed to establish a scientific geodetic system and develop an orderly systematic mapping programme. The main personalities involved during the existence of the Commission, as well as their characteristics and the actions they took, are also presented. To conclude, two major attempts to execute a precise geodetic triangulation network are discussed, the first, which took place in the Rio Grande do Sul State on the Uruguay border, and the second, in the Neutral Municipal District of Rio de Janeiro. Both supported the initial development of the Brazilian geodetic triangulation network.

P.M.L. de Menezes (✉) • A.J.S. Graça
Federal University of Rio de Janeiro, Rio de Janeiro, RJ 21941-970, Brazil
e-mail: pmenezes@acd.ufrj.br; alanjsg@gmail.com

E. Liebenberg and I.J. Demhardt (eds.), *History of Cartography*, Lecture Notes in Geoinformation and Cartography 6,
DOI 10.1007/978-3-642-19088-9_11, 

## Introduction

Until the end of the seventeenth century, the cartography that developed in the coastal areas and in the rural areas of the Colony of Brazil was mainly schematic and qualitative and did not allow for the determination of even the approximate positions of geographical coordinates. Also important at this time was the occupation and exploration of the colony's territory by expeditions called *entradase bandeiras,* which left from São Paulo, São Vicente and places in the south and arrived in the center-west, especially the captaincy of Goiás, the lands of Mato Grosso and Madeira and the Mamoré River basin, closing the Amazon basin to the south. This activity extended the explored areas of the colony and accentuated the need to define Brazil's borders with the land to the west. Some isolated border demarcation movements, such as the Utrecht Treaty, which in 1715 negotiated the Oiapoque's River border line with France, closed the border to the north, but still did not define the border between the Portuguese and Spanish lands. In 1728, Portugal began to develop an extensive program of scientific mapping. The Spaniards, however, remained static with regard to a demarcation process and only started implementing similar solutions in 1734.

The systematic cartography of the Brazilian Empire dates back to 1730 when the Jesuits priests Diogo Soares and Domingos Capacci were specially commissioned by the King of Portugal, Don João V, to undertake the necessary border demarcation and terrestrial positioning. The important work of these so-called "mathematical priests" was characterized by the use of new cartographic positioning techniques in Brazil, such as the calculation of precise longitudes using observed and hourly differences relating to the eclipses of the satellites of Jupiter. These developments signified that Brazil was entering a new cartographic phase (Cassini, 1699). The accurate determination of latitude was followed by the accurate determination of longitude, which made it possible to accurately determine the true position of the Meridian of Tordesilhas, information which was essential for the demarcation of the Brazilian border with the Spanish territories.

The task given to the priests by D. João V was to map the south and the center-west, as well as to define new mapping rules for the whole Colony. The work, which was accomplished during the period from 1730 to 1748, manifested into maps covering the east and south coasts, the area south of Brazil, and the countryside as far as the Paraná River. As many as 197 coordinate lists of Rio de Janeiro, São Paulo, Minas Gerais, and the Goiás captaincies were determined.

To support the mapping that was done, the meridian of Rio de Janeiro, which was established in the observatory of the Morro do Castelo Hill, was chosen as prime meridian. The observed differences amongst other prime meridians varied from 2 to 5°, especially in the south. According to Cortesão (1945), it is possible that the Rio de Janeiro meridian was accepted as the prime meridian to mask the geographical coordinates, mainly longitudes, which related to either Paris or the Iron Island meridians. The priests were instructed to achieve the following:

- Establish and reference a new Colonial Atlas relative to the meridian of Rio de Janeiro;
- Avoid references to the Paris and Iron Island meridians, and
- Determine the true position of places relative to the Meridian of Tordesilhas.

In 1750, Alexandre de Gusmão negotiated the Madrid Treaty, which defined and established the land distribution between the Colony of Brazil and Spain. The basic premises of the negotiation were to establish a balance in the sharing of land: The Amazon River basin went to Portugal, whereas the Prata Tiver basin was allocated to Spain; Portugal retained the Central Plateau Central with its gold and diamonds mines; limits were extended to the south of the colony to tie Minas Gerais to the southern region; an organic visibility was given to all territories explored by Portugal and, finally, the sovereignty of the juridical sanction of the "*utis possidetis*" was stabilized. Apart from presenting supporting documents, two elements were fundamental to the success of the Portuguese negotiations: Alexandre de Gusmão's geographical knowledge and the maps which gave support and sustainance to his strategy.

Spain accepted the so-called "Map of the Courts" as the basis of the negotiations. This map was compiled using several other maps, such as those by Diogo Soares and Capacci of the southern areas, and the chart by D'Anville of the Spanish lands of the Prata River basin, which was drawn on the Spanish Jesuits' maps of Paraguay. Other contributing maps were those by Gomes Freire de Andrade of the center-west and part of Amazonia, and the 1735 map of the Rio Negro Valley by a French scientist and explorer, Charles Marie de La Condamine.

If the Map of the Courts is superimposed on a current map with the meridian of $-51^\circ$ West as starting point, we find that it presents a distortion which allows for an erroneous interpretation. It namely reduces the area destined to the Portuguese, and increases the area shared with the Spaniards. These errors are basically displacements in longitude which, at that time, no longer made sense, as they showed that the mouths of the Madeira and Mamoré rivers had been diverted eastwards with more than $9^\circ$, or in other words, practically 1,000 km or 620 miles.

From Cortesão's studies, it is clear that the present distortions which were adherent to the Portuguese pretensions were perfectly well known by Alexandre de Gusmão. Thus, the Map of the Courts may be considered a thematic document open to a geopolitical application which transmits dissimulated information to a user who is unaware of its power. It is because of this that Brazil's present political image was already delineated in 1750. Although no specific authorship can be claimed for the map, we know that it was designed and drawn in Lisbon under the direction of Alexandre de Gusmão. Two copies were made which were referred to as "primitive maps," to differentiate between them and some later copies that were made in 1751 – three in Lisbon and three in Madrid – which show some variations.

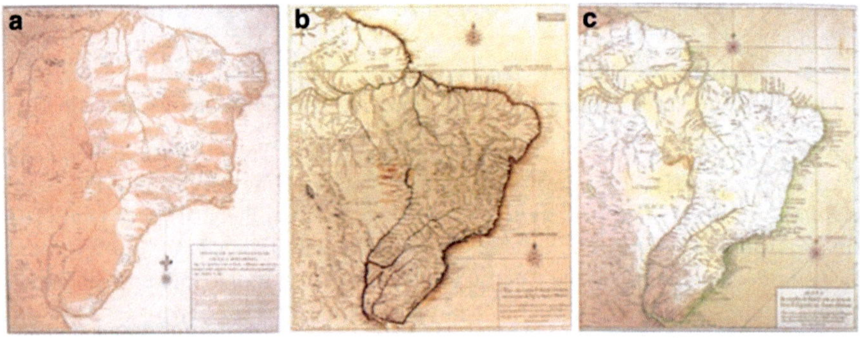

Fig. 11.1 (a) Map of 1746, (b) Primitive map 1749, (c) Copy of 1751

Figure 11.1a depicts the map sent by Gomes Freire de Andrade to support Alexandre de Gusmão's design of the Map of the Courts. This map was compiled in 1746 by an anonymous author. Figures 11.1b and cs show, respectively, the "primitive map" of 1749 and one of its copies of 1751.

On 12 February 1761, the governments of Portugal and Spain signed a new treaty named the Pardo Treaty, which annulled everything that had been agreed upon by the Madrid Treaty of 1750. The main divergences were those which had a bearing on the Mission's territory in the area of the Prata River which had been incorporated into Portuguese territory. Negotiated by Marquês de Pombal, the Santo Ildelfonso's Treaty was signed on 1 October 1777. It maintained the basic agreements of the Madrid Treaty, but imposed a substantial loss of territories in the area of the Prata River on the Portuguese. The occupied lands in Santa Catarina and Rio Grande do Sul were turned to Portugal by the Spaniards during the period after the Madrid Treaty.

All of these treaties imposed a complete knowledge of the controversial territories, taking to the accomplishment of several mapping and demarcations missions, to confrontation and recognition of all areas. After this, many maps were compiled showing positions which had been determined with reasonable accuracy by the astronomical processes that were used.

## Map of New Lusitania

By the end of the seventeenth century, the Portuguese government claimed ownership of a large amount of cartographic information. The result was that D. Rodrigo de Souza Coutinho, Minister of the Portuguese Navy and Foreign Domains, gave an order that a "General Map of Brazil" should be made. The person appointed to organize the project was Dr. Antonio Pires da Silva Pontes.

Headed by Silva Pontes, a group of designers, geographers and cosmographers, all of them in some way linked to Portuguese and Portuguese colonial cartography, compiled the Map of New Lusitânia in 1798. They made use of the best information

**Fig. 11.2** Map of New Lusitania by Silva Pontes

available, especially data that had been used for border demarcations which depicted places by "*their true latitude and longitude positioning*".

Draughtsman Dr. Antonio Pires da Silva Pontes Leme had, at his disposal, the services of José Joaquim Freire and Manoel Tavares da Fonseca. An Italian engineer and astronomer, Miguel António Ciera assisted the team with astronomical observations. Ciera came to Portugal in the mid-eighteenth century to assist the commission entrusted with the demarcation of the borders of Portuguese possessions in southern America. Figure 11.2 shows the Chart of New Lusitania.

This Map was named "*Geographical Map on a Spherical Orthogonal Projection of New Luzitânia or the State of Brazil*" (Carta Geográfica de Projeção Esférica Ortogonal da Nova Luzitânia ou Estado do Brazil), as is shown in its legend. Apart from the legend, which is long and explanatory, it also contains the names of the organizer, the draughtsmen, and 34 individuals and cartographic authorities who supported the venture, amongst them priests, doctors, militarists, scientists and explorers, making up an important list of names linked to Portuguese and Brazilian cartography.

The spherical orthogonal projection, which was used in apparently the same way as the equivalent projection of Sanson-Flamsteed, a projection of which the meridians are curved with the concavity centred on the central meridian. In the case of the latter, the meridian was 315°, which runs through the Ferro Island. The central meridian is projected as a straight line and each parallel is projected as equidistant with all parallels being straight lines. The graphic scale of the map corresponds to 1: 3,865,000.

Only two extant copies of this map are known, one kept in the Tombo Tower in Lisbon, Portugal, and the other is in the fifth Survey Division of the Brazilian Army Geographical Service in Rio de Janeiro. The Brazilian copy was obtained from the royal family in 1808. Another copy, made in the mid-nineteenth century, is held by the Map Library of the Brazilian Foreigners Affair Ministry in Rio de Janeiro.

Noteworthy on this map is the wealth of detail relating mainly to place names of the hydrographic network, as well as to mountains, cities and villages. Another interesting phenomenon on the map is the many quite accurate locations of Indian tribes which once inhabited the rural areas.

It is probably correct to infer that Brazilian systematic cartography had its real beginning with this map and the cartography of the people who demarcated the borders. Amongst them is Lieutenant Colonel Ricardo Franco de Almeida Serra, whose name is linked to the history of Fort de Coimbra in Mato Grosso do Sul and to border demarcations in Brazil's center-west.

## Cartography During the Brazilian Empire

When analyzing the cartography of the Brazilian Empire, it is evident that a considerable period of time elapsed between the last works accomplished by the colony and the first ones accomplished by the Empire.

No negligence was observed with regard to the country's mapping. The territory's mapping processes continued unabated, albeit on a regional basis, in order to provide the necessary topographical maps, boundary maps, hydrographical surveys and geodetic support structures to the provinces, just as it did during the early years of the geodetic triangulation. Notable projects were developed in the rural areas by military engineers, mainly from the Brazilian army, but also by several other civilian organisations.

A large number of topographical plans and hydrographic maps were compiled of several Provinces, including the Amazon and Prata Rivers basins and borders. This includes the maps of the Province of Goiás, which were published in 1836 as the work of Marshal Raimundo José da Cunha Matos, who travelled throughout the territory when he was the provincial governor.

In 1844, the Viscount Beaurepaire Rohan explored the lower Paraguay River basin, and in 1846 he surveyed the connection between the Guarapuava and Paraná Rivers and the navigability of the Iguaçu River to its mouth. Jerônimo Francisco Coelho conducted several recording surveys and maps of the coast of Santa Catarina, and was the author of the "Mappa Topográphico of the South Part of Sta. Catharina's Province," which appeared in1842. Later on he would become the President of the Pará and Rio Grande do Sul Provinces and, finally, the Brazilian Empire Minister.

Army Colonel Jacob Conrado Niemeyer, and Generals José de Souza Soares d'Andréa, Caçapava Baron, and Pedro de Alcântara Bellegarde were active geographers and cartographers of that era who, together, compiled the "Map of the Empire of Brazil," which was first published in 1846. This map was published twice, the first time in 1857, when its publication was ordered by the Marquês of Caxias when he was War Minister, and the second time in 1873, when its republication was organized by the Baron Duarte da Ponte Ribeiro. Figure 11.3 shows a

**Fig. 11.3** Map of the Brazilian Empire, 1873 (1846) and its legend

**Fig. 11.4** Chorographic map of Rio de Janeiro's Province

copy of the 1873 map which is the possession of the Laboratory of Cartography Collection of the Federal University of Rio de Janeiro.

In 1852, Pedro de Alcântara Bellergarde and his uncle, Colonel Niemeyer, worked on a map called the "Chorographic Map of the Province of Rio de Janeiro", which was published in 1865 with the cooperation of Marshall Beaurepaire Rohan (see Fig. 11.4).

In spite of considerable cartographic activity in Brazil, this map was basically a product of scattered and discontinuous mapping projects undertaken throughout the immense Brazilian territory, most of it supported by astronomic positioning only. Because of a lack of coordinated efforts on the geographical and cartographical fronts, no geodetic support was available. Without a proper scientific coordinate and geodetic system, it was not possible to accurately represent the coordinates of any point on the earth's surface.

During this time, approximately 75 maps of several Brazilians regions were compiled: 17 of the northern area, 8 of the western, 6 of the southern, 31 of the eastern and 3 of the central region. Only two regional and eight nationwide maps were developed. It was also at this time that the well-known Baron Duarte da Ponte Ribeiro began to feature prominently in the history of Brazilian cartography.

## General Map of the Empire

In 1862, the Commission for the General Map of the Empire was created under the supervision Dr. Antônio Maria de Oliveira Bulhões. The commission's objective was to organize, design and create a Brazilian geographical map by implementing strict scientific methods, in other words, applying precise geodesy and establishing the necessary triangulation network as had been done in France.

The planning of the project was undertaken in 1864 and the project itself initiated in 1867. Part of the planning constituted of the precise triangulation of the Neutral Municipal district and the measurement of two geodetic bases. Due to a chronic financial crisis in the Empire, the triangulation soon suffered administrative problems, which resulted in the cancellation of the work. The planning of the project was, however, not interrupted and comprised the compilation of 42 sheets to eventually produce a map at a 1:1 million scale, using only astronomical support. By 1871, 31 sheets were ready but, already in 1872, the Commission verified that all sheets had to be changed due to deficient information. Because the information on the maps was not reliable, the execution of the planning sheets also had to be cancelled.

There were several reasons for this slow progress and the unreliable information. One reason was the country's enormous territory; a second, the difficult access to scientific information, and yet another, the lack of preparatory projects that could propitiate an accurate geodetic survey. Last, but not least, a chronic lack of financial resources prohibited the use of methods which had, by then, been used by developed countries in Europe for more than one and a half centuries. To overcome these difficulties, Dr. Antonio Maria decided to apply so-called "expedite" geodetic methods which were not rigorous, but were fast and economical, and could be corrected later. These methods had been successfully applied in Ethiopia where they presented satisfactory results. Implementing these methods would make it possible to produce a reasonable topographical picture of the country, which would be more accurate than any existing map. Unfortunately, these methods, although theoretically feasible, did not provide satisfactory results in practice.

In 1873, Dr João Nunes Campos assumed command of the Empire's General Map Commission, and he immediately identified problems with the existing material. The existing information, which was plotted on the normal Sanson-Flamsteed projection, was in disagreement with the already used map projection, specifically the modified Flamsteed projection. To rectify this, Nunes Campos ordered the compilation of a new Map of Brazil, which was to be drawn on a divided, modified Flamsteed projection and would consist of 30 instead of the original 42 sheets. In 1874, the Commission's command went to Marshall Beaurepaire Rohan who, in 1843, was responsible for the "General Map of Brazil," which was presented at the Vienna's Universal Exhibition. In reality, it was not possible to fully accomplish the envisaged work which had to wait for the cartography of the colonial period under the valuable supervision of Baron Duarte da Ponte Ribeiro.

Ponte Ribeiro, along with the Baron of Rio Branco, devoted many years to the study of Brazilian borders and solved several of the country's border problems with neighboring countries. Throughout this period, he collected maps and geographical material of Brazil and its bordering areas in Portugal, Brazil and several Latin America countries. Aided by his considerable knowledge of the area and the relevant cartographic documentation, he could overcome the difficulties experienced during the necessary preparatory work for the Map. Thanks to him, the Map of the Empire could be published and exhibited at the Philadelphia Exhibition in 1875. Instead of the original 30 sheets, the map was published in 4 sheets after plotting the same information to a scale half of that of the original.

Assisted by Jose Ribeiro da Fonseca Silvares, the map was drawn by Jose Martins Penha and Jose Cupertino do Amaral. The meridian passing through the Sugar Loaf Rock was taken as prime meridian, once it was decided to remove the Castelo Hill where the old Jesuit College was and through which the meridian of Domingos Capacci's and Diogo Soares' Charts passed. The central meridian had a displacement of $0° 00' 57''$ east of the Imperial Astronomical Observatory of Rio de Janeiro.

The Map was compiled using both old and modern maps of various regions of Brazil, most of them collected by Ponte Ribeiro. He used 85 maps from the northern region, 43 from the western region; 45 from the southern region; 116 from the eastern region; 17 from the central area; 15 maps which represented territories in one of the 5 others regions, and 9 maps of Brazil as a whole.

Regarding the border areas, cartographic documents from eighteenth century demarcation expeditions were used, especially those which defined the borders as proposed by the Santo Ildefonso's Treaty. This contribution was particularly noticeable in the Northern region, where 32 of the 85 maps were from border demarcations of that time. Although a large number of modern maps were used for the Eastern region, information was also taken from old maps such as the map of New Lusitânia, which represented a collection of many older cartographic documents. With regard to the Brazilian coast, older hydrographic works were used, such those by Jose Fernandes Portugal and Patricio José de Sousa, both from the eighteenth century and the beginning of the nineteenth centuries.

In 1875, the commission for the General Map of the Empire published the "*Map of the Empire*" and exhibited it at Philadelphia's International Exhibition (see Figs. 11.5 and 11.6). The map was compiled on the modified projection of Flamsteed at a scale of 1: 3 710 220.

**Fig. 11.5** Map of 1875 (reduced)

**Fig. 11.6** 1875s Map legend

In 1876, a new method for editing and updating the General Map of the Empire was introduced by using a so-called "Archive Map" or "*Carta Arquivo*". The idea was that this type of map would be distributed to all provinces so that their deficiencies and errors could afterwards be verified and corrected on the General Map. Unfortunately, the results were not satisfactory. In spite of this situation, the final report of the Commission on the Brazilian Empire General Map, which was introduced in 1878 by Marshal Baurepaire Rohan, renders praise to Ponte Ribeiro's work.

The Commission had, however, a short lifespan and was terminated in 1878. As an initiative of Baron de Capanema, an Itinerary Map Commission was appointed in the same year to survey roads and trails in the rural areas. Likewise, this organisation also did not last long and was terminated 2 years later. It was mostly composed of Austrian engineers who managed to accomplish only a few surveys and some astronomical observations.

## Triangulation Within the Empire

Although not linked to the Empire's General Map Commission, a very important geodetic survey which took place in Brazil was the triangulation undertaken between 1852 and 1857 by Marshal Francisco José Souza Soares d'Andréa, Baron de Caçapava, along the Brazilian-Uruguayan border between the Chuí and Quarai rivers.

The main purpose of this triangulation was to support the relevant border demarcation work. At that time, it was already considered the most important monument left behind by military cartographic engineers in Brazil. Also, it was undoubtedly the first established and calculated triangulation executed not only in Brazil, but also in South America. In 1860, Baron Pedro Alcântara Bellegarde had an identical objective when he carried his triangulation along the entire length of Lagoa Mirim. These two projects are considered the first geodetic networks measured in Brazil.

Figure 11.7 shows the Plane Map of the Border of Chuy, which depicts the border lines projected by the Commission for the Brazil-Uruguay border, as well as the initial structure of the triangulation.

The mapping project initiated in 1866 in the Neutral Municipal district of Rio de Janeiro city owed its origin to a request from the General Finance Bureau, which was interested in establishing the city's cadastre together with a topographical plan to support it. Under the direction of engineer Antônio Maria Oliveira Bulhões, the Imperial Court Inspectorate of Public Works decided to develop a geodetic triangulation. The triangulation consisted of a small-triangle network that covered the urban and surrounding areas. The departure base was located in the fields of Jacarepaguá. It measured 5.994 m and was linked to another base with a length of 3,019 m which had been measured on Arpoador's beach. To improve the density of the network, a third base of 9.423 m was measured in Santa Cruz. Figure 11.8 depicts the outline of the Neutral Municipal district network.

The project was entrusted to engineer José Manoel Silva, who practically accomplished it on his own, even though the general shortage of money made him experience serious financial problems. The work came to a halt in 1868 when the Triangulation Department was terminated because of the same financial problems that were experienced with the General Map. Therefore, the Commission of Public Works Inspectory abandoned the project to develop a new network.

After the creation of the Commission for the General Map of the Empire, it was decided in 1870 to continue with the triangulation network, which was considered of extreme importance for the Neutral Municipal district cadastral map. The triangulation was developed between the Santa Cruz and Magé Channel baselines

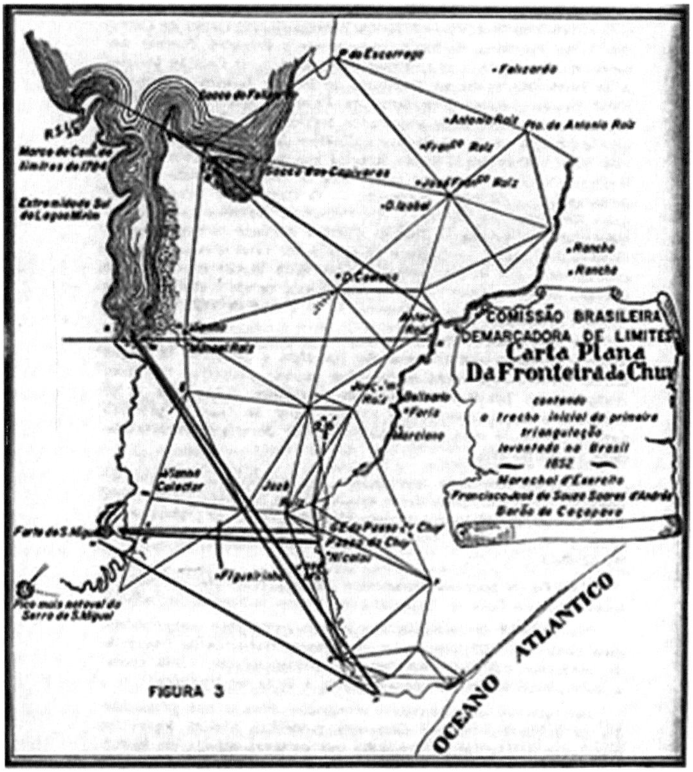


**Fig. 11.7** Plane map of the border of Chuy. Annual report of the management of the geographical service, nr 7.1956

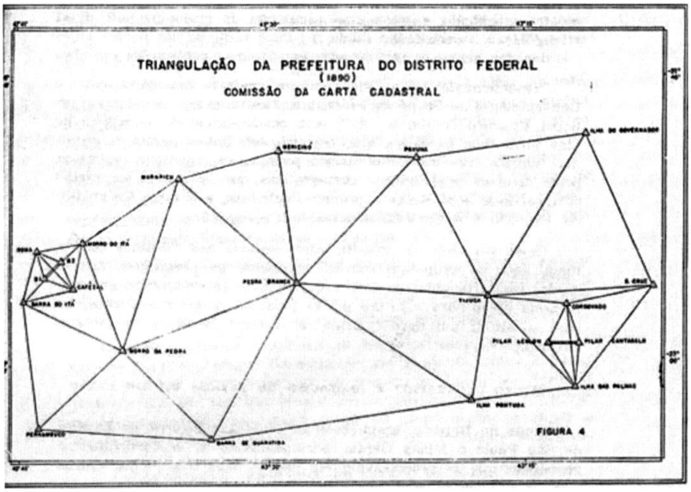


**Fig. 11.8** Triangulation of the neutral municipal district

in the Province of Rio de Janeiro. The first base to be measured was 2.509,82 m in length and was not the same as the original one, but was also located in Santa Cruz. For the first time in Brazil, a French basimeter, *Brunner et Frere,* was used. All first-order vertices were marked with masonry pillars and iron towers. The ellipsoid initially adopted was Bessel's of 1841 which, in the end, was changed to the Clarke 1866.

The initial project planned to establish an astronomical observatory in Santa Cruz, which would provide the initial geographical coordinates and the fundamental starting azimuth of the network. Another project which was considered, was the building of a tidal station in Sepetiba bay.

## Conclusion

With this simplified overview of the history of the cartography of the Colony of Brazil and the Brazilian Empire, it is evident that, although the cartographic output was not large, much has been achieved, especially if one keeps in mind the support needed for the geodetic programme and the financial difficulties which had to be overcome. Practically, all the work that was done served to support the programme which was developed in the Republic since 1903 with the Commission of the General Map of Brazil, and which reinforced the need for a geodetic triangulation network to allow for the necessary mapping of the entire territory.

Many eminent geographers and cartographers have contributed to the history of Brazilian cartography. People such as Colonel Ricardo Franco de Almeida Serra; Colonel Conrado Jacob de Niemeyer; Generals José de Andréa Souza Soares, Baron de Caçapava; Pedro de Alcântara Bellegarde; Baron Duarte de Ponte Ribeiro; and Marshall Beaurepaire Rohan are amongst those who worked tirelessly over a long period for the implementation of a programme of scientific cartography in Brazil.

## Biographical Note

**Paulo Márcio Leal de Menezes** is a surveying engineer who received his MSc in Computational Systems and Automated Cartography from the Brazilian Military Institute of Engineering in 1987. His DSc in Geography was obtained from the Federal University of Rio de Janeiro in 2000. Professor de Menezes is currently Senior Professor of Geography at the Federal University of Rio de Janeiro, and Head of the Cartography Laboratory at the same institution. He has published 1 book, 5 chapters in books, 15 papers in scientific journals, and 128 papers in conference and seminar proceedings. His research interests are Digital Cartography, GIS, Historical Cartography, Web-Cartography, and Thematic Cartography.

# Bibliography

Adonias I (1993) Mapa : Imagens da Formação Territorial Brasileira. Odebrecht, Rio de Janeiro

Adonias I (1945) A Carta Geral do Império do Brasil, de 1875: Trabalhos que a Antecederam. Sua Elaboração e Contribuição do Barão Duarte da Ponte Ribeiro. In: História da Cartografia Política do Brasil. Instituto Rio Branco, Rio de Janeiro, pp 140–196

Coelho PD (1937b) Notícia sobre a Triangulação do Distrito Federal. Serviço Geográfico do Exército, Rio de Janeiro

Coelho PD (1950b) A Obra Cartográfica dos Militares no Passado e no Presente do Brasil. Serviço Geográfico do Exército, Rio de Janeiro

Cortesão J (1944) História da Cartografia do Brasil (Séculos XVI a XVIII). Curso de Mapoteconomia, Instituto Rio Branco, Rio de Janeiro

Cortesão J (1944b) Curso de História da Cartografia e Geografia das Fronteiras do Brasil. Instituto Rio Branco, Rio de Janeiro

Cortesão J (1945b) História da Cartografia Política do Brasil. Instituto Rio Branco, mimeografado, Rio de Janeiro

Furtado SS (1957) Cartas Históricas da Mapoteca da Diretoria do Serviço Geográfico A Cartografia Luso-Brasileira. Anuário da Diretoria do Serviço Geográfico, no. 8. Serviço Geográfico do Exército, Rio de Janeiro, pp 173–214

MJNI (1944) Ministério da Justiça e Negócios Interiores Catálogo dos Mapas Existentes na Biblioteca do Arquivo Nacional, II Reunião Pan Americana de Consulta sobre Geografia e Cartografia, Rio de Janeiro

MRE (1960) Mapas e Planos Manuscritos Relativos ao Brasil Colonial, Conservados no Ministério das Relações Exteriores e Descritos por Isa Adonias para as Comemorações do Quinto Centenário da Morte do Infante Dom Henrique. Serviço de Documentação MRE, Rio de Janeiro

Oliveira Jr JF (1956) As Triangulações Geodésicas no Brasil. In: Anuário da Diretoria do Serviço Geográfico, no 7. Serviço Geográfico do Exército, Rio de Janeiro, pp 73–89

Veríssimo IJ (1956) O Mapa das Cortes. In Anuário da Diretoria do Serviço Geográfico, no 7. Serviço Geográfico do Exército, Rio de Janeiro, pp 95–118

# Chapter 12
# Cartographic Rumors, Brazilian Nationalism, and the Mapping of the Amazon Valley

**Jörn Seemann**

**Abstract** In 1988, J.B. Harley's inspiring paper on silences, secrecy, and hidden agendas in early modern cartography stimulated a debate on power-knowledge relations in the history of cartography. Harley addressed maps as political discourses through which geographic details had been added or left out. Issues such as censorship, intentionality, and strategic interests underlined Harley's argument that maps were "socially constructed perspectives on the world." Despite the recent epistemological debates on authorship, access, and agency in the mapmaking process, the debate is mainly focused on printed maps and does not take into consideration the impact of the increasing number of historical maps that are accessible in the form of digital copies on the internet.

In this context, this paper aims to present the case study of a spurious nineteenth-century map of the Amazon valley that started to circulate in the internet about a decade ago and triggered nationalistic feelings and geopolitical arguments in Brazil. As an inversion of silences and secrecy, the purpose of the map was to exhibit and propagate rather than to hide or leave out information, whereas the anonymous mapmaker sought to link the map hoax to real historical facts and writings such as William Lewis Herndon's exploration of the Amazon (1851–1852) and Matthew Fontaine Maury's reflections on the Atlantic slopes of South America (1853). This example of a cartographic urban legend helps raise questions about the use and abuse of digital historical maps and their impacts in society in the past and the present.

In the 1980s, J.B. Harley wrote several articles about the absences and presences in maps and the relationship between power and knowledge in these paper landscapes

J. Seemann (✉)
Department of Geography and Anthropology, Louisiana State University, Baton Rouge, LA, USA

Departamento de Geociências, Universidade Regional do Cariri, Crato, Brazil
e-mail: jornseemann@gmail.com

E. Liebenberg and I.J. Demhardt (eds.), *History of Cartography*,
Lecture Notes in Geoinformation and Cartography 6,
DOI 10.1007/978-3-642-19088-9_12, 

(Harley 1988a, b). Crafting a "theory of cartographic silence," he affirmed that "which is absent from maps is as much a proper field of enquiry as that which is present," whereas the mission of the historian of cartography should be to establish a dialogue "that arises from the intentional or unintentional suppression of knowledge in maps" (Harley 1988a, pp. 57–58).

Harley's ideas have been widely discussed and critically accessed (for example, Belyea 1992; Andrews 2001; Wood 2002; Edney 2005). In his studies, Harley concentrated on early modern maps, but he also admitted that his framework could equally be used for present-day maps (see, for example, Wood and Fels 2008; Wood 1992, 2010). Surprisingly, the debate is almost exclusively focused on printed maps and does not take into consideration that an increasing number of maps released by map libraries, auction sites, or individuals is now available on the internet. On the one hand, the introduction of new technologies to visualize and disseminate data has resulted in different and more dynamic ways of reading, using or even abusing these "old maps in new clothes." On the other hand, this new virtual nature of cartography sheds a different light on issues such as accessibility, distribution, and authorship in the mapping process and opens up completely new research topics to investigate the relations between maps and society.

In this paper I present the case of a fictitious historical map of the Amazon within the context of political arguments and nationalistic discourse in order to point out that the borders between the present and the past in the history of cartography are blurred and that even map jokes and hoaxes can have a serious impact on how historical facts and fiction are created.

## "Captain Fawry's" Map of South America

On May 10th, 2000, a letter to the editor of the Brazilian electronic scientific bulletin *Ciência Hoje* ("Science Today") revealed upsetting news for the academic community. A young biologist from Rio de Janeiro reported that "in some important American schools" a geography textbook with a map was used in which the Amazon region had been labeled as "area of international control" and not as territory belonging to Brazil. The scientist concluded that "this is only one more proof that the foreigners' idea of intervention in the Amazon has already reached its operative phase" (Almeida 2000a) .

The news spread quickly on the internet – without further verification of its contents. A sample of the supposed American textbook finally appeared as an attachment to e-mails showing a map in which the Amazon was denominated as FINRAF, "former international reserve of Amazon forest," supposedly created in the mid-1980s under the administration of the United Nations and the United States (Fig. 12.1).

A closer look at this digital page, its contents and layout could have easily unmasked the dubious character of the map. However, many Brazilian and other South American internet users, especially the intelligentsia at the universities, kept

An Introduction to Geography

SOUTH AMERICA

in the northern section of South America, forming a land of more than 3.000 square miles.

3.5-5 - THE FORMER INT L RESERVE OF AMAZON FOREST

Since the midde 80's the most important rain forest of the world was passed to the responsability of the United States and the United Nations. It is named as FINRAF (Former International Reserve of Amazon Forest), and its foundation was due to the fact the Amazon is located in South America, one of the poorest regions on earth and surrounded by irresponsable, cruel and authoritary countries. It was part of eight different and strange countries, which are in the majority of cases, kingdoms of violence, drug trade, illiteracy and a unintelligent and primitive people.

The creation of FINRAF were supported by all nations of G-23 and was really a special mission of our country and a gift of all the world, since the possession of these valuable lands to such primitive countries and people should condemn the lungs of the world to disappearance and full destroying in few years.

South America

FORMER INTERNATIONAL RESERVE OF AMAZON FOREST (UN)

BRAZIL

CHILE

ARGENTINA

map 3.5-5.1 - We can see the location of the International Reserve. It took areas of eight South America's countries: Brasil, Bolivia, Peru, Colombia, Venezuela, Guyana, Suriname and F.Guyana. Some of the poorest and miserable countries of the world.

We can consider that this area has the most biodiversity in the planet, with a vast number of species of all types of animals and vegetals. The value of this area is unable to calcule, but the planet can be cert that The United States won't let these Latin American countries explorate and destroy this real ownership of all humanity.

FINRAF is like an international park, with very severe rules of exploration.

76

**Fig. 12.1** False textbook map of the Amazon from the internet (http://www.alerta.inf.br/geral/1165.html)

on believing in these rumors and manifested their opposition to the American imperialism evidenced on the textbook page.

The minister-counselor at the Brazilian embassy in Washington, D.C., Paulo Roberto de Almeida, decided to investigate the source of this rumor which led him to a right-wing website that fervently defended values of Brazilian patriotism and national sovereignty. In his unpublished Amazonia dossiers (Almeida 2000a, b, 2001), the diplomat registered his e-mail correspondence with nationalistic groups, concerned citizens, and scientists. In the attachment of an e-mail sent to him on May 12th, 2000 by a patriotic Brazilian citizen who was concerned about the national integrity of the country and possible imperialistic interests by the United States, Almeida received a digital version of a "historical map" of South America from 1817 which contained the proposal for a new (and rather unusual) territorial division of the continent (Fig. 12.2).

**Fig. 12.2** A false historical map of Latin America (http://www.alerta.inf.br/1165.html). (Translation of English text passages into Portuguese appear in lighter color)

On the digital map that even "preserved" the stains of the "original" paper from the nineteenth century, the following details appeared:

- The areas of Chile and Argentina are clipped in their southern parts that are labeled as "Indian Territory;"
- Colombia and Venezuela are merged into one country named "Gran-Colombia;"
- Uruguay or *Cisplatine* (the lands beyond the River Plate) is supposed to become a British protectorate;
- The Falkland Islands are British;
- Easter Island is French.

Some of these territorial reconfigurations ironize contemporary geopolitical disputes such as the Falkland War and the frequent border hassles between Colombia and Venezuela. However, the most radical changes occur in Brazil whose territory has been split up into smaller units. The southernmost state of Rio Grande do Sul is shown as an independent republic, whereas the state of Santa Catarina appears as a "British occupation" and the state of Bahia as an "autonomous province." Large continental and coastal areas are merged either in the "Empire of Brazil" or the "Republic of Equador" (not to be mistaken for the country with the same name). The vast Amazon watershed is labeled as "soveraigny (*sic*) state of Amazon" whose borders are disputed by the "Empire of Brazil." The Marajó Island at the mouth of the Amazon River and the small island of Fernando de Noronha off the northeastern coast are indicated as "U.S. Naval Stations," while the state of Amapá in Northern Amazonia and at the border of the "Guyanas" is proposed as a "possible French occupation."

Looking at this very unconventional division of South America, the reader quickly realizes that the map is a crude parody of a possible historical map. The texts in English seem like a bad translation of the Portuguese annotations that accompany the map in bold red letters. The map bears the title "MAPE ONE" and is a "GENERAL VIEW" of South America showing the "ideal possible situation at the end of operations." The "document" is classified as the "MOST-SECRET, reference: operation memorandum 157/1816, subject: desestabilization (*sic*) of the cololony (*sic*) of Brazil."

## From "Fawry" to Maury

Despite the loutish character of this fraudulent digital map, the author of the map hoax was at pains to confer authenticity to the map that is signed by a certain "Mathew Fawry US Cap." on April 1st (April Fool's Day), 1817 in Washington, D.C. Despite the fictitious nature of the map and its author, Fawry has even been cited as a real reference in an article on the national border security of Brazil, published by the Woodrow Wilson International Center for Scholars (Costa 2001).

The Fawry map is, in fact, copied from a novel with the title *O Dia em que Napoleão Fugiu de Santa Helena* – "The Day Napoleon Escaped from Santa

Helena" – , written by the Brazilian writer Fernando Sampaio (1994). In the novel, Napoleon flees from his mid-Atlantic exile in a submarine and arrives on the coast of the Brazilian Northeast where he incites a revolution that leads to the creation of the "Republic of Equador." After the successful upheaval, Napoleon plans to continue his maritime journey either to the United States or Mexico.

Behind Fawry's name looms the seriousness of the joke. The allusion to the American naval officer Matthew Fontaine Maury, the so-called "pathfinder of the seas" (Lewis 1927), is not accidental. In the early 1850s, Maury published several newspaper articles signed with the name INCA in which he urged the countries of the Amazon basin to allow free navigation for trade and transport on the river. The texts were compiled in a book that was published in 1853 under the title *The Amazon and the Atlantic Slopes of South America* (Maury 1853).

In his writings, Maury emphasized the economic character of his plea. For him, the American interest was not in the territory itself, but in the trade routes that should be open to international commerce. In his own words, "the policy of commerce, and not the policy of conquest, is the policy of the United States" (Maury 1853, p. 5). Maury underscored his point by alluding to his own calculations of the ocean currents and the favorable winds between North and South America that would facilitate the journeys of commercial vessels: "The winds and currents of the sea are such as to place the Atlantic ports of the United States on the wayside of all nations, either going or coming from the mouth of the Amazon" (Maury 1854, p. 11). Maury considered the free access to the Amazon basin for all countries an issue of national importance since "there is no region of country beyond our own borders, the physical geography of which is more interesting to the people of the United States" (Maury 1854, p. 11).

Commenting on the favorable conditions for colonization, he praised the Amazon as a southern Eden with a climate in which "anything and everything will grow" (Maury 1853, p. 10). For him, the "Amazon [was] one of the most salubrious and delightful of climates" in the world (p. 10). Maury urged Brazil to be more open-minded and discard its "dog-in-the-manger policy" (p. 22) in order to share the Amazon not only with its neighboring countries, but also with "enlightened and commercial nations" (p. 63), taking into account that

> [t]he time will come when the free navigation of the Amazon will be considered by the people of this country as second in importance, by reason of its conservative effects, to the acquisition of Louisiana, if it be second at all; for I believe it is to prove the safety-valve of this Union (Maury 1853, p. 63).

In his discourse, Maury easily shifted from ocean currents and international trade to issues of American national politics. In his address at the annual meeting of the Geographical and Statistical Society of New York on February 16th, 1854, the pathfinder of the seas saw an incredible potential in the Amazon region that any demographer, economist or geographer of the present would judge totally unrealistic:

> The capacities of the country drained by the Amazon to sustain population are thought to be the greatest in the world, and with a population equaling that of Belgium to the square mile – that river basin includes an area that is large enough to sustain a population of *Six*

*Hundred Millions*. Its seed time is perpetual, its harvests never end, and its capacity to sustain population has been computed to the fourfold that of Belgium (Maury 1854, p. 11, emphasis and capitalized letters in original).

Apart from scientific and commercial arguments, Maury also showed his concern with the geopolitical future of the United States, especially the American South with its lack of industry, the decreasing ratio of its commerce in relation to the North, and the high percentage of the enslaved population in the country (Harrison 1955, p. 187). He lamented the incapacity of the Europeans to convert the Amazon into a rich center for agricultural production and highlighted the pioneer spirit of the American settlers: "In the valley of the Amazon the plow is unknown, and the American rifle and axe, the great implements of settlement and civilization, are curiosities" (Maury 1853, p. 6). Maury's argument was that the Amazon could be "an outlet for the increasing slave population of the United States... the Amazon Valley could absorb the excess slave population of the United States" (Bell 1939, p. 494). He contended that the South would not be able to liberate its slaves without losing most of its capital. Simply put,

it must either sell the increase of slaves to other slave territory or await the inevitable struggle for race supremacy. The amount of land in the United States which could be cultivated by slave labor – and therefore the market for slaves – was limited; but if the slave increase could be disposed of to Brazil, the economic and the social problem alike would be solved. Brazil was thus the safety-valve of the American Union and insurance against race conflict (Bell 1939, p. 498).

Maury himself described this solution more directly in a letter to William Lewis Herndon:

The question then comes up who shall people the great valley of this mighty Amazon. Shall it be peopled with an imbecile and an indolent people or by a go ahead race that has energy and enterprise equal to subdue the forest and to develope [sic] and bring forth the vast rescources [sic] that lie hidden there? The latter by all means (Dozer 1948, p. 217).

A general map of South America was included on the first page of Maury's book on the "Amazon and the Atlantic Slopes" (Fig. 12.3). The map is not so different from the contents outlined in the Fawry map. The continent is squeezed into a frame of coordinates and is abruptly cut off at 40° of southern latitude. The map shows the names of the countries and a few place names. The borders are not a major concern for the author who emphasizes the hydrographic features, especially the Amazon River and its innumerous branches that penetrate the continent from east to west like a vigorous blood vessel and only stop at the hachured mountain barrier of the Andes. South America appears as a borderless continent that dismisses the representation of any formal frontiers between its countries.

Maury also searched for empirical proof to strengthen his arguments. Between 1851 and 1852, Maury's cousin, Lieutenant William Lewis Herndon, was sent on an expedition to South America in order to explore the Valley of the Amazon from its sinuous sources in the Andes to the Brazilian plains. Herndon released the findings of his expedition in a two-volume report in which he reaffirmed Maury's discourse, based on the commercial interests of the United States and the favorable ocean currents:

**Fig. 12.3** Map of the Valley of the Amazon and parts of South America

> The citizens of the United States are, of all foreign people, most interested in the free navigation of the Amazon. We, as in comparison with other foreigners, would reap the lion's share of the advantages to be derived from it. We would fear no competition. Our geographical position, the winds of Heaven, and the currents of the ocean, are our potential auxiliaries. Thanks to Maury's investigations of the winds and currents, we know that a chip flung into the sea at the mouth of the Amazon will float close by Cape Hatteras. We know that ships sailing from the mouth of the Amazon, for whatever port of the world, are forced to our very doors by the SE and NE trade winds; that New York is the halfway house between Pará and Europe (Herndon 1854, p. 367).

Herndon's report is accompanied by a map (Fig. 12.4) that is far more detailed than the generalized map in Maury's book. Drawn by a certain H.C. Elliott, the map contains clear international borderlines that are indicated by thick traces in blue and red. Mountain ranges, especially in the Andean region, are indicated by what Erwin Raisz (1962, p. 76) called "hairy caterpillars."

A small, almost invisible legend in the top left corner shows the travel routes of Herndon and his midship-man Lardner Gibbon, whereas small arrows signalize "clear navigation of low water." The Peruvian Andes with its meandering rivers appear far more detailed than the Amazon River and the Madeira River that look like throughways rather than the center of Maury's discourse. A vertical section on

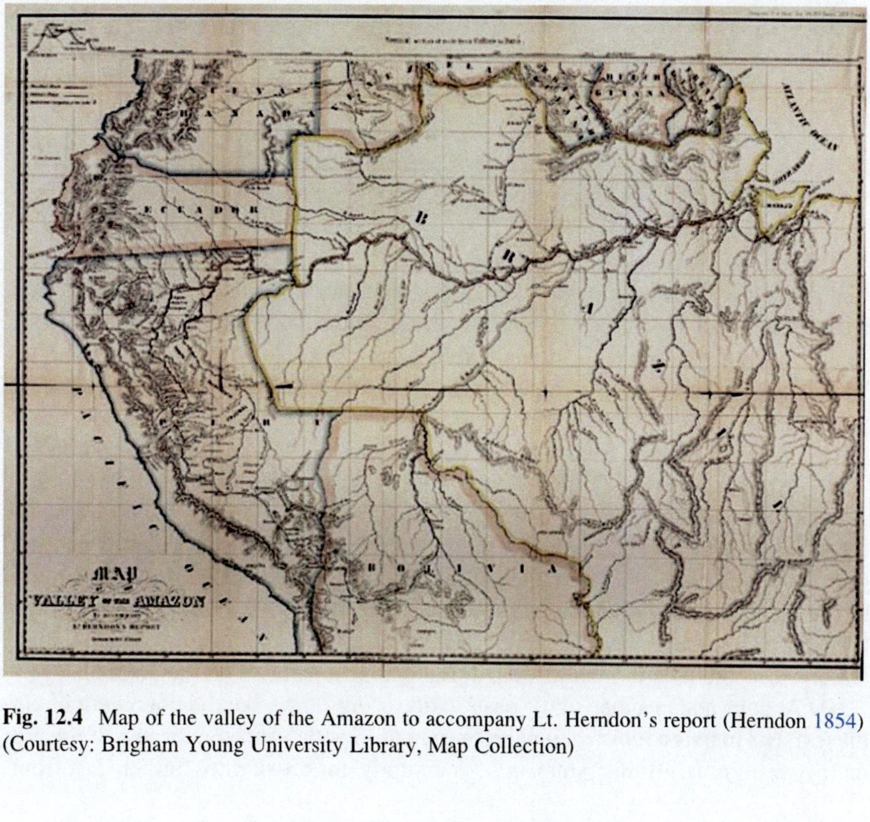


**Fig. 12.4** Map of the valley of the Amazon to accompany Lt. Herndon's report (Herndon 1854) (Courtesy: Brigham Young University Library, Map Collection)

top of the map provides data of the route between Callao in Peru and the mouth of the Amazon River near Belém do Pará.

## South American Reactions

Maury was eager to reveal his proposal to the South American governments. His book about the "Amazon and the Atlantic Slopes" was translated into Spanish and published in Bolivia and Peru in considerably different versions (Pease 1992). The Brazilian journal *Correio Mercantil* published passages of Maury's writings in 1853. By addressing the problem to the neighboring countries of Brazil, Maury attempted to convince these governments to side with the United States and put pressure on Brazil to open Amazonia for the international community. However, Maury's words did not find sympathy, but provoked nationalistic reactions in South America due to their "arrogant and aggressive tone, as well as their gratuitous assumption that the free navigation was already practically accomplished" (Martin 1918, p. 152). Defenders of national autonomy and sovereignty compared the American initiative to a continuation of Manifest Destiny in the southern hemisphere (Harrison 1955; Sternberg 1987). Especially Brazil showed major concerns about the U.S. presence in the Amazon:

> Those Brazilians who had regarded the United States with something akin to fear and dislike now found their worst suspicions confirmed. In many quarters the conviction began to gain ground that the United States harbored sinister designs on certain countries of South America, a conviction which the outcome of the recently concluded Mexican War, our filibustering expeditions to Cuba and Central America and the bombardment of Greytown served to reinforce. It was assumed that once American citizens had established themselves in the southern continent they would introduce their institutions, insist on self-government, and eventually demand annexation to the United States (Martin 1918, p. 152).

The Italian-born scholar Pedro de Angelis (1784–1859), Napolitan by birth, Bonapartist by conviction, and resident in Argentina by choice (Pease 1992), was one of the most passionate defenders of Brazilian interests (De Angelis 1854, 1857). As an international corresponding member of the Brazilian Institute of History and Geography, he counterattacked the geopolitical ideas of the Americans whom he considered "daring and strong, and powerful in men, canons, and warships" (De Angelis 1857, p. 112). De Angelis's arguments were based on "natural rights of use," customs, and sovereignty rather than on international geopolitical interests. For him, the United States did not appeal for equal rights for all nations to access the Amazon, but for "the law of the strongest which the human species has already repudiated forever" (De Angelis 1857, p. 10). His reply to the *marino sureño* ("mariner from the American South"; Pease 1992) reflected the idea of an independent South America that maintained its unity against other nations that intended to manifest territorial claims territories on the continent.

De Angelis added a map to his observations (Fig. 12.5), but did not mention it in his text. His map is a more regional representation of the Amazon dispute. Centered on the western part of the Amazon, it is abruptly cut off slightly further East from

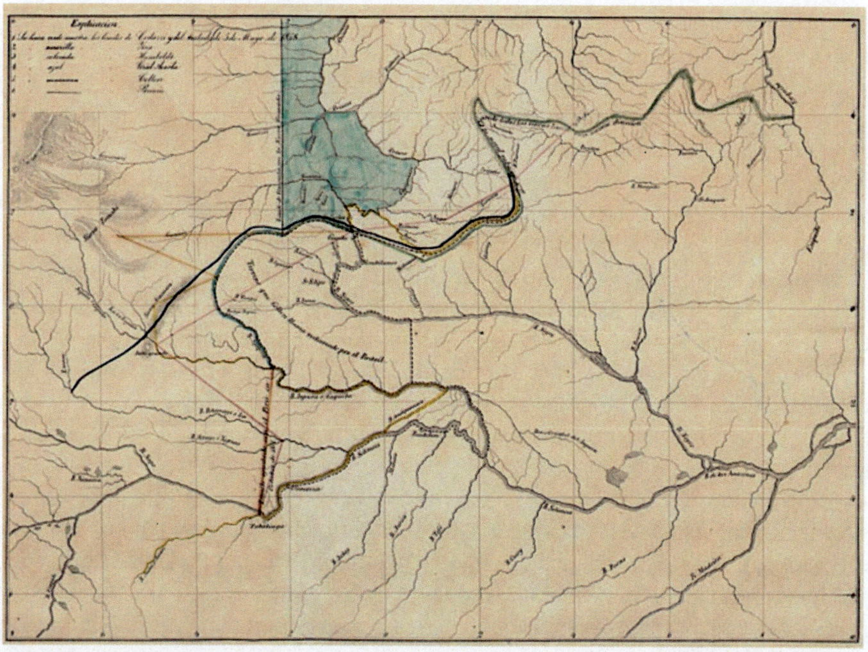

**Fig. 12.5** Untitled map of the western Amazon (De Angelis 1857) (Copyright by Harvard University Open Library Program)

the confluence of the Solimões River and the Rio Negro near Manaus. Previous official border delimitations between the countries proposed by authors such as Alexander von Humboldt, Father Zea, and General Acosta are added to the drawing. The vast Amazon basin is represented as a South American space shared and/or disputed by South American nations, namely Colombia, Peru, and Brazil.

Much paper was wasted to defend either the U.S. American or the Brazilian point of view. However, a few years later, Brazil decided to open up its frontiers for immigrants voluntarily (Tavares Bastos 1866; Hill 1936), including for the Confederate refugees from the Civil War who settled with little success in the Amazon region (Harter 1985) and more significantly and permanently in the Brazilian South East (Dawsey and Dawsey 1995; Griggs 1987) – without the need of political discourse and maps.

## Final Considerations

This paper is an exploratory attempt to dialogue with historical maps, be they facts or fiction. Even map jokes or hoaxes could be used as inspirations, triggers, or gateways for further research. I have only presented a few cartographic moments in the case of the systematic exploration and territorial definition of the Amazon in the mid-1800s. In fact, the maps failed to translate the complex political discourse behind them, but even so, they could stand as "propositions" (Wood and Fels 2008) that invite us to plunge deeper into the debates on international and national geopolitics and cartographic anxiety (Gregory 1994).

The fraudulent historical map that I presented in this paper circulated on the internet and has even been quoted as if it were a real historical document. Even as a piece of fiction, "the document is not the product of pure imagination; it reinterprets historical events" (Mitchell 2010, p. 92), in this case Matthew Fontaine Maury's plan to colonize the Amazon. The false internet maps of the Amazon represent only one case among many others that have had an impact on the media and the public opinion in recent years (another popular example is a map of the U.S.-Mexican border from the times before the Mexican-American War that Absolut Vodka used as a background for their advertisements and that provoked a storm of protests in the United States).

More attention should be paid to the reading between the lines of digital maps. New technologies for visualization and the increasing digital conversion of historical maps call for a new research agenda that takes into account the impact of maps and their use in a specific context and out of context, whereas the borders between facts and myths become blurred and the emphasis is not on the secrecy but on the ways the information is spread. Maps can be extraordinary sources for rumors, and the historians of cartography must not underestimate the tremendous iconicity of maps in the processes of communication and the consequences of reading the past through the lens of the present.

With these examples, I wanted to show that historians of cartography should not only think about silences and secrecy in maps, but also about "a greater pluralism of cartographic expression" (Harley 1989, p. 87). In J.B. Harley's words,

> We need many different map texts that will serve the breadth of social and historical problems that now engage our attention. And we need a narrative cartography that tells a story and portrays at the same time as it is revealing the interconnectedness of humanity in space (Harley 1989, p. 87).

The more different maps about specific places and spaces we have, the more stories can be revealed, be they fact or fiction, testimonies of historical events or reinventions of the past.

## Biographical Note

**Jörn Seemann** is a cultural geographer, cartographic educator, and historian of geography from Hamburg (Germany). He has recently been awarded a Ph.D. in geography at Louisiana State University. In his dissertation with the title "Regional Narratives, Hidden Maps, and Storied Places: Cultural Cartographies of the Cariri Region, Northeast Brazil," he presented four different qualitative mapping strategies in order to investigate the relations between geographical knowledge and representation, society and cartography, culture and maps. He has been an assistant professor in the area of cartography at Universidade Regional do Cariri in Northeast Brazil since 2002. His main research interests include cultural perspectives in cartography, history and histories of cartography and geography, and cartographic education. He has published more than 20 articles on these topics in Brazil and edited a book with the title "A Aventura Cartográfica: Perspectivas, Pesquisas e Reflexões sobre a Cartografia Humana" (2006).

## References


Almeida, Paulo Roberto de (2000a) Dossier Amazônia 1. Primeira Onda de Boatos, Maio-Junho de 2000. Available at http://www.pralmeida.org/04Temas/07Amazonia/01DossierAmazonia1.doc. Accessed 11 Nov 2010

Almeida, Paulo Roberto de (2000b) Dossier Amazônia 2. Segunda Onda de Boatos, Outubro-Novembro de 2000. Available at http://www.pralmeida.org/04Temas/07Amazonia/02DossierAmazonia2.doc. Accessed 11 Nov 2010

Almeida, Paulo Roberto de (2001) Dossier Amazônia 3. Terceira Onda de Boatos, Novembro de 2001. Available at http://www.pralmeida.org/04Temas/07Amazonia/03DossierAmazonia3.doc. Accessed 11 Nov 2010

Andrews JH (2001) Meaning, knowledge and power in the map philosophy of J.B. Harley. In: Harley JB (ed) The new nature of maps: essays in the history of cartography. Johns Hopkins University, Baltimore, pp 1–32

Bell J, Whitfield J (1939) The relation of Herndon and Gibbon's exploration of the Amazon to North American slavery, 1850–1855. Hisp Am Hist Rev 19(4):494–503

Belyea B (1992) Images of power: Derrida/Foucault/Harley. Cartographica 29(2):1–9

da Costa TG (2001) Brazil's SIVAM: As it monitors the Amazon, will it fulfill its human security promise? ECSP Report 7:47–58

Dawsey CB, Dawsey JM (1995) The Confederados: old south immigrants in Brazil. University of Alabama Press, Tuscaloosa

De Angelis P (1854) De la navigation de L'Amazone: réponse a un Mémoire de M. Maury, Officier de la Marine des Etats-Unis. Imprimerie du Rio de la Plata, Montevideo

De Angelis P (1857) De la Navegación del Amazonas: respuesta a una Memoria de M. Maury, Official de la Marina de los Estados Unidos. T. Antero, Caracas

Dozer DM (1948) Matthew Fontaine Maury's letter of instruction to William Lewis Herndon. Hisp Am Hist Rev 28(2):212–28

Edney MH (2005) The origins and development of J. B. Harley's cartographic theories. Cartographica 40(1/2):1–143

Gregory D (1994) Geographical imaginations. Blackwell, Oxford

Griggs WC (1987) The elusive Eden: Frank McMullan's confederate colony in Brazil. University of Texas Press, Austin

Harley JB (1988a) Silences and secrecy: the hidden agenda of cartography in early modern Europe. Imago Mundi 40:57–76

Harley JB (1988b) Maps, knowledge, and power. In: Denis Cosgrove, Daniels S (eds) The iconography of landscape: essays on the symbolic representation, design and use of past environments. Cambridge University Press, Cambridge, pp 277–312

Harley JB (1989) Historical geography and the cartographic illusion. J Hist Geogr 15(1):80–91

Harrison JP (1955) Science and politics: origins and objectives of mid-nineteenth century Government expeditions to Latin America. Hisp Am Hist Rev 35(2):175–202

Harter EC (1985) The lost colony of the confederacy. University Press of Mississippi, Jackson
Lewis Herndon W (1854) Exploration of the valley of the Amazon, made under the direction of the Navy Department. Part 1, 1854th edn. Robert Armstrong, Washington DC
Hill LF (1936) The confederate exodus to Latin America. Texas State Historical Association, Austin
Lewis CL (1927) Matthew Fontaine Maury: the pathfinder of the seas. The United States Naval Institute, Annapolis
Martin PA (1918) The influence of the United States on the opening of the Amazon to the World's Commerce. Hisp Am Hist Rev 1(2):146–62
Matthew Fontaine M (1853) The Amazon, and the Atlantic slopes of South America: revised and corrected by the author. Franck Taylor, Washington DC
Maury MF (1854) Annual address. Bull Am Geograph Stat Soc 1(3):3–31
Mitchell ST (2010) Paranoid styles of Nationalism after the Cold War notes from an Invasion of the Amazon. In: Kelly JD et al (eds) Anthropology and global counterinsurgency. University of Chicago Press, Chicago, pp 89–103
de Pease MM (1992) Historia e traducción: presentación de una Situación Intra-Americanista. Livius 2:189–202
Raisz E (1962) Principles of cartography. McGraw-Hill, New York
Sampaio F (1994) O Dia em que Napoleao Fugiu de Santa Helena. SM&B, Porto Alegre
Sternberg HR (1987) "Manifest destiny" and the Brazilian Amazon: a backdrop to contemporary security and development issues. CLAG Yearbook 13:25–35
Tavares Bastos AC (1866) O Valle do Amazonas. Estudo sobre a Livre Navegação do Amazonas, Estatistica, Producções, Commercio, Questões Fiscaes do valle do Amazonas. B.L. Garnier, Rio de Janeiro
Wood D (1992) The power of maps. Guilford Press, New York
Wood D (2002) The map as a kind of talk: Brian Harley and the confabulation of the inner and outer voice. Vis Commun 1(2):139–61
Wood D (2010) Rethinking the power of maps. Guilford Press, New York
Wood D, Fels J (2008) The natures of maps. University of Chicago Press, Chicago

# Chapter 13
# Theoretical Frameworks for the Study of Journalistic Maps: South American Borders in the Brazilian Press

**André Reyes Novaes**

**Abstract** This paper discusses how the theoretical frameworks proposed by Blakemore and Harley (1980) for the history of cartography can be applied to study of journalistic maps representing South American borders in the Brazilian press. Among the different concepts used by geographers and cartographers to study the influence of cultural meanings on maps, two theoretical frameworks are particularly emphasized: *iconography* and *semiotics*. In the light of these contributions, this paper discusses approaches to explore Brazilian journalistic maps concerning South American borders during the 1970s and 1990s. The role of media maps in the construction of geopolitical imaginations in the continent is still being developed. In this "un-chartered territory", the discussion of Brazilian journalistic maps suggests a persistent representation of borders as threatening and dangerous places.

## Introduction

The study of the use of maps in the media has been influenced by different epistemological models of cartographic knowledge. Different ideas about what a map is, and how maps should be understood, have been indirectly or directly applied to the study of journalistic cartography. This paper discusses how the theoretical frameworks proposed by Blakemore and Harley (1980) for the history of cartography can be applied to the study of journalistic maps representing South American borders in the Brazilian media.

Rather than identifying "deficiencies", "errors" and "inaccuracies" (Balchin 1985:339), or trying to find "solutions" for these maps (Gauthier 1988:5), the aim of this paper is to understand Brazilian journalistic cartography as an historical

---

A.R. Novaes (✉)
Universidade do Estado do Rio de Janeiro, Rio de Janeiro, Brazil
e-mail: andrernovaes@yahoo.com.br

E. Liebenberg and I.J. Demhardt (eds.), *History of Cartography*,
Lecture Notes in Geoinformation and Cartography 6,
DOI 10.1007/978-3-642-19088-9_13, 

document of geopolitical imaginations. Among the different concepts used by geographers and cartographers to study the influence of cultural meanings on historical maps, two theoretical frameworks are particularly emphasized, namely *iconography* and *semiotics*.

Blakemore's and Harley's (1980) monograph on concepts in the history of cartography is a good example of the use of these two theoretical frameworks in cartography. Iconography, as discussed in chapter six of their monograph, is an "immediately applicable theoretical structure" for cartographic history (p. 76). The contributions from the "linguistic model" are discussed in chapter seven, which argues that semiotics "offers the most appropriate underlying structure for the history of cartography" (p. 87). In the light of these contributions, this paper briefly introduces the use of iconography and semiotics as theoretical approaches to explore Brazilian journalistic maps representing South American borders during the 1970s and the 1990s. These two different periods in South America geopolitics have been frequently studied in separate and distinct ways.

The dictatorship period during the 1970s, widely studied by scholars in North America and Europe (e.g. Pittman 1981; Child 1979, 1985; Child and Kelly 1988; Dodds 1993; Kacowicz 2000), has often been discussed as a period characterized by "applied geopolitics". During this decade, geopolitical metaphors played an important legitimizing role in military public policies, thus justifying the occupation and militarization of the country's borders (Hepple 1992). In contrast to this, the 1990s were frequently described as a "borderless" period designated by events such as the end of the Cold War, the transition towards democratic governments in South America, and the development of institutions to promote economic integration (e.g. Mercosul).

The role of media maps in the construction of geopolitical imaginations in the continent has yet to be developed. In this "unchartered territory", the discussion of Brazilian journalistic maps suggests a persistent representation of borders as threatening and dangerous places, stimulating politics for colonization and militarization. Historically, Brazilian borderlands are situated far away from the major centres of political decision-making and far away from the major cities which are concentrated on the Atlantic coast. In an imaginative sense, much of the Brazilian countryside was for centuries perceived in the same way, namely as "a vast and uninhabited swath of *terra incognita*" (Safier 2009:139).

It is obvious that one could not, in the twentieth century, say that any part of the Brazilian borderlands was in a literal sense unknown. However, according to John K. Wright (1947), the qualification of an area as *terra incognita* "depends both on whose knowledge and on what kind of knowledge is taken into account" (p. 3). For that part of the population who wrote and read widely-circulated Brazilian newspapers, the borderlands remained far-way and unfamiliar places, distant from their own living spaces and frequently imagined as empty and potentially dangerous. In this context, journalistic maps can be considered important historical documents on Brazilian geopolitical discourses about South American borders.

I will set the scene by showing how the approach developed here is not commonly used in the study of the use of cartography in the press, and by discussing briefly three tendencies in the study of journalistic maps based on different epistemological points of view: the "Darwinian model", the "map communication model" and the "map as representation model". Having done this, I shall explore some selected maps published in the mainstream Brazilian press during the 1970s and the 1990s[1]. By discussing omissions and highlights in South American borders representations, it is suggested that rather than exhibiting two contrasting periods "from geopolitics to geo-economics", journalistic maps published in the Brazilian press show a kind of "persistence of geopolitics" (e.g. Dalby 1990). While Brazilian borders were during the 1970s often represented in relation to internal colonization and militarization, the representation of borders and neighbouring countries during the 1990s were markedly influenced by the "War on Drugs".

## Tendencies in the Study of Maps in the Press

A large part of the study concerning journalistic maps probably follows what Blakemore and Harley (1980) call the "Darwinian paradigm" which considers stages of development in cartographic "progress" from art to science. This model often classifies journalistic maps as "inaccurate" or "not maps at all", and is clearly based on a normative definition of cartography which disapproves the use of "backward" practices by the media.

Early discussions on popular maps among American scholars often differentiates between a "propagandistic" use where maps are "deprived of their scientific character" (e.g. Speier 1941), and the scientific use of maps which is associated with "objectiveness" and accuracy. This kind of contra-distinction can also be identified in more recent works which emphasize that news maps "reflect an ignorance of map projections or cartographic conventions" (Monmonier 1989:14). According to Harley (1996), maps in the press were, by the end of the twentieth century, still evaluated by many cartographers according to standards of "objectivity" "accuracy", and "truthfulness". Although the idea of "progress" in cartography has been critically discussed (Edney 1993), many cartographers are resistant in accepting that maps in the press can be "cartographically relevant" for the entire discipline (Ristow 1957; Schulten 1998).

By using "pre-scientific" practices, this cartographic genre is frequently associated with "old maps" which were made for "illustrative purposes" at a time of "geographical ignorance" (e.g. Speier 1941:314). Writing in a period of much experimentation and developments in the language of journalistic maps, J. K.

[1] The examples discussed in this paper derive from a previous PhD research on the Brazilian press in which I selected more than 250 journalistic maps depicting South America borders from 1964 to 2008. The selected newspapers (*Folha de São Paulo, O Globo, Jornal do Brasil, Estado de São Paulo*) and magazines (*Veja e Istoé*) represent the mainstream press in the country during the studied period (Silva 1989).

Wright (1942) commented on this relation between "old" and "popular" maps as follows: "medieval maps are adorned with castles, towers, sea monsters, ships, and such things, and the employment of pictorial symbols is coming back today. It has certain advantages, "especially on maps for popular use"" (p. 542).

If we consider the development of "scientific cartography", pictorial symbols necessarily migrated to the borders of the image, where designers were "free of the restraints controlling the cartographers who worked on the maps proper" (Rees 1980:64). Maps in the press, however, keep using pictorial symbols in the center of the image. When commenting on Charles Owen's maps during the Second World War, (Cosgrove and Dell Dora 2005:382) pointed out how pictorial geography frequently "invades" the map in order to create narratives that move over time and space. Although Brazilian newspapers and magazines were relatively poor in their own production of maps until the end of twentieth century, many maps published after the 1970s began to use the cartographic practices associated with "old" maps. The use of pictorial symbols, explicative texts and the exploration of different kinds of "bird-eye-view" projections are examples of cartographic practices often used in the Brazilian press in the late twentieth century (Figs. 13.1 and 13.2).

These two cartographic practices can be useful to represent the other side of the border as a dangerous place. While Fig. 13.1 uses pictorial symbols and a red arrow to show how insurgents in Colombia can attack Brazilian borders, Fig. 13.2 explores an unusual view of the South American continent, using this perspective to emphasize how closely situated Brasil is to many countries with internal conflicts. It is evident that in this kind of map "explanation and drama are more important than cartographic accuracy" (Cosgrove and Dell Dora 2005:382). Using the Darwinian model, scholars can keep studying these maps to identify problems and "misleading aspects" (Balchin 1985:339), and criticize them for their lack of cartographic conventions. The question, however, is what the alternatives to the study this kind of map are when going beyond this normative judgement?

In order to answer this question, it is important to understand that the definition of what is a map and how maps should be constructed has developed different tendencies since the second half of the twentieth century. During this period the diffusion of the so called "communicative model in cartography" brought new ideas to the study of journalistic maps. The exploration of "map effectiveness", taking

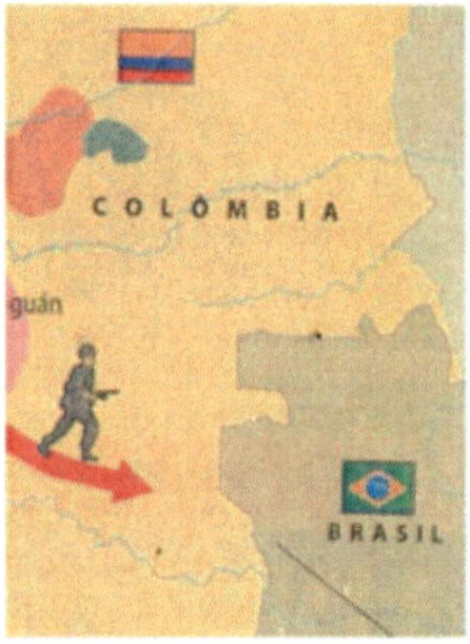


**Fig. 13.1** *Jornal do Brasil*, August 31, 1986 (Copyright for this image by Jornal do Brasil)

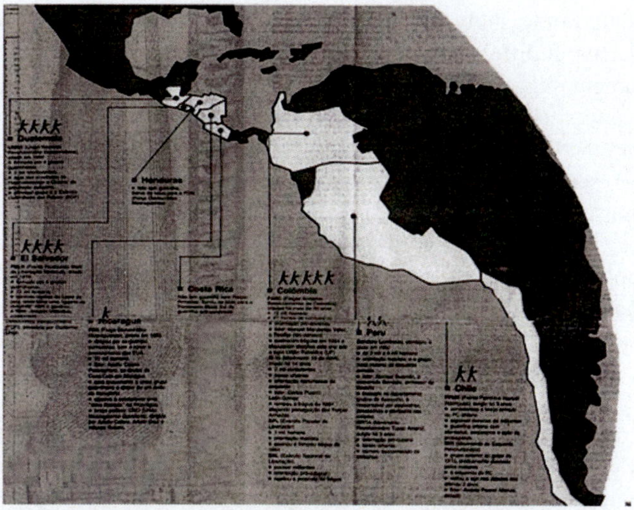


**Fig. 13.2** *Folha de São Paulo*, August 20, 1999 (Copyright for this image by Folha de São Paulo)

into account the capabilities of the map reader, helped the development of new understandings about the use of maps as communication tools. However, according to Crampton (2001:694), under the communicative model the goal is to produce "a single, optimal (best) map" and, following this approach, the study of journalistic maps tends to find "solutions" (Gauthier 1988:5) rather than to discuss the production, diffusion or cultural context of the maps.

These approaches stimulated the idea that graphic artists need courses in scientific cartography to know and understand the rules of graphic communication. Explicitly quoting important cartographers associated with the map as communication model, such as Jaques Bertin and Arthur Robinson, authors like Gauthier (1988), Bodin (1988) and Gronoff (1988), for example, tried to discuss "rules for efficient automation" in journalistic maps by using "graphic semiology". The aim here is to use semiotics to provide a conceptual framework to develop maps that can take advantage of our knowledge of cognitive representations (Maceachen 1995:214).

In contrast to this "applied" use, semiotics can also deal with meaning and power relations and offer a way to integrate approaches to map representations, emphasizing both explicit and implicit meanings in journalistic maps (Maceachen 1995). When Blakemore and Harley (1980) discuss semiotics as a theoretical framework for the study of historical maps, they are turning in this critical way, using the textual metaphor to explore different meanings associated with cartography. In the same way, iconography can also contribute to the understanding of journalistic maps with regard to their cultural meanings and social context.

These theoretical frameworks are clearly related to the "map as representation model" (Crampton 2001) where no one map can be considered "better" than another. The aim here is not to criticize news maps, neither to propose solutions for their design, but to study journalistic cartography as historical documents that can inform us about different ways of seeing places.

This argument suggests that semiotics and iconography can be used to "interpret" and "contextualize" rather than to "construct rules" for journalistic maps. Going beyond the discussion of cartography in the press as a "good or bad" genre (Green 1999), some scholars have been following Harley's suggestions and have considered how these maps can be studied in relation to questions relating to power and knowledge. By discussing the role of media maps in the construction of national identities (Kosonen 1999) and security discourses (Vujakovic 2002), many authors have used the map "as representation model" in the study of journalistic maps, sometimes explicitly relating iconography, semiotics and geopolitical imaginations (e.g. Vujakovic 2002). Although it is important to recognize that Brian Harley actually gives very little advice on the methodological application of these theoretical frameworks, the next section seeks to show how the use of iconography and semiotics can contribute to an "Harleian" research agenda on maps in Brazilian mainstream newspapers.

Interested in overcoming purely formal analysis in artistic images, art historians working during the early twentieth century used terms such as "iconography" and "iconology" to discuss cultural meanings in artistic representations (Heckscher 1994). Nowadays it is possible to say that iconography stands at the "crossroad" of several disciplines, "all of which draw upon images as a source of information" (Cassidy 1993:3). According to Blakemore and Harley (1980), historians of cartography have tended to neglect meanings within maps and iconography, "with its long tradition of studying symbolic meanings within graphic representation" (p. 76), can help to construct an interpretative approach in cartography.

It was also during the early twentieth century that semiotics was developed. Having close relations with linguistics, this theoretical framework was widely applied in the analysis of images in the media in the second half of the century (e.g. Barthes 1977). By seeking to interpret different layers of meanings within an image (denotation and connotation), the tendencies of a semiotic study are clearly similar to an iconographic analysis (Hasenmueller 1978). However, as a "science of signs", semiotics can emphasize the study of the specific language of journalistic maps, describing and interpreting cartographic signs and their meanings. Blakemore and Harley (1980) also highlight the necessity to view maps as a type of language, using semiotics to show how "the language of maps - like other languages - is a living entity and has constantly undergone change through its use and interactions within society" (p. 93).

In the light of these contributions the next sections seek to discuss Brazilian journalistic maps representing borders as a source of information and a specific kind of language. First, I will introduce an iconographic analysis in order to show how Brazilian borders were imagined in relation to exploration and colonization during the 1970s. Having done so, I will explore some semiotic concepts to discuss how the representation of borders during the 1990s was strongly influenced by the new geopolitical factor in the continent, the so-called 'War on Drugs'.

## Colonization and Militarization in Brazilian Borders During the 1970s

Published in 1974 in the *Jornal do Brasil* newspaper, the map below (see Fig. 13.3) demonstrates how the geopolitical interpretation of Brazilian borders as an empty space to be colonized and militarized influenced the construction of journalistic maps during the 1970s. Picturing how the Brazilian military government was occupying the "massive empty space, leading security and civilization to these territories" (*Jornal do Brasil*, August 26 1974), this map depicts the focus on ideas about colonization and the vigilance on border representation during the military dictatorship.

Brazilian journalistic cartography during the 1970s was relatively rare and publications frequently reproduced maps drawn by other institutions. This is the case with this hand-drawn map that does not have pictorial symbols, big arrows, innovative projections or explicative texts, and which was reproduced from a military publication. In order to discuss this simple location map I would like to use two trends related to an iconographic approach. On the one hand it is important to identify different layers of meanings in the map, following Harley (1988)

Fig. 13.3 Jornal do Brasil, August, 26, 1974 (Copyright for this image by Jornal do Brasil)

suggestion to "equate Panofsky's levels of interpretation in painting with similar levels discernible in maps" (p. 279). On the other hand we should consider the historical context in which the image was produced and diffused, articulating image history and cultural history (Woodfield 2001).

The map shows the geographical distribution of the army on the Brazilian borders, localizing with extreme simplicity the military's presence in border cities (using triangle symbols) and in some capitals (using circle symbols). No city or political divisions are shown in the other side of Brazilian borders, and we identify neighbouring countries simply by reading their names. This description of "pure forms" and "natural meaning" of the map can be classified as a preiconographical analysis in Panofsky terms. An iconographical level dealing with "conventional meaning" can discuss how this map has a specific subject matter: the colonization and militarization of the Brazilian border during the 1970s. The third level of interpretation, iconology, seeks to find hidden meanings and symbolic values related to the image, discussing silences and secrecies in order to identify the relation between the image and the "basic attitudes of a nation, a period, a class, a religion or philosophical persuasion" (Panofsky 1972).

By considering these "intrinsic meanings" at a iconological level, it is possible to discuss how symbolic interpretations on South American borders can be identified taking into account the meanings "communicated by the blank spaces on the map" (Harley 2001:88). If the representation of the Brazilian army on the border could be interpreted in relation to the military government's conceptions, the blank space on the other side of the border could equally well demonstrate how the focus of the map is exclusively related to the protection of the Brazilian territory.

The absence of important cities and political delimitations in the other countries can be interpreted as "cartographic silence" (Harley 2001). By omitting the neighbouring cities and highlighting the presence of the army, the map does not recognize the pre-existent local articulations between important cities on both sides of the border. This representation encourages the idea that only the Brazilian state, by means of militarization and infrastructure, can bring development to these places. As Harley (2001) points out, "through these silences, the map becomes a license for the appropriation of the territory depicted" (p. 105), providing a justification for colonization and vigilance policies.

The discussion of this kind of emphasis and omission in journalistic cartography is obviously important if one wants to become familiar with the historical contexts and symbolic values related to the map. Otherwise it would be difficult to "decipher the deeper subconscious structures and values were communicated to the original map maker and users" (Blakemore and Harley 1980:80). In order to discuss what this map could have meant to a reader during the 1970s, it is important to understand how the military government explored ideas about the formation of Brazilian borders and the colonization of the national territory.

The history of Brazilian border delimitation is strongly related to the exploration and the colonization beyond imaginary lines on the maps. During the military dictatorship (1964–1985) the idea that the border should be occupied found strong theoretical support in Brazilian geopolitical ideas of the early twentieth century.

Travassos (1935), for example, argued that Brazilian destiny was to be found in expanding westwards towards the Amazon, whereas Meira Mattos (1960) defined Brazilian borders as a "skin of a growing organism" (Kelly 1984). In this context, Brazilian authors often discussed national border delimitation using terms such as "March to the West" (Ricardo 1942), "Frontier in March" (Mendonça 1956) and "Continental Projection" (Travassos 1935).

The "empty space" was an important theme in Brazilian geopolitical writing (Child 1979) and influenced many important scholars during the dictatorship, such as Couto e Silva. Golbery (1967), highlighting the necessity to occupy "empty hinterlands and unguarded borders". In order to colonize Brazilian "*terra incognita*", the military regime made massive investments to integrate and protect the national territory. By considering this context, the map discussed in Fig. 13.3 can be understood as a graphic expression of the intention related to colonization and the militarization of the border. The presence of the army appears in the entire extension of the Brazilian border, diffusing the idea that the "1964 revolution brought new directions for Brazil and his army, improving the conquest of the 'Green Hell' and colonizing the immense border area" (*Jornal do Brasil*, August 26 1974).

Since these maps were published, there has been a series of transformations at local, national and international levels which have affected the representation of the interior borders of Brazil by the media. Locally, the borderlands have become more occupied every year, with the creation of new cities and increased economic activity. Nationally, the dictatorship was replaced by a new democracy with less emphasis on the language of internal colonization. Internationally, new interpretations about globalization emerged, putting the emphasis on free trade and a borderless world. In the light of these changes, it is necessary to consider the extent to which the representation of borderlands changed during the 1990s.

## Neighbors, Illegal Activities and the Persistence of Geopolitics During the 1990s

The classical geopolitical usage of the terms "frontier" and "border" highlight their military aspects. The term "frontier", for instance, derives from the idea of a battle "front" (Foucher 1991), a place of danger and potential conflict with one's neighbour. In addition, according to Newman (2006), the language associated with borders is defined by binary distinctions between "us" and "them", and the "included" and the "excluded", so that "the traditional function of borders has been to create barriers to movement rather than bridges enabling contact" (p. 150).

In the face of a series of transformations in international and regional geopolitics during the last decades, many theorists have chosen a "cross-boundary" vocabulary for their studies, using terms such as "transition zone" and "points of interaction" (Newman 2006). Since the late 1980s, the increase in the influence of economic blocs like the European Union (Donnan and Wilson 2001) has often been associated with the coming of a so-called "borderless world". This context has been influencing some studies about geopolitics and borders in South America. According to Kacowicz (2000), the emphasis in South American politics during the 1990s was on the role of economics ('geo-economics') rather than on just territory as epitomized in the case of Mercosur (p. 83).

Considering the changes in South America geopolitics between the 1970s and the 1990s, an increase in the circulation of media maps picturing bridges and interactions between South American countries was to be expected. In portraying a South America with "erased" borders, Brazilian newspapers should have followed the tendency of some maps published in the international media (see Fig. 13.4) which commonly highlighted the "borderless" period in the history of the continent.

If we look at maps and articles concerning South American borders in Brazilian newspapers and magazines during the 1990s, we see how geopolitical discourses and practices producing and maintaining the border remained very significant. In contrast with some academic analyses and international representations, questions about military actions, international threats and territorial sovereignty

**Fig. 13.4** The Economist, October, 12, 1996 (Copyright of this image by The Economist)

were often associated with borders in popular geopolitics in the Brazilian press. The new US-South America relations and the increase of international illegal drug traffic were both important elements to understand the persistence of geopolitics in Brazilian imaginative geographies on national borders during this period.

The end of the Cold War changed the relation between South American countries and US foreign policy. With the dissolution of the Soviet Union, the strategy of containing communism lost its driving foundation, but new constructions of containment have since emerged. According to Sharp (2000), the most prominent alternatives to communism include terrorism, drugs dealers and economic threats. Particularly important to South America is the new US emphasis on combating the drug traffic with military action beyond American borders (Olmo 1990). The popularly known "War on Drugs" stimulates many South American countries to militarize their borders in order to combat the "illegal enemy". This tendency can justify a new geopolitical moment in the representation of South American borders in the Brazilian press.

During the 1990s, the Brazilian media increasingly characterized neighbouring states with negative connotations such as "threat" or "insecurity", usually in the context of cross-border drug traffic and other illegal activities. The borderlands themselves were more frequently characterized as "the entrance gate" to the country's misfortunes, focussing on the geopolitics discourse in which the neighbour is the enemy. Consequently the other side of the border started to have a higher representation in maps instead of being merely ignored with "silences" such as was evident during the 1970s.

The map below (Fig. 13.5) which was published in 1997 is a clear example of how Brazil is presented as a victim of activities which originate elsewhere in Latin America and violate the open borders. This discourse is evident in the sub-heading of the map: "this border belt is the origin of the violence which will burst out in several forms in the Brazilian urban centres" (*O Estado de São Paulo*, May 15 1997). In this new context, borders are perhaps, more than never, related to danger and military actions.

Compared with the maps shown previously, this map reflects significant improvements in the production technology used by the media. It uses several colours and symbols to represent the relationships between Brazil and its neighbouring countries. However, not only the technical aspects changed in journalistic cartography during the 1990s, but also the emphasis on the processes taking place in neighbouring countries. In contrast to the blank spaces in the maps produced during the dictatorship, here the threats towards the Brazilian territory come from activities which originate on the other side of the border. The coloured arrows entering Brazil and the signs which signify other South American countries clearly show this renewed geopolitical context.

By considering the different kind of signs used to qualify neighbouring countries and borders, we can use some concepts developed in semiotics to analyse the relation between "signifier" and "signified", cartographic signs and cultural meanings. The "cartographic alphabet" explicitly defined by the legend uses different kind of signs, employing indices and symbols to qualify border spaces. Taking

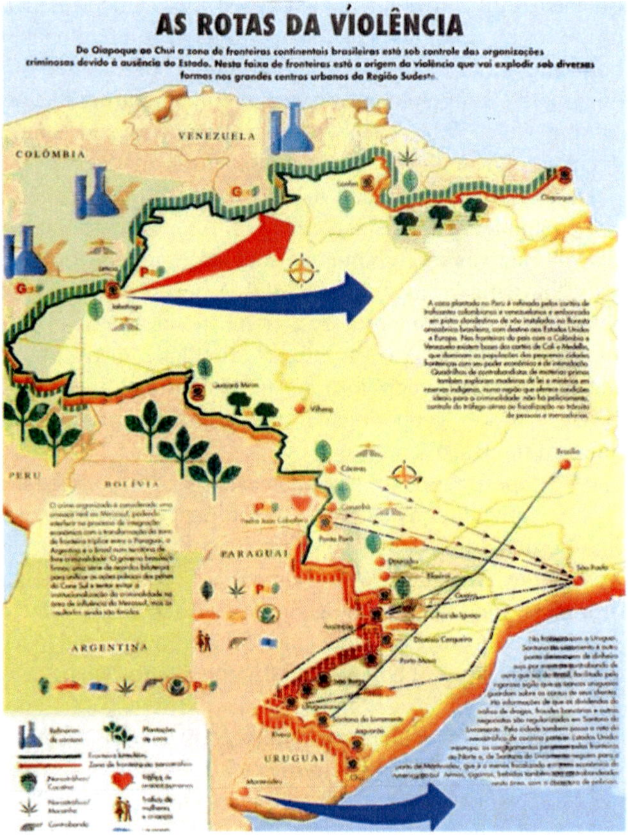


**Fig. 13.5** O *Estado de São Paulo*, May 15, 1997 (Copyright for this image by O Estado de São Paulo)

into account Peirce and Charles Sanders (1932) definitions of signs, the test tubes can be classified as an example of indices which represent laboratories that process illicit drugs by association. The emblems of the Brazilian Federal Police are examples of symbols, identified by conventions and representing the attempt to control illicit activities at the borders.

According to Harley (1988), the association of particular emblems with a territory is a potent means of cartographic representation. This is particularly clear when we consider the test tubes in Colombia and Venezuela, or the coca leaves in Bolivia and Peru. The coca refineries, which actually occupy relatively restricted locations, are here represented as synonymous with entire regions along the border (Machado 2000; Novaes and André Reyes 2005). In turn, coca leaves occupy the larger part of the Peruvian and Bolivian territories. By associating such localized cross-border activities organized in an international network with whole countries and regions (Machado 2000), these cartographic symbols help to produce

**Fig. 13.6** O *Estado de São Paulo*, May 15, 1997 (Copyright for this image by O Estado de São Paulo)

and reproduce specific kinds of imaginative geographies regarding South American borders.

The signs chosen for the small image present on all pages of this special report (see Fig. 13.6) summarize a clear geopolitical message: the illicit drugs (coke and cannabis) are on the other side of the border and the Brazilian army should close the border to protect the country against the "entrance of the evil". Moving from a denotative description to a connotative interpretation, these "signifiers" (plants and arms) can be interpreted as expressions of a renewed geopolitical context in South America. By arguing that "organized crime is considered a real threat to Mercosul", and discussing how it can be an "obstacle for the economic integration" (*O Estado de São Paulo*, May 15 1997), this type of discourse does not stimulate the diffusion of an imaginative geography related to a "borderless" world in South America.

## Conclusion

The relationship between journalistic cartography and geopolitical discourse becomes clear when we consider how the national press uses cartography to represent Brazil's borders and neighbours. Representing spaces unknown to the majority of the population which makes and reads the newspapers, these maps can be an important influence on how people imagine their country's borders. Historically related to empty places and closed systems, Brazilian borders often play a significant role in geopolitical imaginations inside the country. As we saw in the set of examples discussed here, these imaginations regarding South American borders exhibit both changes and continuities.

Journalistic cartography during the two periods referred to – the 1970s and 1990s – presents many contrasts. While the first example was related to the colonization and occupation of the borderlands of the state, the second was more concerned with the drug traffic and related illegal activities; while the first was hand-drawn, the second was produced using computer graphics. In spite of these differences both imaginative geographies about the borderlands can result in the same material practices in the territory: militarization, control and vigilance.

These conclusions suggest that rather than be considered as "backwards" and "not maps at all", maps in newspapers and magazines can be studied as historical documents where both iconography and semiotics can assist the map-user to construct new ways to study these images. By considering journalistic maps as "representation models", it is possible to stimulate new research agendas which use images in the South American media as sources of information on different historical subjects.

## Biographical Note

**André Reyes Novaes** works at the State University of Rio de Janeiro since 2006 and received his PhD in 2010 from the Federal University of Rio de Janeiro. The author was an academic visitor at the Royal Holloway University of London and worked for the Brazilian government in South American borderlands. His research interests include visual methods, popular geopolitics, journalistic cartography and geographical education.

## References

Balchin WGV (1985) Media map watch: a report. Geography 70:339
Barthes R (1977) Rhetoric of the image. Image, music, text. Hill and Wang, New York
Blakemore MJ, Harley JB (1980) Concepts in the history of cartography: a review and perspective. University of Toronto Press, Toronto, Published in *Cartographica* 17, no. 4
Bodin S (1988) Un bilan dês images graphiques (diagrame et cartes) dans la presse francese, 1980–1986. In: Gauthier M-J (ed) Cartography in the media. Presses de l'Université du Québec, Quebec
Cassidy B (1993) Iconography at the crossroads. Princeton University Press, Princeton
Child J (1979) Geopolitical thinking in Latin America. Lat Am Res Rev 14:89
Child J (1985) *Geopolitics and Conflict in South America: Quarrels Among Neighbors*. New York: Praeger
Child J, Kelly P (eds) (1988) Geopolitics of the Southern Cone and Antarctica. Lynne Rienner, Boulder
Cosgrove D, Dell Dora V (2005) Mapping global war: Los Angeles, the Pacific, and Charles Owens's pictorial cartography. Ann Assoc Am Geographers 95:373
Couto e Silva G (1967) Geopolitica do Brasil. Jose Olympio, Rio de Janeiro
Crampton JW (2001) Maps as social constructions: power, communication and visualization. Prog Hum Geogr 25(2):235–252
Dalby S (1990) American security discourse: the persistence of geopolitics. Pol Geogr Q 9:171–188
de Meira Mattos C (1960) Projeção wundial do Brasil. Grafica Leal, São Paulo
del Olmo R (1990) A face oculta da droga. Revan, Rio de Janeiro
Dodds K (1993) Geopolitics, cartography, and the state in South America. Pol Geogr 12(4):361–81
Donnan H, Wilson TM (2001) Borders: frontiers of identity, nation and state. Oxford Press, Oxford/New York
Edney MH (1993) Cartography without progress: reinterpreting the nature and historical development of mapmaking. Cartographica 30(2):54–68
Foucher M (1991) Fronts et frontières – un tour du monde géopolitique. Fayard, Paris

Gauthier M-J (1988) Les cartes et Les diagrammes das les médias: reflexions générales et quelques exemples procédant de la graphique. In: Gauthier M-J (ed) Cartography in the media. Presses de l'Université du Québec, Quebec
Green DR (1999) Journalistic cartography: good or bad? A debatable point. Cartographic J 362:141–153
Gronoff J-D (1988) Approche théorique dl un savoir-faire graphique: étude de quelques produits graphiques des hedomanaires nordamericans, newaweek et time. In: Gauthier M-J (ed) Cartography in the media. Presses de l'Université du Québec, Quebec
Hasenmueller C (1978) Panofsky, iconography, and semiotics. J Aesthetics Art Criticism 36 (3):289–301
Harley BJ (1988) Maps, knowledge and power. In: Denis C, Stephen D (eds) The iconography of landscape. University of Cambridge Press, Cambridge
Harley BJ (2001) Silences and secrecy: the hidden agenda of cartography in early modern Europe. In: Paul L (ed) The New nature of maps. Essays in the history of cartography. The Johns Hopkins University Press, Baltimore
Harley BJ (1996) Deconstructing the map. In: John A (ed) Human geography: an essential anthology. Blackwell Publishers, Oxford
Heckscher WS (1985) The genesis of iconology. In: Egon V (ed) Art and literature, study in relationship. Duke University Press, Durham
Hepple LW (1992) Metaphor, geopolitical discourse and the military in South America. In: James D, Trevor B (eds) Writing worlds. Discourse, text e metaphor in the representation of landscape. Routledge, London
Kacowicz AM (2000) Geopolitics and territorial issues: relevance for South America. Geopolitics 5(1):81–100
Kelly P (1984) Geopolitical themes in the writings of general Carlos de meira mattos. J Lat Am Stud 16:439–461
Kosonen K (1999) Maps, newspapers and nationalism: the Finnish historical experience. Geo J 48:91–100
Maceachen AM (1995) How maps work. Representation, visualization and design. The Guilford Press, New York
Machado LO (2000) Limites e fronteiras. Da Alta diplomacia aos circuitos da ilegalidade. Rev Território 8:7–23
Machado LO, Novaes AR, Monteiro LC (2009) Building walls, breaking barriers: territory, integration and the rule of law in frontier zones. J Borderlands Stud 24(3):97–144
Novaes AR (2005). A iconografia das drogas na pmprensa (1975–2002). Masters Thesis, Universidade Federal do Rio de Janeiro
Novaes AR (2010) Fronteiras mapeadas: geografia imaginativa das fronteiras Sul-Americanas na cartografia da imprensa Brasileira. PhD Dissertation, Universidade Federal do Rio de Janeiro
Monmonier M (1989) Maps with the news: the development of American journalistic cartography. University of Chicago Press, Chicago
Newman D (2006) The lines that continue to separate us: borders in our 'borderless' world. Prog Hum Geogr 30(2):143–161
Panofsky E (1972) Studies in iconology: humanistic themes in the art of the renaissance. Harper Torchbooks, New York
Peirce CS (1932) In: Charles H, Paul W (eds) The collected papers of Charles Sanders Peirce, vol II. The Belkmap Press of Harverd University Press, Cambridge, 1960
Rees R (1980) Historical links between cartography and art. Geogr Rev 70:60–78
Ricardo CL (1970) Marcha Para oeste: a influência da bandeira na formação social do Brasil, 4oth edn. José Olympio, Rio de Janeiro
Ristow WW (1957) Journalistic cartography. Surv Mapp 17(4):369–390
Safier N (2009) The confines of the colony: boundaries, ethnographic landscapes, and imperial cartography in iberoamerica. In: James RA (ed) The imperial map. Cartography and the mastery of empire. University of Chicago Press, Chicago

Sharp J (2000) Condensing the cold war: reader's digest and American identity. University of Minnesota Press, Minneapolis
Schulten S (1998) Richard edes Harrison and the challenge to American cartography. Imago Mundi 5:174–88
Silva CEL (1989) La influencia Americana en el periodismo brasilenõ. Dialogos Comuicacion 24:7–18, Junho
Speier H (1941) Magic geography. Soc Res 8:310–330
Travassos M (1935) Projeção continental do Brasil. Companhia editora nacional, São Paulo
Vujakovic P (2002) Mapping the war zone: cartography, geopolitics and security discourse in the UK press. Journal Stud 3(2):187–202
Woodfield R (2001) Art history as Cultural history. Warburg's Projects, Biel London University Press, Amsterdam
Wright JK (1947) Terrae incognitae: the place of imagination in geography. Ann Assoc Am Geographers 37:1–15

## Newspapers

*Jornal do Brasil*, August 26, 1974
*Jornal do Brasil*, August 31, 1986
*O Estado de São Paulo*, May 15, 1997
*Folha de São Paulo*, August 20, 1999
*The Economist*, October 12, 1996

# Part VI
# Southern Africa

# Chapter 14
# Unveiling the Geography of the Cape of Good Hope: Selected VOC Maps of the Interior of South Africa

**Elri Liebenberg**

**Abstract** The **United Dutch East India Company** (*Vereenigde Oost-Indische Compagnie* or **VOC** in Dutch, literally meaning "United East Indian Company") was a chartered company which was active in the Asian trade from 1602 until 1795. During this period, cartographers in the service of the VOC produced numerous maps of the coasts of southern Africa and of the VOC settlement at the Cape. Amongst map historians, the majority of these charts, maps and drawings are well-known and well documented. What is perhaps less known, is that the Dutch National Archives holds an important collection of pioneering VOC manuscript maps of the *interior* of the Cape Colony on which very little has been published.

When the Cape of Good Hope fell into British hands for the first time in 1795, Britain believed that no Dutch cartography of the Cape or of the interior of the country existed. This belief held until 1950 when a large number of manuscript maps of the Cape was discovered in the archives of the Dutch Topographic Service at Delft in the Netherlands. This paper deals with a selection of the topographical maps found in 1950 which is here discussed with regard to their origin, style, accuracy and completeness.

## Introduction

The **United Dutch East India Company** (*Vereenigde Oost-Indische Compagnie* or VOC in Dutch, literally "United East Indian Company") was a chartered company which was established on 20 March 1602 when the States-General of the Netherlands granted it an initial 21 year monopoly to undertake trading between the Cape of Good Hope and the Strait of Magellan. At the height of its power in the seventeenth and eighteenth centuries, the VOC was the greatest commercial concern

E. Liebenberg (✉)
Department of Geography, University of South Africa, Pretoria, South Africa
e-mail: elri@worldonline.co.za

E. Liebenberg and I.J. Demhardt (eds.), *History of Cartography*,
Lecture Notes in Geoinformation and Cartography 6,
DOI 10.1007/978-3-642-19088-9_14, © Springer-Verlag Berlin Heidelberg 2012

in the world. It was also arguably the world's first megacorporation, possessing quasi-governmental powers, including the ability to wage war, negotiate treaties, coin money and establish colonies. Between 1602 and 1796 the VOC eclipsed all its rivals in the Asian trade and enjoyed huge profits from its spice monopoly in the East. The VOC's territories in the East which became the Dutch East Indies were expanded over the course of the nineteenth century to include the whole of the Indonesian archipelago, and in the twentieth century would become Indonesia.

In 1652 the VOC founded a refreshment station at the Cape of Good Hope for the scurvy-ridden crews of its fleets plying between Europe and Asia. The officials making decisions for the VOC, the so-called "Lords Seventeen" or Heren XVII, did not intend for the settlement at Table Bay to be an independent economic or political entity. The Company was a commercial firm and the ownership of overseas territories was by no means the object of the VOC. The first persons of European extraction at the Cape were all employed in the direct service of the VOC. The Company, however, soon decided that it would be more lucrative to grant some of the VOC servants the status of free citizens (*free burghers*) and offer them tracts of land. These farmers could then produce agricultural goods at their own cost and sell their products to the VOC which was the only buyer. As stock-farming proved to be both easier and more economical than tilling the land, many free burghers soon switched to cattle and sheep farming. Known as *trekboers* (migrating farmers), these pastoralists gradually settled further and further away from the settlement at Table Bay in search of better grazing for their herds.

Due to the Mediterranean type semi-arid climate resulting in poor grazing, the expansion of the settlement followed two main directions: northward between the west coast and the mountain ranges running parallel to the Atlantic Ocean and eastward along the foothills of the east–west mountain ranges. In the 1680s the European population of the Cape was boosted when some 180 French Huguenots arrived. To provide in the need for manual labour, slaves were imported from the Dutch East Indies, Madagascar and Mozambique.[1] When the VOC period officially came to a close in 1795, the frontier of the white pastoral settlement had shifted as far as the Buffels River in the north and the Fish River in the east (see Fig. 14.1).

## Cartographical Knowledge

Dutch cartographical knowledge of the coast of southern Africa was initially based on Portuguese sources (see Fig. 14.2). In 1596 this body of knowledge was considerably strengthened by the publication of Van Linschoten's *Itinerario*[2]

[1]Elphick, R. and Giliomee, H. (ed.). 1980. *The shaping of South African society*. Cape Town: Longman.

[2]Varley, D.H. 1955. "Jan Huygen van Linschoten and his *Itinerario*". *Quarterly Bulletin South African Library*, Dec. 1955.

**Fig. 14.1** Expansion of stockfarmers or *trekboers* from 1700 until 1780 (Source: Giliomee, H. & Mbenga, B. 2007. *A new history of South Africa*. Cape Town: Tafelberg. Copyright for this image by Tafelberg Publishers)

which contained a wealth of navigational information. In 1597–1598, the first Dutch fleet under Cornelis Houtman reached India via the Cape. In the first half of the seventeenth century the publication of various Dutch coastal charts and maps of the coastline of southern Africa indicated that the cartographers of the VOC were steadily building up their own cartographic database to assist sailors attempting the treacherous waters of the south Atlantic and Indian Oceans (see Fig. 14.3).

Knowledge of the coastline did not imply knowledge of the interior, and the early settlers at the Cape had no knowledge of what could not be seen from the coast. Throughout the late sixteenth and early seventeenth century, influential Dutch cartographers such as Ortelius, Mercator, Blaeu, Hondius, etc., published maps of Africa and southern Africa which depicted mythical and imaginary features such as the gold mines of the Empire of Monomotapa, the river Spirito Sancto, and the cities of Vigiti Magna, Samot and Camissa. The early settlers considered these places to be real and as late the 1680s, regular expeditions were sent out to locate them. It was only by the end of the seventeenth century when all these expeditions had returned empty-handed, that the VOC gradually realized that these places did not exist.[3]

[3]Paver, F.R. 1955. "A puzzle in antique cartography: Vigita Magna", and A.H.S. 1955. "Further Notes on Vigiti Magna". *Africana Notes and News* 11(6), 186–196.

**Fig. 14.2** (*above*): Manuscript map of the south-western coast of Africa by João Albernaz II, c. 1680 (Groote Schuur Collection no 50, Cape Town. Size: 64 × 44 cm. Copyright for this image by the South African Heritage Resource Agency)

Early maps of the Cape either depict the area around Table Bay, or the stretch along the west coast between the Cape and Saldanha Bay. Maps of the interior would only follow later when the country beyond the mountains surrounding the settlement became better known. Throughout the VOC period expeditions were sent out both northward and eastward to explore the interior and investigate the possibility of trade with the indigenous peoples. Not all these expeditions produced maps but some did and, for the purposes of this paper, seven VOC maps which could be considered "overview" maps of the interior or parts of thereof, are discussed. Before doing so it might however be appropriate to explain why the VOC cartography of the Cape, as opposed to the Company's maps of for instance the East Indies or Australia, remained unknown for long.

From the very beginning of the settlement at the Cape, the VOC made the allocation of land to free burghers subject to cadastral registration (see Fig. 14.4). The official position of Land Surveyor was created in 1657 and land surveyors were employed throughout the company era. It is not clear whether map-makers also held an official position at the Cape. It is more likely that this function was combined with other work according to need. Apart from surveyors and map-makers, the VOC also employed military engineers who were active in the compilation of engineering drawings, building plans and topographic maps. Although cartographic

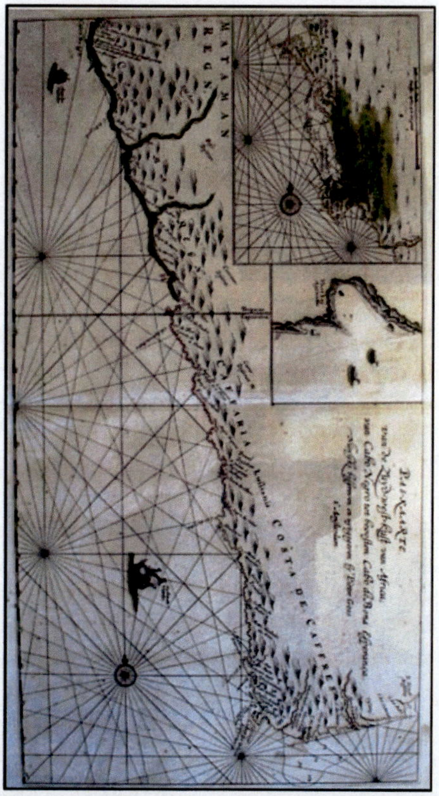

**Fig. 14.3** (*right*): *Paskaarte van de Zuyd-west-kust van Africa; van Cabo Negro tot beoosten Cabo de Bona Esperança*, by Pieter Goos, c. 1670. Size: 29 × 52 cm

**Fig. 14.4** Eighteenth century Cape land diagram (Copyright for this image by L.F. Braun)

work was carried out throughout the eighteenth century, it was especially during the governorship of Jacob Cornelis van de Graaff that Cape cartography reached a level of professional excellence comparable to that of the motherland. Van de Graaff was an expert in the field of military fortifications and during the Napoleonic wars of the 1780s, he was ordered to the Cape to improve the military defence systems. He implemented extensive new cartographic projects at the Cape which resulted in a large collection of maps and plans. To prevent them from falling into the hands of the English, he took many maps, charts and plans produced during his term of office with him when he left in 1791. When Britain took possession of the Cape in 1795, the British authorities found no maps amongst the remaining Dutch documents, a situation which led to the belief that the interior had been left unmapped by the VOC. This erroneous conclusion remained in place until 1950 when Cornelis Koeman discovered Van de Graaff's collection of manuscript maps of the Cape in the Archives of the Topografische Dienst in Delft.[4]

Compiled by specialist cartographers, many of the VOC maps of the Cape are of an excellent quality. As they are manuscript maps and were unknown for so long, they have not as yet received the attention they merit.[5] The fact that most of them are of an exceptionally large size and therefore difficult to photograph, also prevented them from becoming more widely known. The recent publication of a well-produced Dutch atlas in which digital images of these maps are produced[6] should do much to improve this situation.

## Claudius' Map of Governor van der Stel's Journey to the "Copper Mountains", 1685

Rumours that an indigenous Khoisan tribe, the Namaqua, who allegedly lived to "the North and West of the Cape", was rich in copper, first reached the first commander of the Cape, Jan van Riebeeck, in 1657.[7] Between 1659 and 1663 seven

---

[4]Koeman, C. 1988. "Nieuwe bijdragen tot de kennis van Zuid-Afrika's historische kartografie", in G. Schilder and Peter van der Krogt (ed.), *Miscellanea Cartographica. Contributions to the History of Cartography*. Utrecht: H&S Publishers.

[5]The only publications to date dealing with (some of) the VOC maps of the Cape are Koeman, C. 1952. *Tabulae Geographicae Quibus Colonia Bonae Spei Antiqua Depingitur*. Kaapstad and Amsterdam: NV Hollandsch-Afrikaansche Uitgevers; Forbes, V.S. 1952. "An atlas of eighteenth century cartography of the Cape", *Africana Notes and News*, 9(3), 97–103; Koeman, C. 1955. "Beschrijving van een belangrijke 18e eeuwse manuscriptkaart van de Kaapkolonie", *Africana Notes and News* (11(7), 276–287; and Forbes, V.S. 1965. *Pioneer Travellers in South Africa*. Cape Town and Amsterdam: Balkema, 7–9.

[6]Brommer, B. 2009. *Grote atlas van de Verenigde Oost-Indische Compagnie, Vol . 5: Afrika*. Amsterdam: Asia Maior.

[7]Moodie, D. 1960[1838]. *The Record, or a series of official papers relative to the condition and treatment of the native tribes of South Africa. A photostatic reprint of the original book of 1838*, 116. Cape Town: Balkema.

successive expeditions were dispatched to the country of the Namaqua to look for riches such as copper, gold and pearls, none of them successful.[8] In April 1682 the interest in copper flared up again when it was reported that some Namaqua, to whom "the river of Vigiti Magna... was known",[9] visited the Fort at Table Bay and brought pieces of good quality copper with them. In October of that year an expedition was sent out under Ensign Olof Bergh, and another the following year, also under his command.[10] Both expeditions penetrated far into Namaqualand, but still could not reach the source of the copper.

In 1685 the Commander, Simon van der Stel,[11] decided to lead an expedition to the so-called "Copper Mountains" himself. The expedition, consisting of 60 men, ventured out on 25 August 1685 and reached the mountains near the present-day Springbok on 21 October where they dug ore and surveyed the surrounding area[12] (see Fig. 14.5). The expedition returned to the Cape on 26 January 1686. The ore samples were sent to the Netherlands to be assayed, but were found to be of low grade, yielding as little as 7.5–11.5% of copper.[13] This, together with the mountains' distance from the Cape and the difficulty of transporting the ore, caused the possibility of mining the copper to be abandoned. Although not an economic success, the expedition nevertheless gathered valuable information on the natural environment as described in Van der Stel's Journal, a copy of which was sent to the Netherlands. The drawings of plants, insects, reptiles and animals made by the apothecary and artist Heinrich Claudius, who was a member of the expedition, still count amongst South Africa's most valuable historical artifacts.[14] To geographers and cartographers it is, however, his map of Van der Stel's journey[15] which is important (see Fig. 14.6a). This map depicts the most important rivers and mountain ranges between the Cape and the Copper Mountains. The route outwards is

---

[8]Mossop, E.E. 1927. *Old Cape Highways*, 122 ff. Cape: Maskew Miller.

[9]Moodie, D. *Op cit.*, 386.

[10]Mossop, E.E. (ed.). 1931. *Journals of the expeditions of the Honourable Ensign Olof Bergh (1682 and 1683) and Isaq Schrijver (1689)*. Cape Town: Van Riebeeck Society, no. 12.

[11]Simon van der Stel (1639–1712) was appointed Commander at the Cape of Good Hope in 1679, and was promoted to the rank of Governor in 1691.

[12]Waterhouse, G. 1932. *Simon van der Stel's Journal of his expedition to Namaqualand, 1685–6*. Dublin: Hodges, Figgis, and Co., Folios 3 and 5. Waterhouse translated and published the only extant MS of the Van der Stel Journal which is kept in the Library of Trinity College, Dublin. Prior to this the *Journal* was published in 1726 by Francois Valentijn in his *Oud and Nieuw Oost-Indien*, Vyfde deel, Thiende Boek, 54 ff.

[13]Rogers, A.W. 1917. "An Old Report on the Copper Field of Namaqualand (written by Friederich Mathias von Werlinhof, dated 18.3.1686)". *South African Journal of Science* 14: 146–151. Von Werlinhof was the VOC Superintendent of Mines at the Cape. See also Smalberger, J.M. 1975. *Aspects of the history of copper mining in Namaqualand 1846–1931*. Cape Town and Johannesburg: C. Struik, pp. 12–13, and Waterhouse's 1953 *Supplement* to his original book of 1932, 11–12.

[14]Wilson, M.L., Toussaint van Hove-Exalto, T. and Van Rijssen, W.J.J. (ed.). 2002. *Codex Witsenii*. Cape Town and Amsterdam: Iziko Museums.

[15]Dutch National Archives, 4. VEL. 851

**Fig. 14.5** Simon van der Stel's journey to the Copper Mountains, 1685 (Source: Wilson, M.L. (ed.). 2002. Codex Witsenii. Cape Town: IzikoMuseums. Copyright for this image by Iziko Museums, Cape Town)

indicated by a red line, and a yellow line marks the way back. A total of 26 important topographic features are numbered and named in the legend and a red monogram situated close to Table Bay certifies Claudius as the author. A similar map which seems to be a copy of Claudius map, was recently found in the Bibliothèque National in Paris.[16] Although it is doubtful whether these.maps were ever used for topographical purposes, they marked important steps forward in the gathering of local geographical information.

[16]Bibliothèque Nationale, Paris, Pf. 114, Division 2, P.1. The existence of this copy of Van der Stel's map is reported in Maré, E.A. and Dubourg Glatigny, P. 2006. "A map and its copy of Governor Simon van der Stel's expedition to Namaqualand (1685): an enquiry into their visual values". *South African Journal of Art History* 21(1), 104–113. The map is entitled "Plan de la route suivie en 1688 par le gouverneur du Cap de Bonne Esperance dans les terres du Nord de le Cap". The authors purport that the map was compiled by the Jesuit scientist and missionary Guy Tachard (1651–1712) who visited the Cape four times during the 1780s.

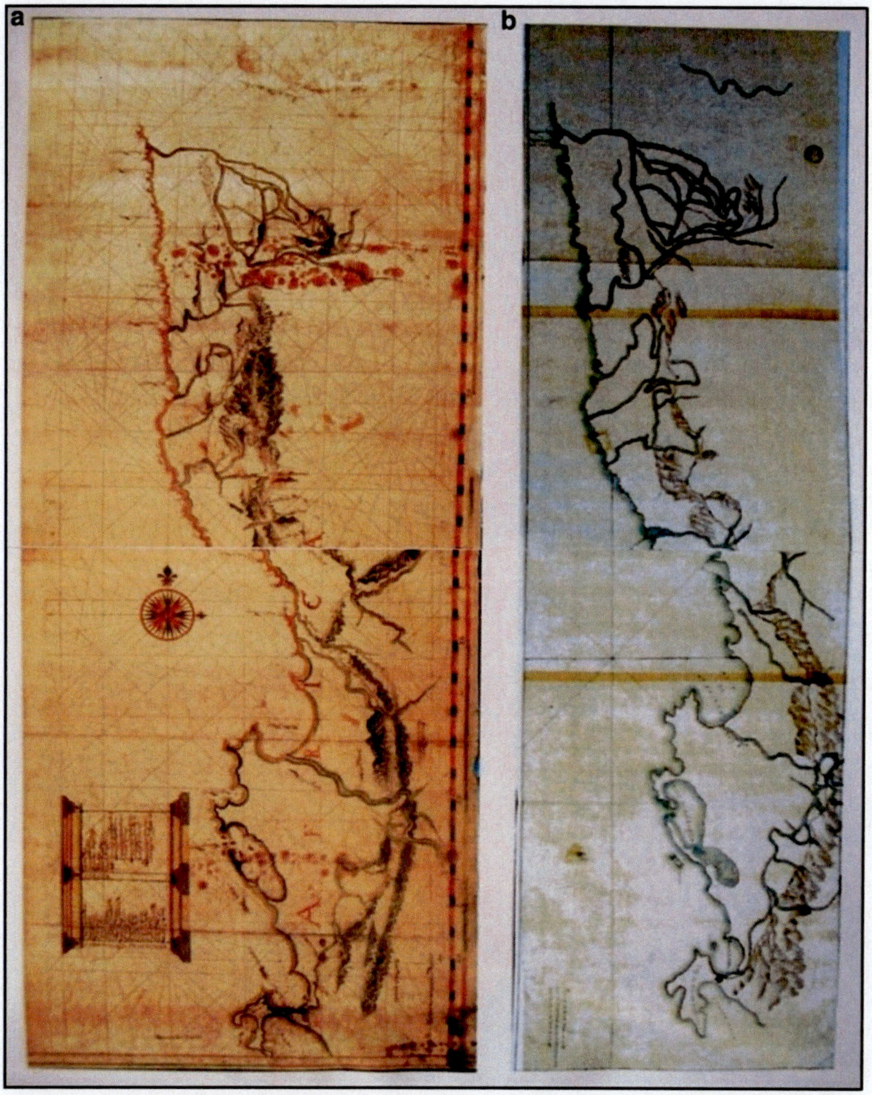

**Fig. 14.6** (**a**) (*left*) Hendrik Claudius' map of Van der Stel's journey. Copyright for this image by the Dutch National Archives, 4 Vel 851). (**b**) (*right*) A copy of Claudius' map, presumably drawn by the Jesuit priest and astronomer Guy Tachard (Copyright for this image by the Bibliotheque Nationale, Paris. Pf. 114, Division 2, P.1)

## Wentzel's Map of Beutler's Expedition 1752

On 19 February 1752 the first VOC-sponsored expedition, ordered by Governor Ryk Tulbagh, left the castle at the Cape for the eastern part of the country to investigate this region's possibility for development. The expedition consisting of

71 people, was led by Ensign August Frederik Beutler with Carel Albregt Haupt as official diarist, Carel David Wentzel as surveyor and cartographer, and Pieter Clement as assistant surveyor.[17] Beutler's company travelled past the present Port Elizabeth and East London and penetrated the area beyond the Kei River as far as the present Butterworth. This alone was a remarkable achievement considering that *trekboer* settlement had by then only reached the vicinity of Mossel Bay. The expedition lasted more than 8 months and the carefully maintained journal provided valuable information on the topography, climate, vegetation and inhabitants of the Eastern Cape. What was also important, was that the journal recorded in writing much that had previously depended on hearsay.

The map of Beutler's route as produced by Wentzel[18] depicts a compass-traverse of over 1,000 km outward-bound and is drawn on a sheet of paper 2.5 m in length (see Fig. 14.7). It shows the route of the expedition with extraordinary detail between a considerable number of points which can today be identified with fair certainty. According to Forbes[19] there are few points the position of which is likely to be in error further than 8 km, a remarkable feat considering the existing knowledge and instruments of the day. What is obvious from a close inspection of the map, is that points in the far eastern part are generally mapped west and north of their correct positions (see Fig. 14.8). These errors were initially ascribed to an

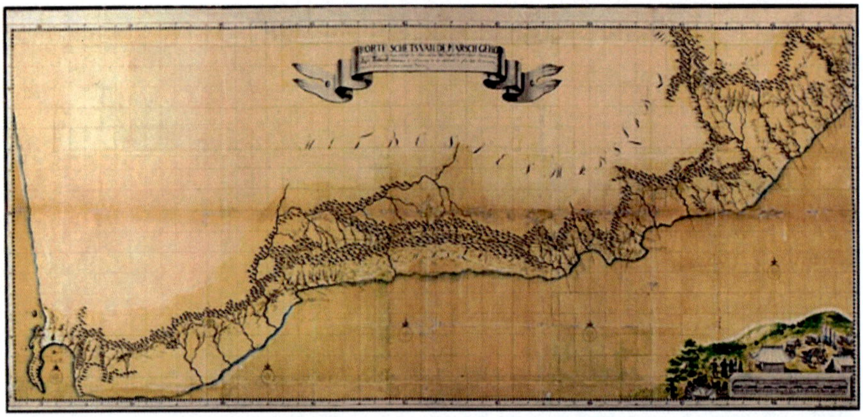

**Fig. 14.7** C.D. Wentzel's map of Beutler's journey, 1752 (Copyright for this image by the Museum Africa, Johannesburg)

[17]Forbes, V.S. 1965. *Pioneer travellers in South Africa*, 7. Cape Town and Amsterdam: Balkema.

[18]Dutch National Archives, 4. VEL 852. According to Koeman (see his "Nieuwe bijdragen…", 83) this is the "original drawing of Wentzel (de originele tekening van Wentzel)". Apart from this map, three equivalent copies exist: one in the Dutch National Archives 4.TOPO.15.119; one in the Africana Museum, Johannesburg, and one in the Cape Archives, M. 79.

[19]Forbes, V.S. *Op. cit.*, 9.

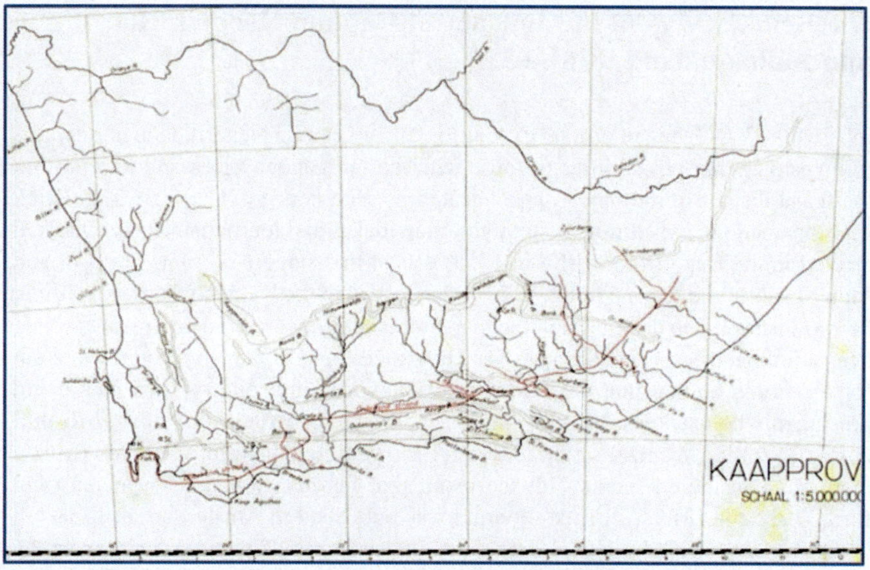


**Fig. 14.8** Wentzel's map (*in red*) superimposed on a modern map of South Africa (Source: Koeman: *Tabulae Geographicae*, p. 44. Copyright for this image by URU-Explokart Research Group, University Utrecht)

excess correction for magnetic variation,[20] but according to Forbes[21] this phenomenon should be ascribed to a much simpler reason. After studying both the map and Haupt's journal, he concluded that Wentzel purposefully compressed the detail towards the eastern margin of his map so as not to challenge the authority of the established cartographers of his time. All standard maps of that period showed southern Africa much narrower than it actually was,[22] which meant that the solar observations made by his assistant Clement put the farthest point from the Cape, reached by the expedition, out at sea. Unwilling to contest the authority of the existing small-scale maps of Africa, Wentzel decided to ignore Clement's values and to plot the route to the west of the observed coordinates.

Although Beutler's expedition cannot be seen as a necessary prelude to the settlement of the Eastern Cape, it certainly contributed to the extension of the official administration of these territories. The fact that Beutler planted a beacon on the shores of Algoa Bay bearing the VOC monogram, established the Company's dominion at least as far east as that point. This action in itself, might have prevented the French from establishing a colony there during the 1790s.

[20]*Molsbergen*, E.C. Godée. 1932. *Reizen in Zuid-Afrika*, vol. IV. The Hague: Linschoten Vereeniging.

[21]Forbes, V.C. *op. cit.*, 7–9.

[22]This interesting point could be the reason for the occurrence of numerous shipwrecks on the Eastern Cape Coast over a period of three centuries.

## C F Brink's Map of the Journey of Captain Hendrik Hop into Namaqualand, 1761–62

In 1761 the VOC sent an official expedition under Captain Hendrik Hop northwards to investigate the possibilities of trade with the indigenous tribes and to report on the feasibility of mining the copper located by Van der Stel. Carl Friedrich Brink accompanied the expedition as surveyor, map maker and journalist. Brink, who had arrived in the Cape from Berlin in 1758, was at that stage Assistant-Surveyor and Map Maker in the service of the VOC. With regard to the Namaqualand expedition, he was instructed to draw a "suitable map of the undiscovered unknown country and keep a journal of daily occurrences".[23] The expedition also had to be on the lookout for a "certain animal that has never been known in this country" [24] which could presumably be used as a beast of burden. The journey lasted from July 1761 until April 1762. Like Wentzel's map of Beutler's expedition, the information on Brink's map was considered secret with the result that his map remained in manuscript form[25] (see Fig. 14.9a,). Brink's journal was published in Amsterdam in 1758.

On 16 August 1761 the exploring party, consisting of 85 people, set off from the Olifants River (situated 150 km north of the Cape), crossed the Orange River (now called the Gariep River) into what is now Namibia, and moved northwards towards the present Keetmanshoop in Namibia. They met with various Namaqua tribes, but by 6 December 1761 the constant lack of water and the difficult terrain and heat made them decide to turn back. Although Hop had no existing map to follow, Brink mapped the route of the expedition with sufficient precision to enable the present reader to identify most of its details with confidence today. The map measures 190 × 98 cm on a scale of approximately 1:350,000 and the route of the expedition is indicated by a red line. Because the expedition did not travel along the coast, the coastline initially plotted was presumably taken from an existing small-scale chart and is therefore altogether wrong.[26] The mouth of the Orange River is for instance shown as a deep bay called "Angra das Voltas". An interesting feature of the drawing of the coastline is that various corrections were made at a later stage. These corrections which agree more accurately with the actual situation also appear on the map. Obviously being on the lookout for the "unknown animal", Brink also included drawings of a giraffe, a zebra, a lion and a horse on his map.

---

[23]Mossop, E.E. 1947. *The Journals of Brink and Rhenius*. Cape Town: Van Riebeeck Society.

[24]Brommer, B. 2009. *Grote atlas van de Verenigde Oost-Indische Compagnie, Vol . 5: Afrika*. Amsterdam: Asia Maior, 248.

[25]Dutch National Archives, 4.JSF.2. A digital image is published in Brommer, B. *op. cit.*, 248–9. Another facsimile, together with a 1685 copy which was lithographically produced, appear in Koeman, C. "*Tabulae Geographicae*. . ..", Plates III and IV.

[26]Forbes, V.S. 1965. "Some early maps of South Africa 1595-1795". *Journal for Geography*, 11(6), 14.

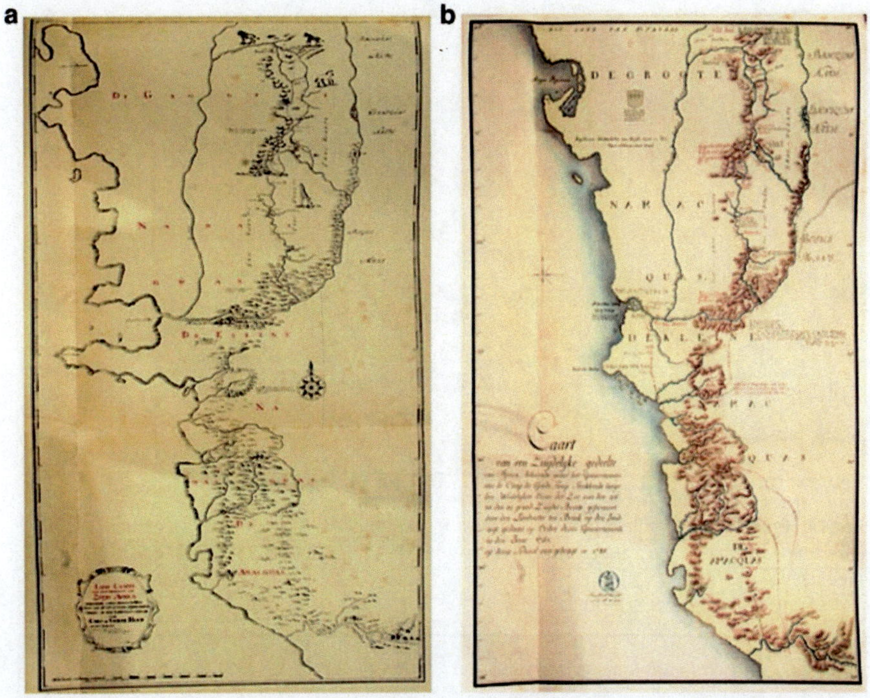

**Fig. 14.9** (**a**) (*left*): Brink's map of Hop's journey into Namaqualand, 1761–62. (Copyright for this image by Dutch National Archives, map 4.JSF.2.) **(b)** (*right*). The same map as corrected by Van de Graaff in 1785 (Source: Koeman, *Tabulae Geographicae*..., PL. III. Copyright for this image by URU-Explokart Research Group, University Utrecht)

## Map of the Route Along Which Swellengrebel Travelled, 1876–77

Hendrik Swellengrebel, the son of a former governor of the Cape, belonged to an aristocratic Dutch family. When visiting the Cape in 1776–77, he undertook several journeys into the interior. His first journey was to Saldanha and St Helena Bay on the west coast. His second journey took him eastwards from Stellenbosch across the Great Karroo to the Great Fish River, whereas his third journey was towards the north and the estuary of the Olifants River.[27] The remarkably accurate map which depicts Swellengrebel's routes[28] is considered to be the "first reasonably accurate general map"[29] of the entire spread of the area settled by white pastoralists in the second half of the eighteenth century (see Fig. 14.10). The courses of the main

[27]Forbes, V.S. *Pioneer Travellers, 59–80.*

[28]Koeman, C. *Tabulae Geographicae*..., Plate V.

[29]Forbes, V.S. "Some early maps of South Africa 1595–1795". *Journal for Geography* 11(6), 14.

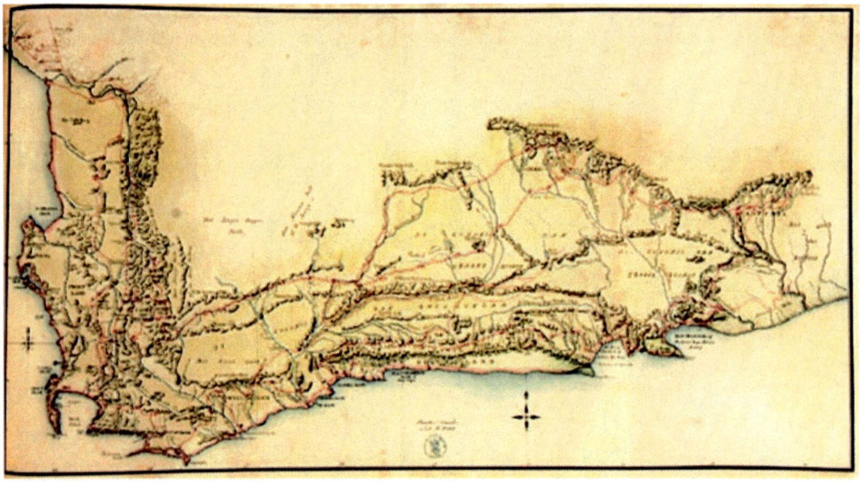

**Fig. 14.10** Map of Swellengrebel's journey to the Eastern Cape, 1779 (Koeman: *Tabulae Geographicae. . .*, PL. V. Copyright for this image by URU-Explokart Research Group, University Utrecht)

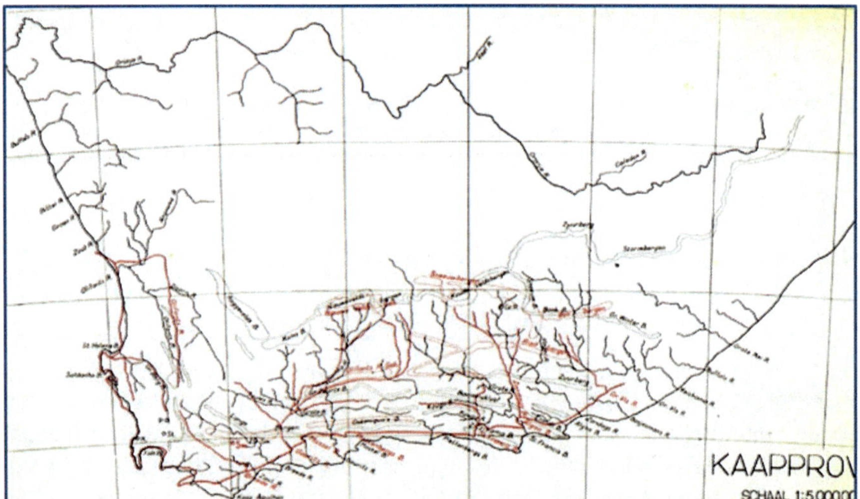

**Fig. 14.11** Map of Swellengrebel's journey maps (*in red*) superimposed on a modern map of South Africa (Source: Koeman: *Tabulae Geographicae*, p. 52. Copyright for this image by URU-Explokart Research Group, University Utrecht)

rivers and mountain ranges are depicted relatively correctly and are placed within the graticule with as much accuracy as could be expected from an exploratory survey conducted at that time. Although the south coast shows the correct number of bays, their size and the distance between them need correction (see Fig. 14.11).

Both Koeman[30] and Brommer[31] accredit the original map depicting Swellengrebel's travels to Pieter Cloete (1756–1787) who accompanied Swellengrebel on his journey to the eastern frontier in 1779, and to Europe in 1779. Forbes initially accepted Koeman's claim, but rejected its validity after he had examined the map himself. Likewise, Forbes also contested Godée Molsbergen's conclusion that Cloete was the author of both the journal of the expedition, and the map.[32] According to Forbes it is unlikely that Cloete played any significant part in the expedition either as he was, at the time of Swellengrebel's eastward journey, still very young and unqualified. Cloete only became an unpaid assistant surveyor in the service of the VOC in 1779 and his promotion to engineer with the rank of Ensign did not take effect until 1781. Forbes is of the opinion that Swellengrebel wrote the journal himself and that the map of the journey is probably a compilation of Surveyor C F Brink of his own work and that of other official surveyors. It was probably given to Swellengrebel by Governor Van Plettenberg for use on his journey. The fact that the artist Johannes Schumacher was a member of Swellengrebel's party which travelled to the eastern part of the country, also opens possibilities as regards the true origin of the map.

## R. J. Gordon's Maps and Drawings 1770–85

The culmination of VOC cartography at the Cape was the remarkable Map no 3 of the Gordon Collection in the *Prentenkabinet* of the Rijksmuseum, Amsterdam (see Fig. 14.12). Robert Jacob Gordon, a Dutchman of Scottish descent, was Commander of the Dutch garrison at the Cape from 1780 to 1795. He undertook four notable expeditions into the interior from 1777 to 1786 which he recorded in journals. [33] He also left behind a portfolio of important maps and charts,[34] as well as exquisitely executed topographical drawings and sketches of the inhabitants of the country and its fauna and flora.[35]

---

[30]Koeman, C. "Nieuwe bijdragen...", 84.

[31]Brommer, B. *op. cit.*, 242. The digital image reproduced here is of Map 221 of the Van de Graaff Collection which is ascribed to Brink and which is similar, but not identical , to Map 222 which is ascribed to Cloete.

[32]Forbes, V.S. *Pioneer Travellers, 59–63; Molsbergen*, E.C. Godée. 1932. *Reizen in Zuid-Afrika*, vol. IV. The Hague: Linschoten Vereeniging.

[33]Forbes, V.S. *Ibid., 93–116.*

[34]The maps and plans in the Gordon Collection (Prentenkabinet, Rijksmuseum) are listed in Forbes, V.S. *Pioneer Travellers*, 113–114.

[35]Raper, P.E. and Boucher, M. (ed.). 1988. *Robert Jacob Gordon Cape Travels, 1777 to1786*, vols. I and 2. Johannesburg: Brenthurst; Cullinan, P. 1992. *Robert Jacob Gordon 1743–1795. The man and his travels at the Cape*. Cape Town: Struik Winchester. After his death, Gordon's maps, drawings and manuscripts were taken to England by his wife and sold to the Marquess of Stafford. When put up for sale in 1914, they were bought by the Dutch government.

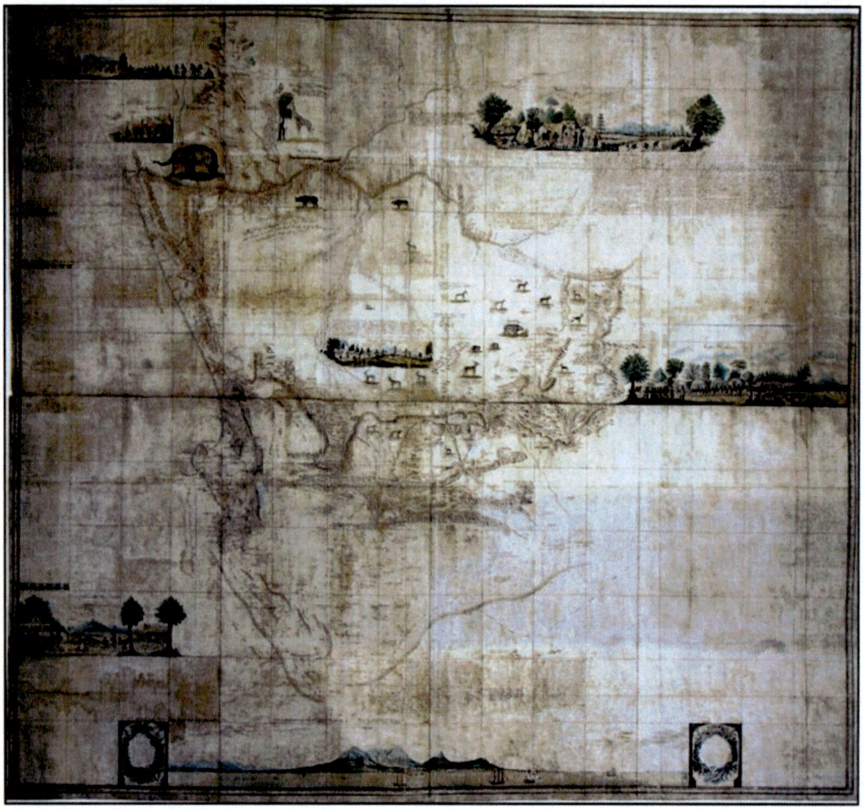

**Fig. 14.12** Gordon's Map no 3 of the Cape Colony (Copyright for this image by the Rijksmuseum Prentenkabinet, Amsterdam, map RP-T-1914-17-3)

Gordon's Map no 3, also known as Gordon's "Great Map", measures 180 × 196.5 cm on a scale of approximately 1:750,000. It covers Southern Africa eastwards to the Great Fish River and northwards to the Orange River and is drawn within the graticule of a modified cylindrical projection. In the west there is an extension north of the Orange River into the present Namibia which was copied from C.F. Brink's map of Hops' expedition referred to above. The map depicts a large amount of finely drawn and accurately positioned topographical detail and contains numerous illustrations of the indigenous peoples and animals that are commented upon in copious notes. A few small areas are incomplete and although the southern coastline is only lightly drawn, it is a faithful representation. The names of numerous farms and their owners are depicted and the barometrically

Until 1964 Gordon's travel journals were considered lost. In that year they were found in the Staffordshire county archives and bought by the Brenthurst Library in Johannesburg where they are now lodged.

ascertained elevations of over 36 points are shown. The fact that Koeman refers to this map as one which "exhibits all the characteristics of an amateurish product",[36] probably stems from the fact that he never made a study of it himself. Forbes, who did, is of the opinion that Gordon's Map no 3 " bears everywhere the stamp of authenticity complemented with the greatest accuracy obtainable under the circumstances in which it was compiled".[37]

Gordon's life came to a tragic end when he, weakened by illness, committed suicide shortly after the surrender of the Cape to Britain in 1795. Forbes states that Gordon was as instrumental as Van de Graaff and maybe even more so, in preventing the complete cartographic archives of the Cape from falling into British hands.[38] In January 1794 Gordon listed 143 maps, plans and topographic drawings in the castle at the Cape. It is highly probable that of the 95 cartographic documents taken to England by his wife in 1797, many if not all belonged to the 143 listed 3 years earlier.[39] Unfortunately for later generations, the larger part of Mrs Gordon's consignment to England has not survived as there are now only 15 maps and one plan in the Gordon Collection in Amsterdam.

It is not entirely clear whether Gordon drew Map no 3 himself, or whether the actual drawing was done by his German batman, the artist Johannes Schumacher, who accompanied him on all four his journeys into the interior. Certainly Schumacher was responsible for the execution of many of the beautiful panoramic drawings (some up to 4 m in length) which are associated with Gordon and his travels.

## J. C. Frederici's Map of the Southern Coast of Africa, 1789–90

Johann Christian Frederici was a military officer who arrived in the Cape in 1784 after he had been recruited by the VOC in Germany. Governor Van de Graaff employed him as military surveyor and in 1791 he was promoted to the position of Lieutenant in the Cape Artillery. Frederici was responsible for many beautifully executed maps of the Cape, the most important being his map of the coastal area between Cape Agulhas and Algoa Bay, entitled *Caerte van den Zuydelijken oever van Afrika* (see Fig. 14.13). In the execution of this map he was assisted by

[36]Koeman, C. 1952. "Landmeters in dienst van de VOC in de Kaapkolonie". *Tijdschrift voor Kadaster en Landmeetkunde*, 01.04.1952, p. 96. The Dutch sentence reads" "Deze kaart vertoont . . . al de kenmerken van een amateurproduct".

[37]Forbes, V.S. *Pioneer Travellers*, 112.

[38]Forbes, V.S. *Ibid.*, 112.

[39]Forbes V.S. 1952. "Further notes on Colonel R.J. Gordon". *Africana Notes and News* 9(3), 88-89.

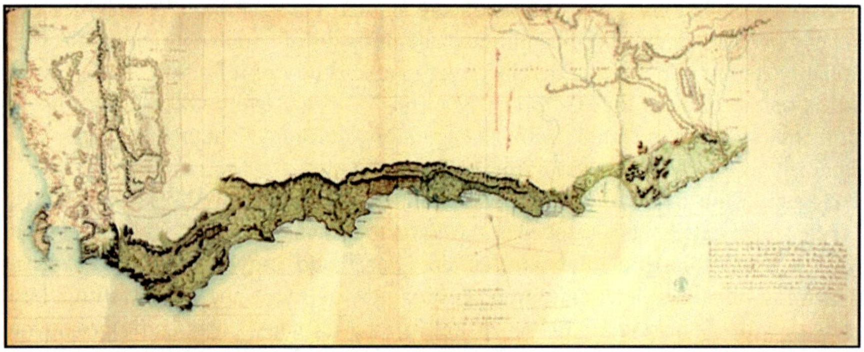

**Fig. 14.13** J. C. Frederici's map of the south coast of southern Africa, 1789–90 (Copyright for this image by the Dutch National Archives, 4, TOPO 15 11)

Bombardier Josephus Jones.[40] The fact that this map is one of only two[41] to which Koeman dedicated separate, stand-alone articles[42] after his discovery of the collection of VOC maps in Delft, emphasises its importance.

Frederici's map of the south coast of southern Africa is drawn on a scale of approximately 1:120,000; measures 100 × 400 cm, and depicts an area more than 400 km long and 40–50 km wide with considerable accuracy and detail (see Fig. 14.14). The coastline is well drawn and well placed in latitude and is better represented than on any earlier printed map or chart. An important attribute of this map is the depiction of all existing homesteads, the names of the relevant farms and owners, and the connecting roads between these settlements. No other map provides such a detailed picture of the landscape at the time and it is not strange that Koeman refers to it as "the best map dating from the eighteenth and beginning of the nineteenth centuries of the south coast of the Cape Colony (sic) known to us." (my translation).[43]

[40]Jones (1769–1811), who was of French descent, joined the Cape artillery in 1784. In 1981 he left to serve in the corps of engineers where he worked as draughtsman and cartographer. He was a competent artist and his most famous work is the 1808 panorama of Cape Town which gives a 360° view of the city. The original drawing is kept by the Rembrandt van Rijn Art Foundation in Stellenbosch, South Africa.

[41]The other article discusses a 1687 hydrographical chart of False Bay by Governor Simon van der Stel. See Koeman, C. 1951. "De eerste kaart van de hydrografische opneming van de Baai Fals en De Houtbaai aan die kust van Z-Afrika". *Tijdschrift van het Koninklijk Nederlandsch Aardrijkskundig Genootschap* 68, 448–451.

[42]Koeman, C. 1955. "Beschrijving van een belangrijke 18e eeuwse manuscriptkaart van de Kaapkolonie". *Africana Notes and News* 11(7), 276–287.

[43]*Ibid.*, p. 277: "*. . .de beste kaart van de Zuidkust van de Kaapkolonie vervaardigd welke wij uit de 18e en 't begin van de 19e eeuw kennen*".

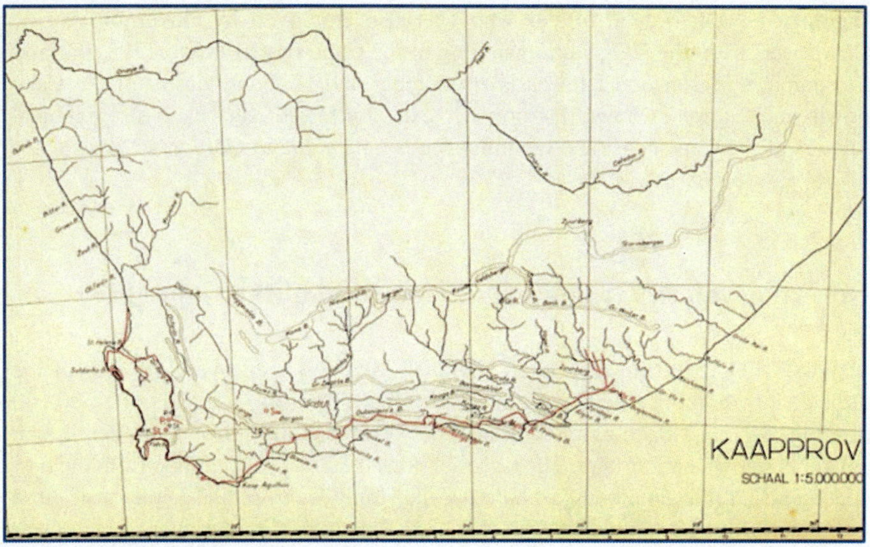


**Fig. 14.14** Frederici's map (*red*) superimposed on a modern map of South Africa (Source: Koeman: *Tabulae Geographicae*..., p. 56. Copyright for this image by URU-Explokart Research Group, University Utrecht)

An almost equivalent copy of Frederici and Jones' map, copied and translated into English, is kept in the British National Archives.[44] This map, which is of the same scale as the one in the Dutch Archives, was signed by the Commanding Royal Engineer at the Cape, Captain Henry Smart, on 4 April 1816. It differs from the Dutch map in that the names of the farms and their owners do not appear on the map face but are listed according to a numbering system on either side of the map.

Various derived copies of Frederici and Jones' map exist. One example is the almost similar map on a smaller scale drawn at the Cape by the military engineer Domenicus Michel Barbier, in which existing material relating to the western and eastern parts of the country was added.[45] A copy of this map which was translated into English and brought to England by Capt Bridges, the Commanding Officer Royal Engineers at the Cape in 1801, is kept in the British Archives.[46] A second derived copy of the original Frederici map is a map/chart in which the hydrographical surveys of the western and southern coasts executed by François Renier Duminy are combined with Frederici's and Jones's survey of the interior.[47]

[44]British Archives, MR 1/1305.

[45]Dutch National Archives, 4.TOPO.15.12.

[46]British National Archives, MR 1/1292.

[47]Cape Archives, Map 3/36 (copy).

Duminy, a French naval officer who was also employed by Governor Van de Graaff, served in the VOC administration at the Cape from 1786 to 1795, initially as Equipage Master and later as Hydrographer and Harbour Master of the Cape Town and Simon's Town Harbours.[48] He completed various hydrographical surveys of the Cape coast and his influence is noticeable in Gordon's Map no 3.

## Governor van de Graaff's Overview Map of the Cape, 1790

Map 4.TOPO 15.116 in the Dutch National Archives is an overview map of the Cape Colony compiled by Governor van de Graaff himself on a scale of 1:500,000 (see Fig. 14.15). This drawing measures 60 × 150 cm and shows more detail than any other known earlier map. The title mentions that it "was compiled from the best documents, at that time being in loco, by the Hon. Governor and Director of these Colonies C.J. van de Graaff 1785 to 1784".[49] What these sources were, is not clear especially as the map bears on its reverse an inscription that it was made during the expedition with Governor van Plettenberg and copied from the map by Secretary De Wet. Olof Gotlieb de Wet, who wrote the journal of the Van Plettenberg expedition, was not a surveyor and must therefore have obtained his map from someone else. It is also obvious that additions to it were made after Van de Graaff's

**Fig. 14.15** Governor van de Graaff's overview map of the Cape Colony, 1790 (Copyright for this image by the Dutch National Archives, map 4. TOPO 15 116)

[48]Duminy, A. 2005. *François Renier Duminy 1747–1811. French mariner and South African Pioneer*. Pretoria: Protea.

[49]Koeman, C. 1952. *Tabulae Geographicae...*, 63.

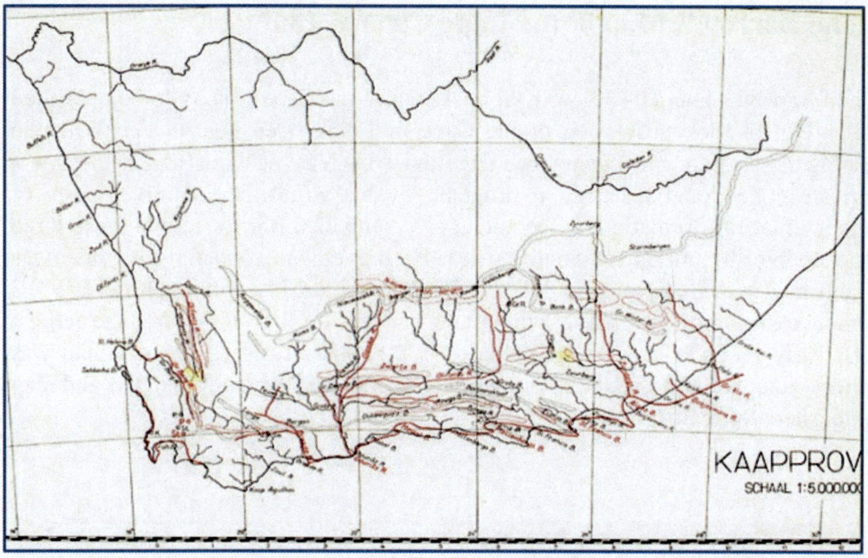


**Fig. 14.16** Van de Graaff's overview map (*in red*) superimposed on a modern map of South Africa (Source: Koeman: *Tabulae Geographicae*..., p. 64. Copyright for this image by URU-Explokart Research Group, University Utrecht)

death in 1812 as Theopolis, a mission station established 1814 in the Eastern Cape by the London Missionary Society, appears on the map (see Fig. 14.15).

Koeman refers to Van de Graaff's map as the "first topographical map of the Colony"[50] and believes that it was meant to serve all the governmental and defensive measures Van de Graaff had in mind. Brommer considers it Van de Graaff's "magnum opus" and "the culmination of over one-and-a half centuries of mapping of the Cape".[51] Although the map is certainly impressive, Forbes considers its cartographic merits "questionable", especially when compared with Gordon's Map no 3.[52] Facts which support this claim and also point to the Van de Graaff map's composite origin, are that the length of a degree of longitude varies across the map; that a major town such as Graaff-Reinet appears in two different places, and that the Kariega River in the Eastern Cape is shown in four localities.

[50]*Koeman, C.* 1952, "Landmeters in dienst van de VOC in de Kaapkolonie." *Tijdscrift voor Kadaster en Landmeetkunde* 01.01.1952, 96.

[51]Brommer, B. 2009. *Op cit.*, 288.

[52]Forbes, V.S. 1952. "An atlas of eighteenth century cartography of the Cape". *Africana Notes and News* 9(3), 102.

## John Barrow's Map of the Cape Colony, 1801

John Barrow's map of 1801 was the first map of the Cape Colony to be published after Britain took possession of the Cape in 1795[53] (see Fig. 14.17). With the cartographical archive having been removed by Van de Graaff, and Gordon's private collection transferred to England by his widow, the complete lack of topographical information on the country [54] compelled Barrow to undertake three travels into the interior of South Africa. His travelogue containing his map was published in two volumes in 1801 and 1804 respectively. Although not a VOC product, his map is nevertheless important as it was for long considered the earliest relatively accurate map of the colony. As a published map, Barrow's map was widely disseminated and the information on it used by other cartographers and map publishers for many years.

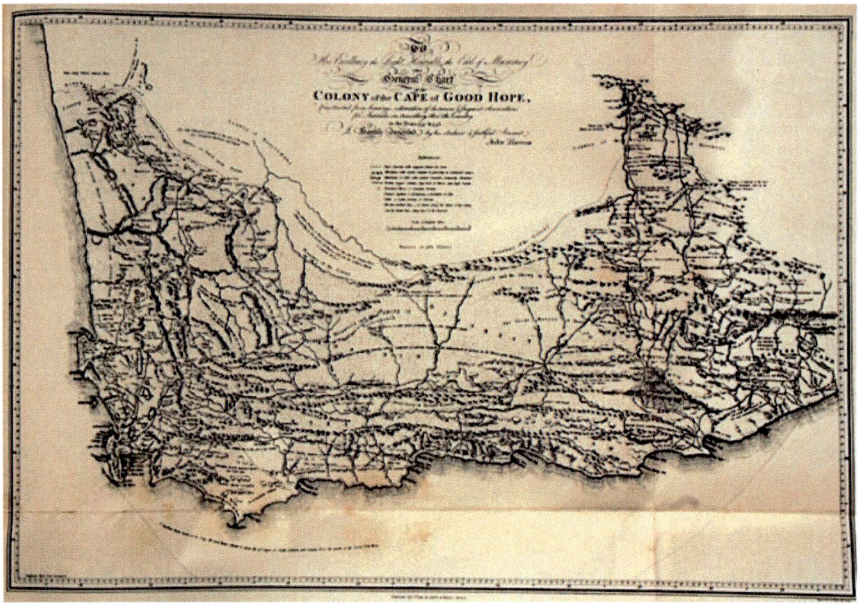

**Fig. 14.17** John Barrow's map of the Cape Colony, 1801

[53]*General chart of the Colony of the Cape of Good Hope*, in Barrow, John. 1801. *Travels into the Interior of Southern Africa...*, vol. 2. London: Cadell and Davies.

[54]Barrow wrote as follows about the intelligence situation at the takeover: "at the capture...we were entirely ignorant of all the points which were most material to be known. There was not a survey of one of the bays that could be depended on.....not a single map that took in one tenth part of the colony." See Barrow. J. *Op. cit.*, 180.

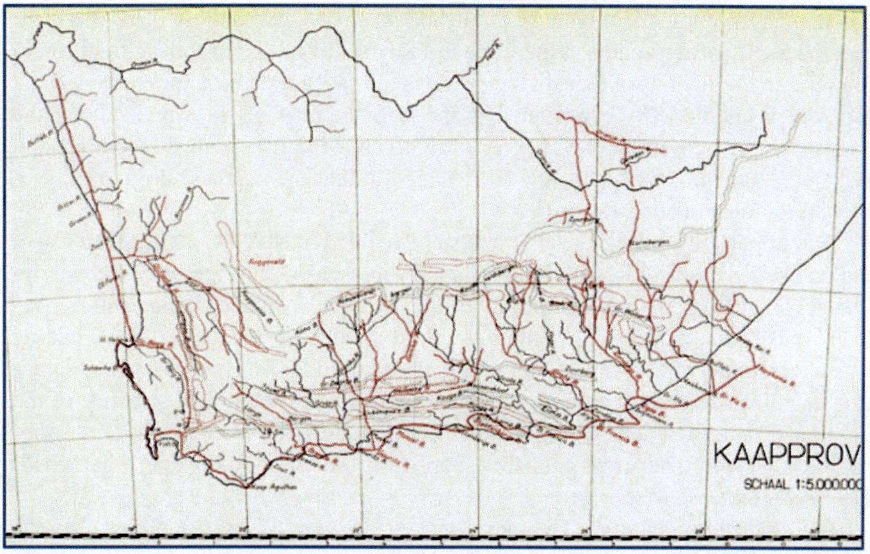


**Fig. 14.18** Barrow's map (*in red*) superimposed on a modern map of South Africa (Source: Koeman: *Tabulae Geographicae*..., p. 60. Copyright for this image by URU-Explokart Research Group, University Utrecht)

Barrow was almost entirely ignorant of the extensive surveys that had been undertaken by trained surveyors during the Dutch period, and his only mention of a professional map drawn under Dutch auspices is of Frederici's Map of the south coast.[55] His own mapping method was that of the compass traverse, with distances estimated by the average speed of an ox-waggon over varying types of country. His latitudes were determined by solar observations at intervals of about 35 km, and his longitudes were estimated by dead-reckoning.[56] In spite of this rather inaccurate fixing of points, his map contains no major inaccuracies (see Fig. 14.18).

## Conclusion

The maps referred to in this paper show a surprising degree of professionalism given the scarcity of topographical information on the interior of southern Africa in the seventeenth and eighteenth centuries. They also show a gradual increase in the amount of detail and accuracy with which the country was mapped as is shown by

[55]Forbes, V.S. *Pioneer Travellers, 135.*

[56]*Ibid.*

Koeman in the comparison map he published for each.[57] From these maps it is clear that the VOC cartographers at the Cape initially followed the European tradition to map the southern and southeastern coastlines too far to the north and to the west, a fact which undoubtedly contributed to the large number of shipwrecks along the coast. This error was, however, gradually amended and in Van de Graaff's and Barrow's maps this situation is even reversed in that the southeastern coastline is plotted too far to the east.

Two important stylistic developments are visible on the VOC maps of the Cape; the one being the way in which the relief is represented, and the other the way in which colour is used. The topography of the Cape shows marked relief differences over short distances and representing the third dimension by means of hill-shading was a technique which was, at that stage, hardly used in Europe.[58] The representation of relief was further enhanced by the creative use of colour. Koeman is of the opinion that the relevant cartographers, knowing that the maps were manuscript maps which would never be published, applied colour more freely than what would have been the case otherwise.

The eighteenth century VOC maps of the Cape merit more attention. Cornelis Koeman was an historical cartographer of international repute and his opinion that the Cape of Good Hope was by the end of the eighteenth century "better mapped than any other area outside Europe", [59] should not be taken lightly.

[57]In the comparison maps, the outline, and a few marked characteristics of the relevant VOC map are superimposed on a copy of map no 86 of South Africa in *Stieler's Grand Atlas de Géographie Moderne*, edited by Jutus Perthes, Gotha, 1934.

[58]Koeman, C. "Landmeters in dienst van de VOC...", 97.

[59]Koeman, C. "Nieuwe bijdragen...", 81.

# Chapter 15
# László Magyar's Cartography of Angola and the Discovery of his 1858 Manuscript Map in the Cholnoky Collection in Romania

**Zsombor Nemerkényi and Zsombor Bartos-Elekes**

**Abstract** With the exception of Livingstone, a Hungarian explorer, László Magyar, was the first European to travel in large areas of the present-day Angola. He lived in Angola from 1848 to 1864 and produced relatively accurate maps during this period. His manuscript map of 1857 which shows the coastline of Angola and a large area of the interior, is stored in Budapest. The other manuscript map from 1858 depicts inner Angola, an area which was at that time almost completely unknown to Europe. This map was known only through a copy which had been published in the German periodical *Petermanns Geographische Mitteilungen* in 1860. The manuscript version was considered lost for a long time, but was in 2007 rediscovered by the authors in the Cholnoky Map Collection in Cluj-Napoca, Romania.

In this paper the authors explain how Magyar's two manuscript maps (both drawn in Angola) ended up in Hungary and Romania respectively. They also analyse the content and planimetric accuracy of the two maps using digital methods (GIS-techniques and statistical index numbers), and some toponymical tools. In the third instance they want to introduce the reader to the Cholnoky Map Collection.

The Cholnoky collection was established at the beginning of the twentieth century by the Hungarian geographer Jenő Cholnoky who was professor at the University of Cluj-Napoca (Kolozsvár), at that time part of the Austro-Hungarian Monarchy. The collection disappeared in the 1950s to be rediscovered and

Z. Nemerkényi (✉)
Department of Cartography and Geoinformatics, Eötvös Loránd University, Budapest, Hungary
e-mail: zsombor76@gmail.com

Z. Bartos-Elekes (✉)
Faculty of Geography, Babeş–Bolyai University, Cluj-Napoca, Romania
e-mail: bezsombor@yahoo.com

E. Liebenberg and I.J. Demhardt (eds.), *History of Cartography*,
Lecture Notes in Geoinformation and Cartography 6,
DOI 10.1007/978-3-642-19088-9_15, 

catalogued between 2001 and 2008. The Cholnoky map collection is presently considered the second largest map collection in Romania.

## Introduction

This paper had its origin in two seemingly independent cartographic research projects which were later found to be closely related. The first of these was the discovery in 2007 of a hitherto forgotten map collection at the Babeş–Bolyai University in Cluj-Napoca, Romania. Known as the Cholnoky Jenő Map Collection, this body of maps has since been catalogued and is now considered to be the second largest map collection in Romania. The other project involved an investigation into the cartographic legacy of the Hungarian explorer László Magyar. That these two projects were related, came to the authors' attention when the newly-found Cholnoky Collection surprisingly yielded a known, but long considered lost, 1858 manuscript map by Magyar. Prior to this discovery, the only extant manuscript map by Magyar was his 1857 map which shows the coastline of Angola and a large area of the interior. This map is now kept in the Library of the Hungarian Academy, Budapest.

The rediscovery of the 1858 map is a significant cartographic event which is also of interest to ethnographers. With the exception of David Livingstone, László Magyar was the first European to travel in large areas of the present-day Angola. He lived in this part of the continent from 1848 to 1864 and produced relatively accurate maps based on his own geographical and ethnographical observations. His findings were regularly sent back to Europe where two of his maps, compiled in 1857–1858 respectively, were subsequently published – his 1857 map in Hungary in 1859 (Hunfalvy 1859), and his 1858 map in Germany in 1860 (Magyar 1860b). After its publication, the manuscript of the 1857 map was retained but, since 1888, the manuscript of the 1858 map has been considered lost. What makes Magyar's 1858 map important to map historians, is that it is a depiction of inner Angola, an area which was in the mid-nineteenth century almost completely unknown to Europe. Another significant feature which Magyar indicated on this map, is the route Livingstone followed in 1854 when he traversed central Angola from the Zambezi to Luanda.

The authors' main objective with this paper was to clarify how these two extant manuscript maps of 1857–1858 (both drawn in Angola) ended up in Hungary and Romania respectively. As a secondary objective, they aimed to analyse the content and planimetric accuracy of the two maps in question.

To achieve these goals, two different methodologies were adopted. As a first step, a comprehensive bibliography consisting of nearly three thousand entries related to Magyar and his contemporary South-Central African travellers, was compiled (Nemerkényi 2008). The relevant material was not only found in libraries and archives in Hungary and Portugal, but also in Germany and Austria. For the analysis of the two manuscript maps, digital methods such as GIS-techniques and statistical index numbers, as well as toponymical tools, were employed (Nemerkényi 2008).

## The Cholnoky Map Collection in Cluj-Napoca, Romania

The discovery of a large number of historical maps and photographs in a store room at the Babeş–Bolyai University of Cluj-Napoca in 2001 came as a surprise to everybody concerned. Viewing the collection initially, the origin of the maps and photographs was difficult to identify. Inspecting the dusty and in some places moulded papers carefully did, however, yield dividends in that it became apparent that the collection had once belonged to Jenő Cholnoky.

Jenő Cholnoky (1870–1950) was one of Hungary's most famous geographers. He graduated at Budapest, travelled in China (1896–1898), and taught at the universities of Cluj-Napoca / Kolozsvár (1905–1919) and Budapest (1920–1940). His research interests were Physical and Regional Geography.

To fully understand the history of the collection, one should relate it to the history of the region. In the Middle Ages the present city of Cluj-Napoca (its Hungarian name is Kolozsvár) was part of the Kingdom of Hungary. The city then became part of the Principality of Transylvania, then of the Habsburg Empire, and eventually of the Austro-Hungarian Monarchy. The university was founded in the Austro-Hungarian Monarchy in 1872 when it was given the name Francis Joseph University. Between 1905 and 1919 Cholnoky headed the Department of Geography at this university (see Fig. 15.1). After the Monarchy was defeated in World War I, the eastern part of Hungary (also named Transylvania) where Kolozsvár was situated, became part of Romania. The university lost its Hungarian character and Cholnoky had to leave the city on short notice, leaving the collection of maps and photographs behind. The official name of the city was changed from Kolozsvár to the Romanian name of Cluj (later Cluj-Napoca), and a Romanian university was established on the same site and occupying the same buildings as the former university. During the communist regime of the 1950s, it was considered illegal to own historical maps and the authorities would have destroyed the collection if they had known about it. To preserve the map and photo collection for posterity, it was hidden in a store

**Fig. 15.1** The University of Kolozsvár in 1909 as it appears on an old postcard

room with the result that, decades later, nobody knew about its existence. When the communist regime came to an end in 1989, the university assumed a multicultural identity. Its present name is the Babeş–Bolyai University of Cluj-Napoca, with Romanian, Hungarian and German as languages of education.

The photo collection which contains more than 5,000 individual photos dating from the beginning of the twentieth century, was initially catalogued by Zoltán Imecs. Most of the photographs were taken by Cholnoky in Transylvania, but hundreds of photos also derive from travels he undertook in other continents such as China and the United States (Imecs 2004). Between 2006 and 2008, the Cholnoky map collection was catalogued under the supervision of Bartos-Elekes. Of the 6,400 items, 3,050 are chorographic and thematic maps, 3,200 (mostly Austrian) topographic map sheets, and 150 atlases. The oldest maps date from the seventeenth century, but the majority of maps is from the nineteenth century. This collection of historical maps which is larger than the map collection of the Romanian National Library, is considered the second largest collection of historical maps in Romania after the collection of the Romanian Academy in Bucharest.

The cataloguing process also yielded some valuable manuscript maps, the most important map being László Magyar 1858 cartographic compilation of inner Angola (Bartos-Elekes 2008).

## László Magyar, a Hungarian Explorer in Angola

László Magyar was born in Szombathely in Hungary on November 13, 1818 (see Fig. 15.2). He was well-educated and spoke five European languages except Hungarian. He left Europe in 1843 and arrived in the present-day Angola in Africa in 1848 where he married a local woman, the daughter of the King of the Bié territory. The fact that the indigenous peoples accepted him as an "insider", greatly aided him in his travels and provided him with the opportunity to intimately study the local people, their habits, and their social organization. Unlike other European travellers who were always on the move, Magyar would stay at a place for a long time where he would record valuable geographical, and especially ethnographical, data. He lived in Angola for 17 years and died in Ponto do Cuio on November 9, 1864 (Nemerkényi 2008).

Magyar was also interested in the geography of the country and undertook several journeys into the interior of the present-day Angola. Most of his travels were in the borderland between the present Angola and the Democratic Republic of Congo (1850–1851), and in the south-eastern part of Angola (1852–1854) along the watershed of the Zambezi and Congo rivers (see Fig. 15.3). He was often the first European explorer to set foot in a specific area and, whilst travelling, he indulged in an extended correspondence with various scientists and scientific institutions in Europe. He used to send his ethnographic-geographical observations and descriptions home to the Hungarian Academy of Science and was elected a corresponding member of the Hungarian Academy in 1858 (Nemerkényi 2008).

**Fig. 15.2** Portrait of Magyar's brother in law which is believed by many to be László Magyar himself as their faces were quite similar – even the squint of their eyes

## Magyar's Manuscript Map of 1857

In 1857 Magyar sent the first volume of his reports, together with a map (see Fig. 15.4), to János Hunfalvy, a member of the Hungarian Academy of Science in Budapest. The despatch of the documents was took place with the aid of the Portuguese government. As the first and primary Hungarian patron of Magyar's travels, Hunfalvy wrote a preface to Magyar's text which he supplemented with notes. In 1859 this text was published by the Hungarian Academy under the editorship of Hunfalvy (Hunfalvy 1859). After publication, Magyar's manuscript map of 1857 became part of the collection of the manuscript department of the Hungarian Academy of Science in Budapest where it has been stored ever since (Magyar 1857). In 1861 Magyar was informed of the successful delivery of the parcel and the subsequent publication of his work, as well as his election as a member of the Hungarian Academy of Science.

Similar to the way in which he treated Magyar's earlier letters and reports, Hunfalvy forwarded the German translation of Magyar's material to August Petermann, the editor of *Petermanns Geographische Mitteilungen* in Gotha, as well as to the officers of the Royal Geographical Society in London (Rónay 1854). Although it cannot be verified that Hunfalvy also sent Magyar's original map of 1857 to other institutions abroad, he most probably forwarded a copy of it, together with Magyar's textual descriptions, to members of other international scientific communities in Europe.

**Fig. 15.3** Travels by Magyar and his contemporaries and the areas covered by his manuscript maps (Copyright for this image by Nemerkényi)

## Manuscript Map of 1858

In a letter to Hunfalvy in 1858 (Magyar 1858a), Magyar appended a description of the countries of Moluva, *alias* Moropuu, and Lobal, as well as a map depicting the territory concerned. In 1859 the main part of this study (without the vocabulary list) was read by Hunfalvy as Magyar's inaugural lecture at the Hungarian Academy of

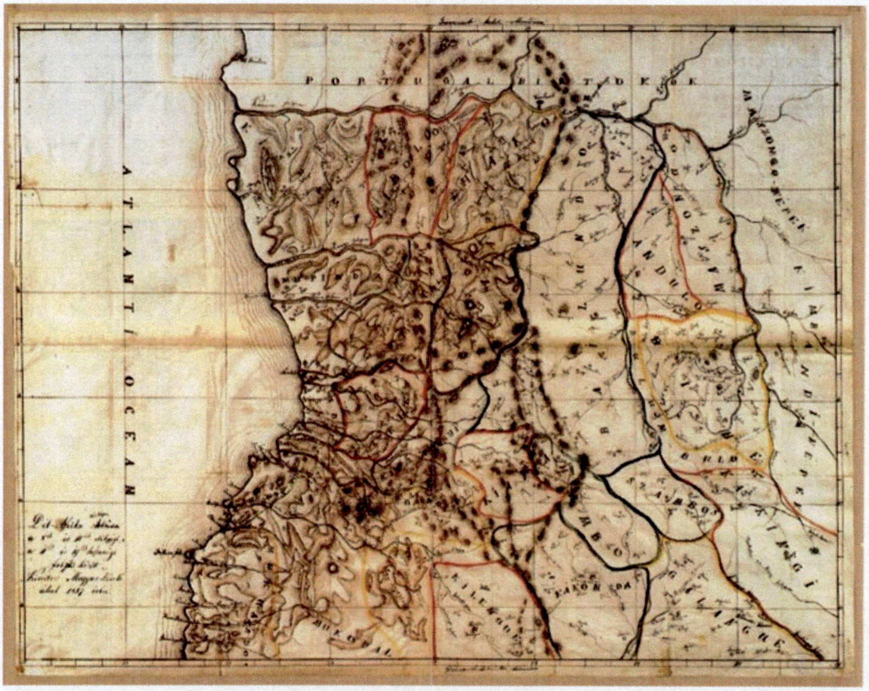

**Fig. 15.4** Facsimile of Magyar's manuscript of 1857 (copyright for this image by the Hungarian Academy of Science, Budapest, and Cartographia Ltd., Budapest)

Sciences (Magyar 1859). With some minor changes, the same paper also appeared in *Petermanns Geographische Mitteilungen* (Magyar 1860a). Petermann also published the redrawn version of the appended map together with the Moluva list of approximately 200 words from Magyar's original manuscript (Magyar 1860b). This was the last European publication which would refer directly to Magyar's original and unaltered manuscript map of 1858. From then onwards, for almost a century and a half, various authors would refer to the edited copy published by Petermann only (see Fig. 15.5).

In his biographical study of Magyar published in 1937, Gusztáv Thirring refers to "maps" (plural) that had been compiled by Magyar, but provided no any accurate indication of his sources (Thirring 1937). Using the plural form was indeed correct, since the afore-mentioned correspondence and reports obviously implied the editing of multiple maps of which Magyar sent at least two to Hungary via Portuguese mediation. In 1888 Thirring visited the editor-in-chief of *Petermann's Geographische Mittheilungen* at the Justus Perthes Geographical Institute in Gotha, Alexander Supan, to ask for the 1858 manuscript map of Magyar (Thirring 1888). The map was, however, no longer in the Institute at the time.

The authors realised that access to the original version of the 1858 map would be imperative to provide a summary of Magyar's scientific achievements and an

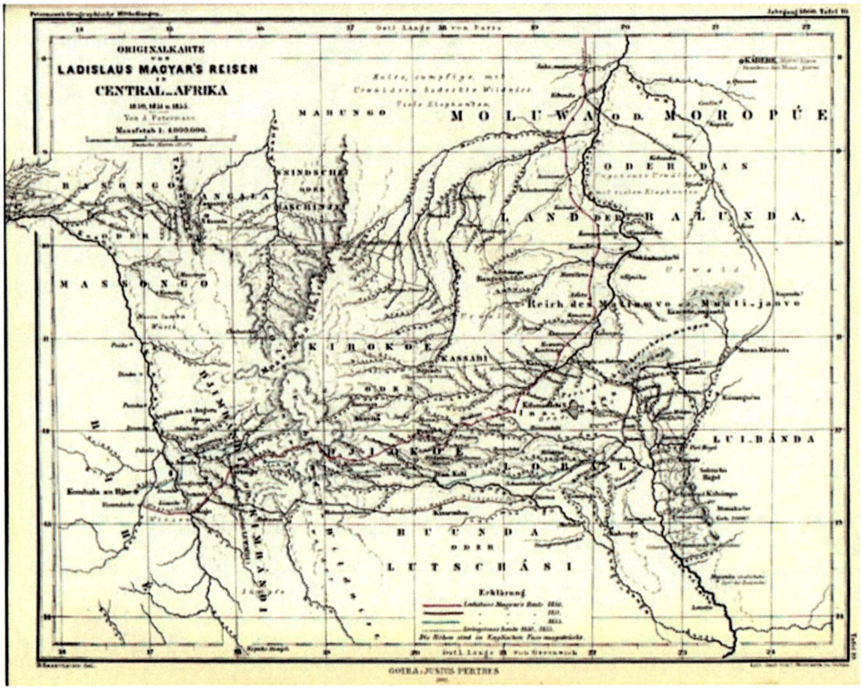


**Fig. 15.5** Magyar's map published in PGM, in 1860 (With permission of National Széchényi Library, Budapest)

analysis of the quality of his cartographic work. In an attempt to locate the 1858 manuscript map for the purposes of this research, the map collections of the Justus Perthes Geographical Institute in Gotha, the Lisbon Geographical Society, and the Royal Geographical Society in London were searched. Only references to the map were found, but not the map itself. Shortly afterwards, however, in April 2007, a manuscript map (Magyar 1858b) with the title *Outline of the Map of South Africa, edited in 1858 by Magyar* (see Fig. 15.6) was discovered during the cataloguing of the Cholnoky Map Collection. With his reference to "South Africa" (instead of southwestern Africa or Angola) Magyar obviously followed a convention established by earlier travellers. It has to remark that beside this map from Cholnoky Map Collection could be other examples of Magyar's manuscript maps (e.g. in Gotha).

The map in question apparently reached Hunfalvy in Hungary as the appendix to Magyar's treatise, who forwarded it to August Petermann in Gotha for publication in *Petermanns Geographische Mitteilungen*. Knowing that the map appeared in the *Mitteilungen* (Magyar 1860b), it was logical that Thirring (in 1888) expected the manuscript version of the map to be still in the archives of the Justus Perthes Geographical Institute in Gotha. Today, however, it can be accepted that Supan was not mistaken when he was unable to locate the map. What probably happened

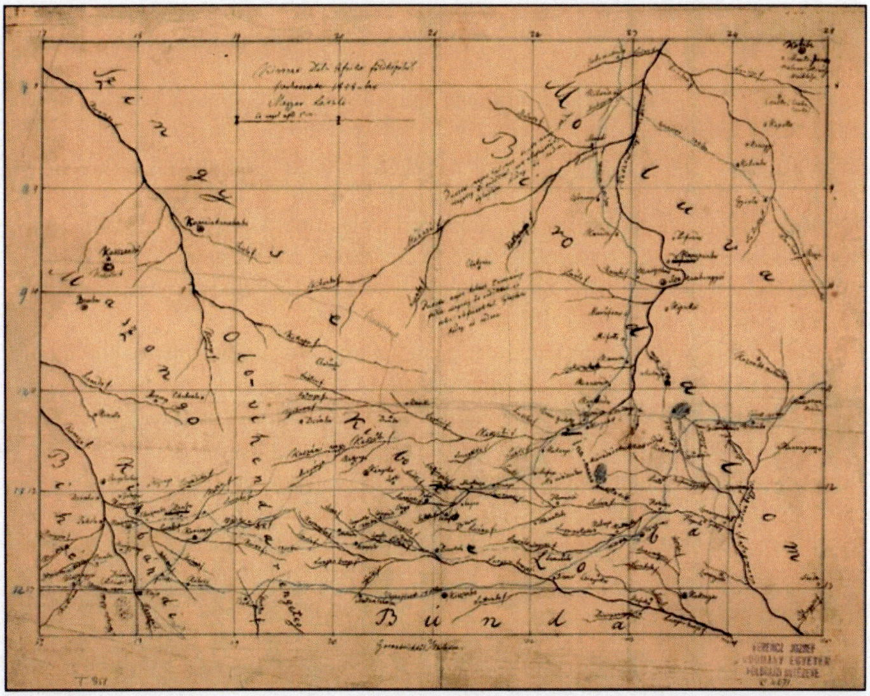

**Fig. 15.6** Manuscript map of 1858 (Copyright for this image by the Babeş–Bolyai University, Cluj-Napoca and Cholnoky Geographical Society, Cluj-Napoca)

was that Petermann, after editing and publishing Magyar's text and the appended map in 1860, by 1888 had returned the manuscript map to Hunfalvy in Hungary. Instead of donating the map to the Hungarian Academy of Sciences, Hunfalvy added the map to the collection of the Hungarian Geographical Society which he had founded himself in 1872. This conclusion is supported by the fact that the map eventually emerged from Cluj-Napoca where it probably ended up through the intervention of Cholnoky. Cholnoky served as secretary of the Hungarian Geographical Society from 1905 to 1910 and as president from 1914 to 1945. It is highly improbable that he would have transferred the map to Cluj-Napoca from the Hungarian Academy of Sciences, but much more likely that he added it to the departmental map collection during his tenure as a professor of Geography at Francis Joseph University. When the Romanian authorities forced him to leave at short notice in 1919, the map remained behind. Today the map bears the seal of the *Ferencz József Tudományegyetem Földrajzi Intézete* (Geographical Institute of Francis Joseph University) in its the lower right corner with the catalog number C 4071 underneath. From this we can infer that Magyar's map became part of the collection in Cluj after 1905, but before 1919.

## Accuracy of the Manuscript Maps

Together Magyar's two manuscript maps represent Central Angola. The 1857 map depicts the western part (plus the coastline), whereas the 1858 map covers the inner part, showing territories which presently belong to the Democratic Republic of the Congo. The two maps overlap and the areas depicted on them reveal a part of the world which was almost totally unknown to Europe. Until that time, the only Europeans who had ever travelled here, were David Livingstone in 1853–1855, António Francisco Ferreira da Silva Porto in 1841–1854, and Alexander Alberto de la Rocha Serpa Pinto in 1846–1900.

Before subjecting Magyar's maps to a digital analysis, a few general comments about his cartography might be appropriate. On the whole his maps reveal both a lack of precision in determining geographical position, and an overestimation of the distances covered during his travels (Hunfalvy 1859): 14,461. The first shortcoming could probably be ascribed to a lack of experience in using astronomical instruments, and the second to the practical difficulties in moving a large expedition party consisting of several hundred people. Whatever the reasons, Magyar's maps were distorted and Petermann, keeping the actual information intact, had to shorten the distances on the maps proportionally before publishing them.

Both the 1857–1858 maps are provided with geographic coordinate systems. The grids are quadratics, which means that the maps were compiled on the Equirectangular Cylindrical Projection which is true to scale along the Equator. Using the dimensions of the grid, it was possible to calculate the theoretical scale of the maps. Both the 1857–1858 maps are provided with geographic coordinate systems. The grids are quadratics, which means that the maps were compiled on the Equirectangular Cylindrical Projection which is true to scale along the Equator. Because this projection is true to scale along the Equator, so we can calculate the scale of the maps: measuring the distance along the Equator and dividing by the real length. Using the dimensions of the grid: on the 1857 map, 1° (111 km) corresponds to 91 mm, which means that the scale is approximately 1:1,220,000; on the map of 1858, 1° (111 km) corresponds to 54 mm, defining the scale as approximately 1:2,040,000.

For the purposes of this article, Magyar's maps were analyzed using the MapAnalyst software which was developed in Switzerland (Jenny 2007). This software calculates the accuracy of an historical map by comparing the content of the old map with the content of a new map using control points. For reasons of comparison, recent map sheets of the Soviet topographic map of Africa on a scale of 1:500,000 were used after it had been converted to an Equirectangular Cylindrical Projection (the same projection as used by Magyar) by means of GIS methods.

Eighteen control points were identified on Magyar's map of 1857 (e.g. Cape Santa Maria; cities such as Benguela, Kuito, Luanda, Lucira, Pungo Andongo, and Quilengues; as well as the mouths of rivers and the points where rivers converge).

The local scale factors of the map (compared to the content of the map and not to the grid of the map) are 1:1,050,000 horizontally and 1:1,170,000 vertically. This means that the map is horizontally 16% and vertically 43% larger than what Magyar anticipated. Expressed differently: the error in latitude is only 4%, while the error in longitude is 16%. The error in longitude is larger because for Magyar the measurement of longitude (measuring difference in time) was more complicated than the measurement of latitude (measuring the angle of the sun or a star above the horizon).

The standard deviation of the displacement vectors is 37 km, whereas the mean positioning error is 52 km. Most of the displacement vectors have a horizontal direction as the larger error was in the longitude and not in latitude. Places on the coastline were drawn further to the west than their actual position, whereas places close to Kuito (Magyar's residence) were drawn further east than where they are in reality (see Fig. 15.7).

On the map dating from 1858 it was more difficult to establish control points as the area had been poorly mapped. Only nine control points could be found (mostly confluences of rivers). In this case the scale factor of the map is also 25% larger than what Magyar had anticipated. The displacement vectors here are between 20 and 30 km. The vectors have varied directions as this map is less accurate than the 1857 map (see Fig. 15.8).

Magyar used astronomical positioning measurements for his latitudes, which allow much more accurate coordinates than the estimation of longitude with larger errors. The map compiled in 1857 is fairly accurate as he probably made use of older maps when compiling it. The map of 1858 which represents inner Angola, is by comparison more inaccurate as it covers an area that had not been mapped before.

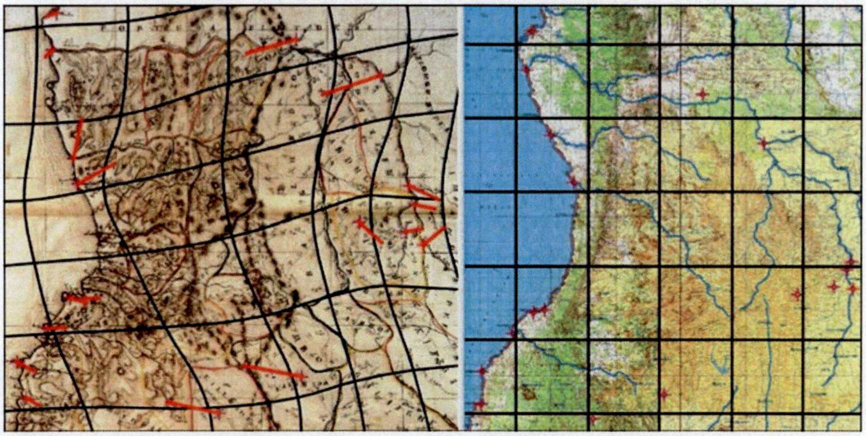

**Fig. 15.7** Geographic grid and distortion vectors on the map of 1857 (Copyright for this image by Bartos-Elekes)

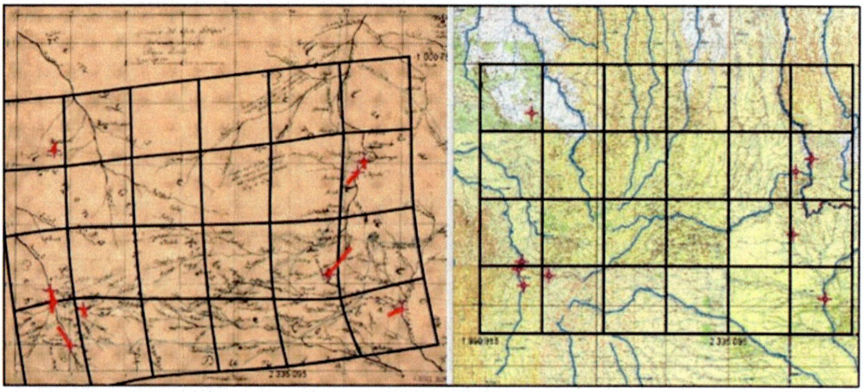

**Fig. 15.8** Geographic grid and distortion vectors on the map of 1858 (Copyright for this image by Bartos-Elekes)

## Content of the Manuscript Maps

Both maps are provided with a geographic grid and both depict hydrographical features and settlements. The map of 1857 also depicts relief and boundaries, whereas the map of 1858 shows the route Livingstone followed from Sesheke to Luanda. Mentioning should be made of the fact that the detail on the maps do not pertain to certain areas only, but are evenly spread. The reason for this was that Magyar was integrated in the local society and that he was knowledgeable about the entire area and not only about the areas in which he himself had travelled. Livingstone's map of 1854 of the same area covers only the narrow zone situated along the explorer's route. From his correspondence with his Hungarian patrons, it can be inferred that Magyar knew about Livingstone's successful journey from the Zambesi to Luanda. It is therefore not strange that Magyar wrote in one of his letters (Magyar 1856) that he intended to personally contact Livingstone. This meeting unfortunately never took place and there does not exist any documentary evidence that Livingstone knew about Magyar.

The number of names on the map of 1857 is 450, with 150 appearing on the map of 1858. The names are mostly toponyms or place names, and hydronyms, with some descriptive notes in Hungarian. Both Magyar's maps are detailed maps when compared to the maps of other travellers such as Livingstone (Livingstone 1854) (see Fig. 15.9).

## Lost Documents

The Hungarian Academy of Sciences only learnt about Magyar's death in 1868, 4 years after he had passed away. When receiving the news, they asked permission from the Portuguese government to transport his papers and personal belongings to

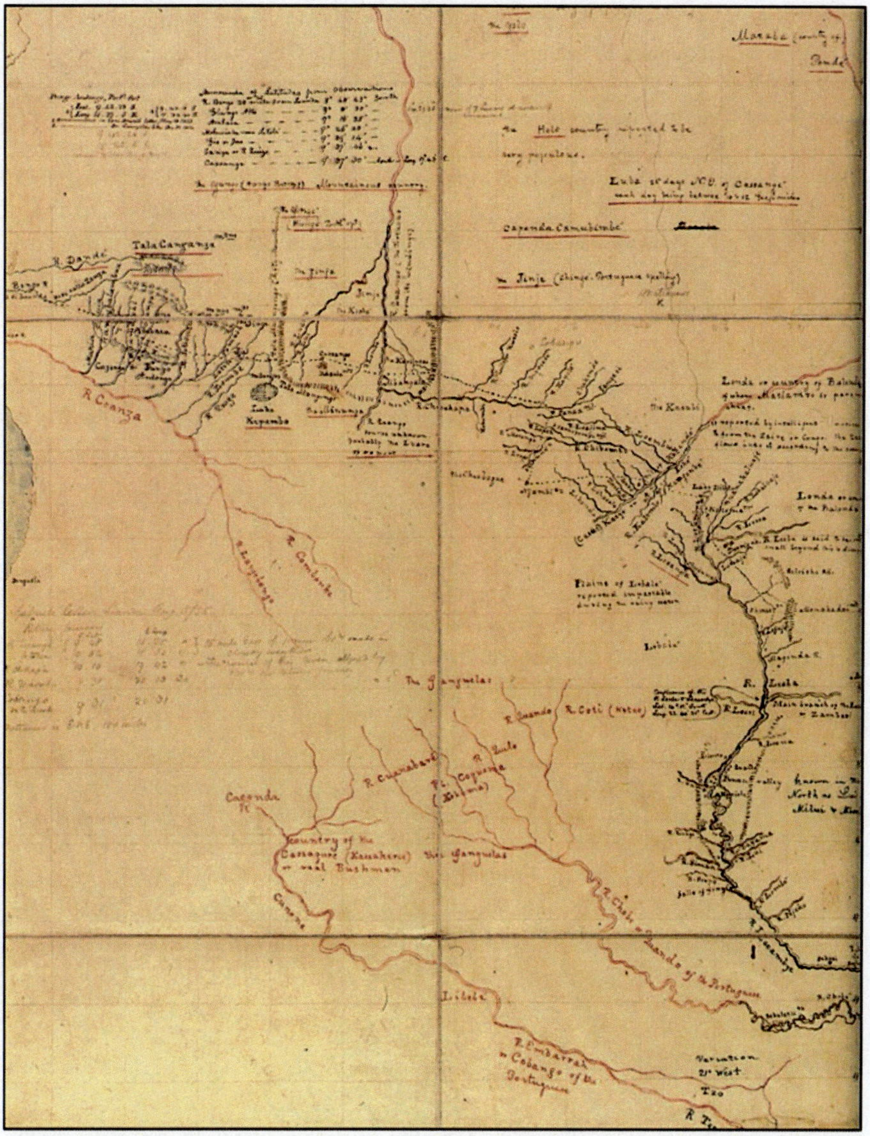


**Fig. 15.9** Livingstone's manuscript map indicating his route from Sesheke to Luanda. 1854 (Copyright for this image by the Royal Geographical Society, London)

Hungary (Arany 1868). Four years later, in 1872, the Portuguese authorities reported that the crate containing Magyar's papers had been destroyed by fire (Anonymus 1873). Hungarian scientific circles found this explanation difficult to accept, but the period of 9 years which had already passed since Magyar's death was too long to allow for effective action.

We can accept with certainty that some of Magyar's manuscripts were destroyed, but it would be premature to conclude that this was the case with all his documents. The research undertaken by the present authors does indicate that artifacts and documents belonging to Magyar which may still be found could provide valuable further insights into his discoveries. At this stage one cannot entirely exclude the possibility that Magyar's second and third volumes, together with the relevant maps, may still be found in either Angola or the Arquivo Histórico Ultramarino in Lisbon, Portugal, or maybe elsewhere in Europe. Judging from the example of the manuscript map of 1858, the most logical archival or library collection to research is not necessarily the correct one.

## Conclusion

In the second half of the nineteenth century Magyar's travels received much attention in scientific circles in Europe. It was especially his ethnographical work which was deemed important and he became a celebrity in the world of ethnographical researchers and geographical explorers. Unfortunately his second and third volumes never reached Europe with the result that his later travels only became known through fragments of his diary and the letters he had forwarded to Europe. Amongst scientists this lack of accurate reporting soon led to a diminished interest in his achievements. New research was only to be instigated in the 1980s when a Hungarian cultural anthropologist, Éva Sebestyén, uncovered previously unknown documents related to Magyar's death in Angola (Sebestyén 1998). The finding of Magyar's 1858 manuscript map in Cluj-Napoca in 2007 by the present authors, and their subsequent biographical and cartographical analysis of his work, were further steps in unravelling the life and work of this remarkable Hungarian explorer. Last, but not least, the analysis of his 1857–1858 maps contribute to the history of the exploration of Angola in the nineteenth century.

## Biographical Note

**Zsombor Nemerkényi** received his M.Sc. in Cartography from the Eötvös Loránd University in Budapest in 1999 and his Ph.D. in the History of Cartography in 2009 from the same university in 2009. His research interests are the European colonization of Africa in the nineteenth century and the exploring of the Arctic and Antarctic. His doctoral research was about the history of the colonization era in Southwestern Africa. He has been associate editor of the official magazine of the Hungarian Geographical Society since 2006. He also edits maps for scholarly periodicals, books and atlases.

**Zsombor Bartos-Elekes** followed a GIS-course at the University of Utrecht in 1998; received his M.Sc. in Cartography from the Eötvös Loránd University in

Budapest in 1999, and his Ph.D. in the History of Cartography and Toponymy in 2006 from the same university. The theme of his doctorate was the language of maps, especially Transylvanian place-names on maps since the nineteenth century. Since 2000 he has been teaching Cartography and Toponymy at Babeş–Bolyai University (Cluj-Napoca) as senior lecturer. His main research interests lie in Cartography (the territory of Romania on old maps, the Cholnoky Map Collection, the projections of the old maps, and ethnic maps) and Toponymy (the standardization of place names, exonyms, and the language of the maps). His edited maps and papers are published mostly by Hungarian and Romanian publishers and periodicals.

## References

Anonymus (1873) Magyar László iratairól irja a VKM. hogy valószinüleg elégtek. [Brief report about the perished letters of László Magyar.] Unpublished manuscript, Archive of the Hungarian Academy of Sciences (38/1873)

Arany J, (1868) Magyar László hagyatéka felkutatása tb. az elnökség levelezése: Arany fogal. Eötvöshöz [Correspondence about the research of László Magyar's bequest]. Unpublished manuscript, Archive of the Hungarian Academy of Sciences (1400/583)

Bartos-Elekes Zs (2008) A kolozsvári Cholnoky Jenő Térképtár. [The Cholnoky Map Collection of Cluj.]. Földrajzi Közlemények 4:489–494

Hunfalvy J (1859) Magyar László délafrikai utazásai 1849–57 években. [The South African Travels of Magyar László in the Years 1849–1857.], 1st edn. Eggenberger, Budapest, p 204

Imecs Z (2004) Cholnoky Jenő fényképi hagyatéka a kolozsvári egyetemen. [The Photos of Cholnoky Jenő at the University of Cluj.]. Földtani Kutatás 3(4):18–24

Jenny B (2007) Planimetric analysis of historical maps with MapAnalyst. In: Oehrli M (ed) Paper and Poster Abstracts of the 22nd International Conference on the History of Cartography. ICHC, Bern, pp 62–63

Livingstone D (1854) Manuscript map of Livingstone's route from Sesheke to Luanda by David Livingstone. Ca: 1:2,900,000. Royal Geographic Society, London

Magyar L, (1856). Levél Magyar Imréhez (Bihé, 1856. augusztus 20). [Letter to Amerigo Magyar – Bié, August 20, 1856.] In: Thirring G (ed), Magyar László élete és tudományos működése. Kritikai adalék a magyar földrajzi kutatások történetéhez Magyar László kiadatlan írásaival [The Life and Scientific activity of Magyar: Critical Addition to the History of Hungarian Geographical Research – with Unpublished Writings of Magyar.]. Kilián, Budapest, p 133

Magyar L, (1857). Dél-Afrika térképe, a 8dik és 15dik szélességi, s a 11dik és 19dik hoszasági fokok között. Késztve Magyar László által 1857 évben. [Map of South Africa, between the 8th and the 15th degrees of latitude, and the 11th and 19th degrees of longitude. Produced by Magyar László in the year of 1857.] (Manuscript map) Ca: 1:1,170,000. Budapest: Library of the Hungarian Academy

Magyar L (1858a) Levél Hunfalvy Jánoshoz (Lucira, 1858. november 16). [Letter to Hunfalvy – Lucira, November 16, 1858.] In: Thirring G (ed), Magyar László élete és tudományos működése. Kritikai adalék a magyar földrajzi kutatások történetéhez Magyar László kiadatlan írásaival [The Life and Scientific activity of Magyar: Critical Addition to the History of Hungarian Geographical Research – with Unpublished Writings of Magyar.]. Kilián, Budapest, pp 145–149

Magyar L (1858b) Kivonat Dél-Afrika földképéből szerkesztette 1858-ban Magyar László. [Outline of the Map of South Africa, edited in 1858 by Magyar]. (Manuscript map) Ca: 1:2,020,000. Cholnoky Map Collection, Cluj-Napoca

Magyar L (1859) Rövid tudósítás a Moluva vagy Moropuu és Lobál országokról. [Short Report on the Countries of Moluva or Moropuu and Lobal.]. Akadémiai Értesítő 11:921–941

Magyar L, (1860a) Ladislaus Magyar's Erforschung von Inner-Afrika. Nachrichten über die von ihm in den Jahren 1850, 1851 und 1855 bereisten Länder Moluwa, Morupu und Lobal. Petermann's Geographische Mittheilungen 1860:227–237

Magyar L, (1860b) Originalkarte von Ladislaus Magyar' Reisen in Central Afrika 1850, 1851 und 1855. In: Ladislaus Magyar's Erforschung von Inner-Afrika. Nachrichten über die von ihm in den Jahren 1850, 1851 und 1855 bereisten Länder Moluwa, Morupu und Lobal. By László Magyar, Plate 10, Gotha: Petermann's Geographische Mittheilungen, 1860

Nemerkényi Zs (2008). Magyar László térképészeti munkájának összehasonlító elemzése. [Comparative Analyses of the Cartographical Work of Magyar László.] Ph.D. dissertation, Eötvös Loránd University, Budapest

Rónay J (1854) Extracts from the letters of an Hungarian traveler in Central Africa. J R Geogr Soc 24:271–275

Sebestyén AÉ (1998) Levéltári kutatástörténet: Magyar László. [History of an archive research: László Magyar.]. Africana Hungarica 2:303–327

Thirring G (1888) Újabb adalékok Magyar László életrajzához. [Additional facts for biography of László Magyar.]. Földrajzi Közlemények 1:333–344

Thirring G (1937) Magyar László élete és tudományos működése. Kritikai adalék a magyar földrajzi kutatások történetéhez Magyar László kiadatlan írásaival. [The Life and Scientific activity of Magyar: Critical Addition to the History of Hungarian Geographical Research – with Unpublished Writings of Magyar.]. Kilián, Budapest

# Chapter 16
# Missionary Cartography in Colonial Africa: Cases from South Africa

**Lindsay Frederick Braun**

**Abstract** Among the first agents of European cultural exchange in Africa before the 1880s were missionaries, who were especially active in the region of modern South Africa. These missionaries came from a wide variety of societies and nations, including French Protestants in Lesotho, the Berlin Missionaries in the Transvaal, the London Mission Society, Wesleyans, and others in the Cape Colony and Natal, to name only a few. These missionaries were not only interested in spreading the Gospel but also in disseminating information about their African surroundings, which included a variety of scientific and literary data. One of the most prominent forms of data collected was geographical, and it expressed through a variety of maps produced or used by missionaries. Mission stations enjoyed generally static locations, often beyond the writ of colonial law, and close relationships between missionaries and local informants. These qualities worked together to produce maps with interesting mixes of topographical, social, political, and cultural content. This essay explores a few examples of 19th-century South African missionary and mission-based cartography, in particular that of London and Wesleyan missionaries in the Transkei, the Berlin missionary Alexander Merensky in Mpumalanga, and the Swiss Berthouds (Paul and Henri) in Limpopo, to discuss the relationship between particular southern African missionaries, maps, and the history of the region's colonization.

Among the first agents of European cultural exchange were Christian missionaries of many denominations, who collected not only ethnological and linguistic data, but geographical and other scientific information as well. This essay explores a few nineteenth century South African examples of missionary and mission geography expressed cartographically, in order to divine some characteristics and key roles of

L.F. Braun (✉)
Department of History, University of Oregon, Eugene, OR, USA
e-mail: lfbraun@uoregon.edu

E. Liebenberg and I.J. Demhardt (eds.), *History of Cartography*, Lecture Notes in Geoinformation and Cartography 6,
DOI 10.1007/978-3-642-19088-9_16, 

missionaries' mapping activity in the history of the colony. Of particular interest are the degree that output contributed to the broader geographical archive and the evident contribution of non-European sources of knowledge to missionaries' map output. Indeed mapping by missionaries straddled the notional boundary between topographical and topological mapping, and sought to merge people's relational knowledge with some sort of scale representation of physical geography; ultimately this mapping became colonial knowledge. However, missionaries' map output tended to reflect networks of information (topologies) much more than accurate or precise survey, much like their other ethnological and linguistic work, even though the forms these maps took suggested greater certainty.

Missionary cartography understandably follows in some ways the breakdown of imperial cartography that Jeffrey Stone offered in his *Short History of the Cartography of Africa* (1995, 47–61) although Stone tended to include the work of missionaries within the first "exploratory" phase and only considers their work within broader phases and processes, not as a distinct genre. However, the cartography produced by missionaries changed over time and did not uniformly follow the program of administration and rule, so the evolution and characteristics of maps produced by and for missionaries were at times quite different. By looking at a few examples from southern Africa, we can see the great variability in missionaries' maps, from the route maps that were also part of the precursor cartography of empire to thematic maps supporting mission activity and even compilation maps produced for commercial purposes. All, however, reflected particular needs and certain blind spots that were more than simply information selectivity.

The questions of what a "missionary" map is, and how one imparts that adjective, are more open than they appear at first blush. A few missionaries, such as Alexander Merensky, David Livingstone, and brothers Paul and Henri Berthoud, produced decidedly secular maps in addition to plans and charts connected in some direct way with mission work (Kayser 1937; Driver 2001, 68–89; Harries 2007, 113–116). Other maps produced by non-missionaries for missionary accounts, such as the 1840s engravings of Wyld for Robert Moffat and Thierry for Arbousset and Daumas, purported to show travel routes but may have been creative with geographical data from the missionaries themselves and relied heavily on existing compilations (Moffat 1842; Arbousset and Daumas 1842; Cartwright 1976).[1] This ambiguity replicates in many ways the complex relationship between missions, knowledge, and empire, whereby a religious figure might represent all three at various times or even simultaneously. It is therefore imperative to bear in mind there are no clear occupational boundaries when considering the terminology around missionary maps. Because missionaries operated within the context of the

[1]See also freestanding examples such as Heinrich Berghaus, "Map of South Africa Showing the Mission Stations of the Various Missionary Societies . . ." (1855), M1/1217, Cape Archives Depot, Cape Town (hereafter KAB). Etherington (2004) points to statements by early cartographers that the missionaries were key sources of early geographical information on the interior, but the exact materials they furnished and how cartographers used them remain uncertain.

era of imperial expansion, their influence and agency effaced many lines; because missionaries were often among the earliest agents of European cultural influence in a given region, geographers and cartographers in Europe much desired whatever geographical knowledge they might possess (Comaroff and Comaroff 1991, 86–125; Driver 2001, 59, 63; Etherington 2008).

The intensity of mission activity in and around the settler colonies of southern Africa meant that the religious practitioners involved in cartography and geography had a larger local venue for their output than those in areas farther from those centers who transmitted knowledge for more distant European consumption. Local consumption of geographical information was a factor even in areas without large European populations, whether among the small ecumenical communities, their African congregants, or other African neighbors generally, but colonists changed the calculus of meaning and necessity (Kokkonen 1993; Hunt 1994, 32–35; Bassett 1998, 24–25).[2] Even before the South African mineral discoveries of the 1870s and 1880s, the movement of colonists across the land and the conflicts that movement engendered meant that knowing the local geography in places like the South African subcontinent took on an imperative quality. At the same time, a local survey profession took part in the compilation of that knowledge and served as the primary vehicle for the colonial state's efforts to impose order in the area where its legal writ carried weight.

Outside of such areas, the collection of information fell to those who had access to local knowledge and those who traveled further afield. Missionaries fell into the latter category together with an eclectic group with varying motivations, but in the former the missionaries were usually the only Europeans with the time to develop access to African social networks and travel freely in those areas. Therefore, in the kingdoms of southeastern Africa and on the Highveld, missionary geography was often instrumental to cartography if not actually the only source of information. Although not all missionaries produced maps themselves, a few took special interest in charting routes and mapping more widely, and the history of their cartography gives us a useful window into the relationship between missionaries, local people, the production of knowledge, and colonial domination in the second half of the nineteenth century.

## Useful Data: Transkeian Missions

Few if any of the missionaries on the eastern Cape produced finished (published) cartography. This lack was especially evident in the region between the Kei and Mzimkhulu rivers, but its reason had less to do with a lack of mission stations and

[2]Kokkonen (1993) suggests the use of this information to promote cartographic literacy or give instruction, but that suggests that symbolic cartography was more alien than was likely the case, as Stone (1995, 6–8) also suggests.

more with the well-worn paths of knowledge transfer that already existed in the area. Those mechanisms involved a number of parties that did produce finished compilation cartography, most notably the Surveyor General in Cape Town (and the Deputy Surveyor General in King William's Town from 1847 to 1866), the British Royal Engineers, and occasionally the Royal Geographical Society in London or other mapmakers who could draw from materials at the War Office and the Colonial Office. Of course the interest of these bodies derived from the nature of the particular colonial environment, which was one of repeated confrontation that pitted colonial domination and political conflict against humanitarian impulses, to the detriment of the latter and the disillusionment of the missionaries (Price 2008, 127–33). Military and administrative cartography thus tended to eclipse what a missionary might produce, even as such maps included missionary data. It is tempting to speculate that alienation at the missions combined with the valorization of geographical science helped to push some sons of Eastern Cape missionaries and clergy to become land surveyors, including S. B. Melvill, Frank Skead, John X. Merriman, and the brothers Joseph and Francis Orpen, but there is insufficient data to be certain.

Two of the missionary societies most active in the Transkei before 1900 were the London Missionary Society and the Wesleyan Methodist Mission Society. They were hardly alone; there were over a dozen Protestant missionary groups with stations on the Eastern Cape (including several different Lutheran bodies), along with a French presence primarily in Lesotho (Scriba and Lislerud 1997, 173–80). Those two societies were however much more strongly connected to the colonial apparatus than missionary societies from outside the British Empire. This connection was not a simple one of instrumentality; rather, it was a social and cultural network of connections between people of widely varying viewpoints but who nevertheless shared a wide range of information (Lester 2001, 7, 31–35). Route maps, reports on land occupation and use, political observations, and the like were almost certainly shared between missionaries and colonial administrators, and the missionaries' intents had little bearing on the way their topological knowledge might be used once it entered the colonial archive.

The maps of the Methodist missionary James Stewart Thomas (1852) and the Anglican missionary D. W. Dodd (ca. 1865) give us a window into this variable process.[3] Both are partial maps of areas showing points of personal interest to the missionaries – stations, trading posts, a few key great places and affiliations – and each clearly takes some data from personal communications if not from other printed charts. Each entered the broader geographical archive in some way, for the maps known today are copies of originals the missionaries retained and that reproduction attests to their importance as interventions in the map compilation process. Beyond that, however, their journeys were somewhat different.

[3]"Copy of sketch made by Mr. Thomas, Missionary Clarkbury, 1852," DSGBK 56, KAB; "Copy of Map Drawn by Rev D Dodd 1868," MP 1453, G. E. Cory Library, Rhodes University, Grahamstown (hereafter "Cory Library").

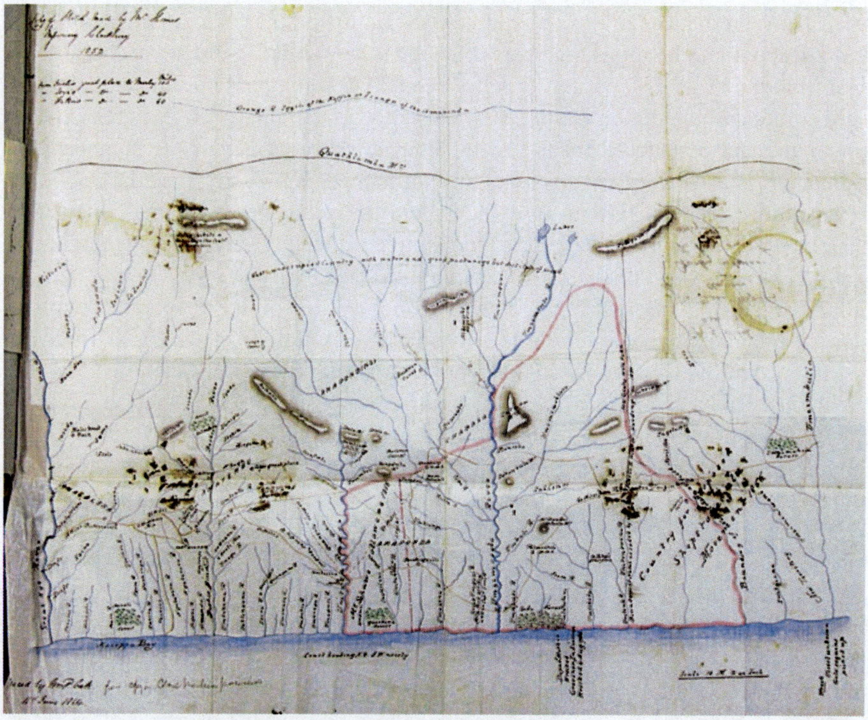

**Fig. 16.1** The 1864 second-hand copy of Rev. J. S. Thomas's route map of the Transkei, originally 1852 but modified by colonial administrators for their needs in the interim (DSGBK 56, KAB)

Thomas's map (Fig. 16.1) is known from its presence among the maps inherited from the Deputy Surveyor General, British Kaffraria (King William's Town), when the Cape Town office subsumed it in 1866. It is a sketch of the Transkei made by William Oak, the DSG's draughtsman, in 1864 that includes a tremendous amount of information that postdates its declared original date (1852) and is of uncertain origin. In fact it was not copied from the original map Thomas had, but from a copy of that original carried by Colonel John Maclean, the Commissioner of the colony of British Kaffraria. When Maclean obtained his copy from Thomas, and where that copy is today, are unknown. It is however clear that even before Oak copied it, Maclean had made a number of notes about potential threats (e.g. the movements of the amaGcaleka paramount Sarhili), policy particulars such as treaty lines, and important locales including mission stations and forts across the Kei from the colony proper. Such active usage suggests strongly that the map was the best schematic available for some time, even after 1858 when military reconnaissance surveys began on the eastern banks of the Kei River (Braun 2009).

From the appearance of this second-generation copy, we can tell that the original was almost certainly a standard route map connected to Thomas's trips across the roads between the Kei and Mzimkhulu rivers. He linked his river crossings directly to the sea, with undulations that were intended to be symbolic and not accurate. It is read from the Cape Colony, left to right, even though compass north is in the upper right corner and the region is not nearly as square as depicted. Few features exist at any significant distance from the road Thomas indicated across the southeastern region. Tributaries very far upstream are not named as they are speculative, but a few features do exist farther away and there are some generic labels about the character of the country. The latter could well have been observed and generalized, but the former very likely came from informants. In the map's upper left, a mountain labeled "Qalaga" has a textual description of its appearance with a name ("Intanjaikala a large hole thru centre of mountain") that suggests relayed testimony; similar are features in the upper left, including fictive "lakes" feeding the Mzimvubu River, the positions of Nehemiah Moshoeshoe's Sotho faction, reputed caves, and a few mountains in the foothills of the large Qathlamba (Drakensberg) range. Unless it provides a relevant marker to the routes and sites, the map eschews showing the region's sometimes remarkable topography and convoluted hydrography.

The simplicity of the geography Thomas drew clearly provided a useful schematic for Maclean, and the rivers and roads gave him markers for locating positions. A scale appears in the lower right, which is a moment of farcical practice on a map so clearly drawn with an eye to relative position and without any absolute positions (indeed none would be fixed until the 1860s, and none reliably until the 1880s). But for Maclean, obsessed with the question of security and trying to enumerate "native" law and custom for the first time through the observations of missionaries and officials appointed from among their progeny, the political notes at various points and their relation to one another were the most important things. In his compendium of Xhosa laws, customs, and politics (Maclean 1858), however, he ironically did not include a map! This combined route map and noting plan in Maclean's possession was therefore doubly precious to the Deputy Surveyor General and worth reproducing for that office. Although it was not itself redrawn in Cape Town, the map was among the variety of plans kept separately and used to compile maps of the Transkei by the Surveyor General's chief compiler, Charles Neumann Thomas (no relation), in the 1870s (Braun 2009, 4–6).

Dodd's map (Fig. 16.2), on the other hand, represents the broader networking of relational information among the missionaries themselves. The extant copy of this map was made in 1868 by the Methodist Rev. E. J. Barrett, who had since 1865 taken up a succession of posts centering on the Kei River (Methodist Church of South Africa 1931, 10) and who later bequeathed his entire library, including many maps, to the G. E. Cory Library at Rhodes University in Grahamstown. Barrett came into contact with a variety of traveling missionaries, many of different denominations or societies, and some of them had charts of the areas to the northeast. Dodd, too, was in contact with other mission stations through travelers, for the map Barrett copied shows them prominently as well as the roads between. However, the accuracy of the non-Wesleyan stations' positions is questionable;

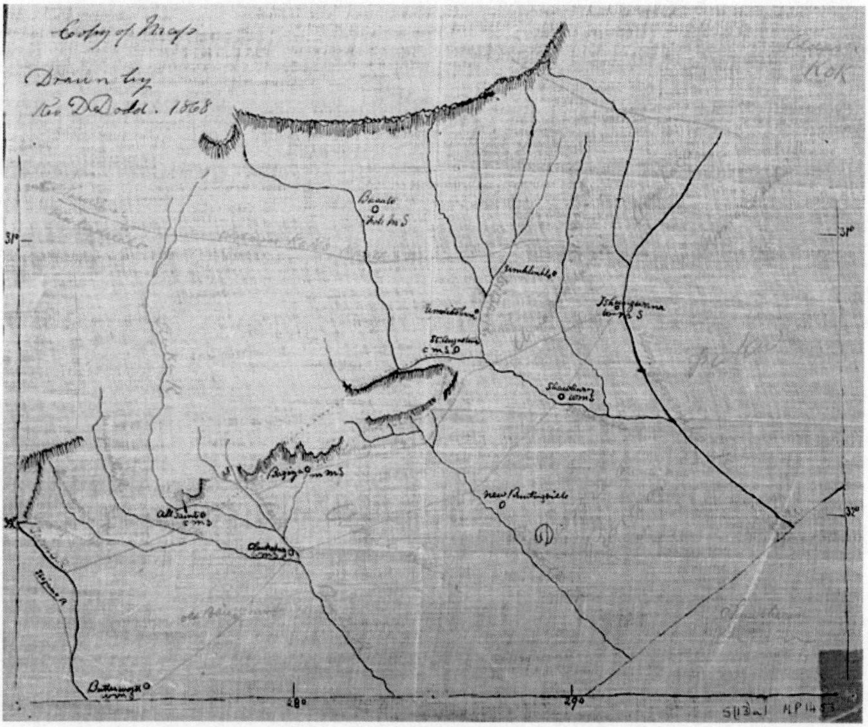


**Fig. 16.2** Rev. E. J. Barrett's 1868 copy of Rev. D. W. Dodd's map of the area between the Mbashe and Mzimvubu rivers (MP 1453, Cory Library)

they bear unusual names, and the presence of French missions in Lesotho (Basutoland) only appears as "Basuto." To a degree it indicates the flux of missionary activities in the region between the Kei and Mzimkhulu rivers, but it also suggests that much of the information was transmitted from intermediary to intermediary. The positions of key chiefs, including Mhlonto and Mditshwa, are approximated relative to the roads and the rivers. Only the most prominent topography appears, and seems to serve to bound a number of ethnonyms that were probably copied from the original. The map also shows the division of Pondoland as well as, in the far upper right, the route and position of the eastern Griqua people who trekked across the mountains in the 1860s. Like Thomas's map, this one is schematic more than it is geographically reliable, although unlike Thomas's this map (or at least its copy) places north at the top and seeks to convey a sense of absolute spatiality.

It is uncertain if Dodd's map figured into any colonial cartography itself. Barrett appears to have cobbled that map together with a copy of an 1864 map of the eastern banks of the Kei, the latter area clearly being based on the military surveys of J. M. Grant and G. P. Colley because of the inclusion of its characteristic

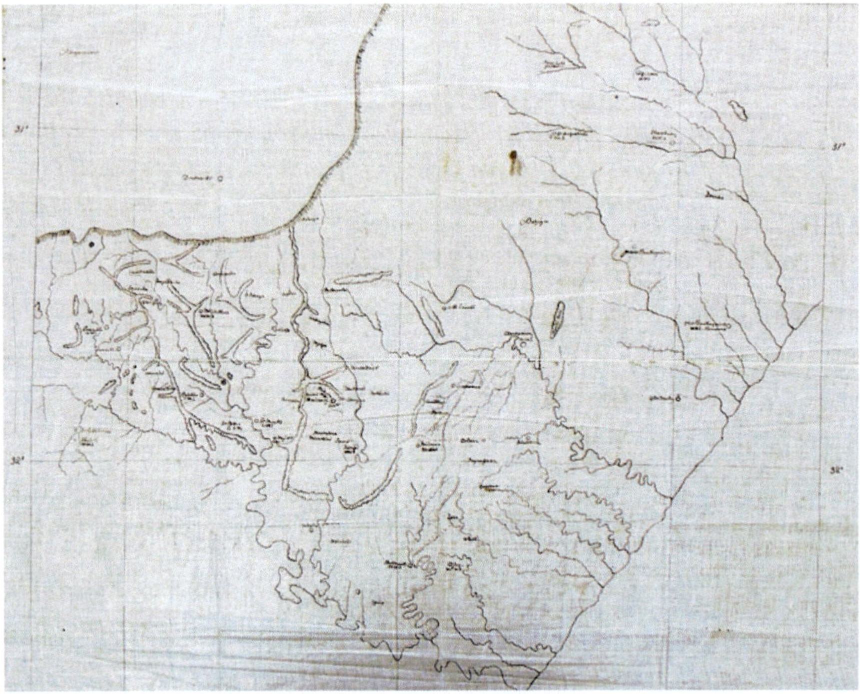

**Fig. 16.3** Barrett's compilation of multiple sources, ca. 1870 (MP 1452, Cory Library)

topography and boundaries.[4] The product (Fig. 16.3) was a fusion that attempted to marry the military survey and its markers with a reorientation of the Dodd map to fit the coastline as known from hydrographic surveys.[5] At the same time, a number of mission stations in the surveyed area were added, likely from Barrett's own knowledge, and uncertain information vanished as did points indicating chiefs' Great Places (capitals) and the ethnonyms. The mission stations of questionable existence remained, but lost their attributions to societies. This map appears to have been an intermediate state towards a chart of mission stations between the Kei and Mzimkhulu rivers, but whether it was carried forward or shared with colonial authorities is uncertain. Barrett absolutely had the connections necessary to put forward his map, although it is possible that mapmaking out of Cape Town may have superseded his efforts or he simply intended this map for his own limited use as Dodd and Thomas originally intended theirs. The records of the bequest from

[4]Barrett's copy of this map is "Outline Map of the Country Between the Kei and Bashee," n.d. [1864], MP 1457, Cory Library. The only known manuscript version of the original survey plan was destroyed by the Surveyor General's office in Cape Town in 1921 (owing to damage) after an identical copy (Plan 5614) had been made; that one matches Barrett's.

[5]Untitled compilation map from Barrett bequest, n.d. [1870], MP 1452, Cory Library.

Barrett to the Cory Library are silent on the point, but regardless of the eventual use of Barrett's map, it is a testament to the collection, processing, and sharing of cartographic information that the missionaries facilitated.

## Colonial Agents: Alexander Merensky and the Pedi Polity

Far to the north, in the creole state historians colloquially call "the Transvaal" but which has gone under a variety of names (most often the South African Republic), cartographic practice and geographical knowledge were not centralized in a meaningful way until the end of the nineteenth century. The information and labors of missionary cartographers there were on the whole more likely to see print or other distribution directly, all the more because popular and scholarly interest in the interior geography of Africa in North America and Western Europe waxed in the decades leading up to 1900 as Europe laid claims to the continent (Butlin 2009, 277). The idea of a civilizing mission was not incompatible with this interest, but the "civilizers" who sent back maps relied upon the same incomplete and relational data and produced similar topological maps, even though their appearance was increasingly professional and precise.

The Lutheran Berlin Missionary Society was one of the most active in the Transvaal, certainly in the areas most densely populated and under the effective control of African kingdoms. One such area was the boPedi, or the Pedi heartland, site of a powerful state under the Maroteng king Sekwati. In that endeavor they in 1861 received the acquiescence, if not necessarily the blessing, of the Boer state headquartered in Pretoria that nominally claimed control of the area after 1845 although the baPedi always claimed otherwise until the 1870s when the Republic pressed the issue (Delius 1983, 32–33). The BMS and its handful of converts were always kept at arm's length by the Maroteng and ultimately the society was ejected by new king Sekhukhune I around the end of 1865. One of the Berlin missionaries, Alexander Merensky, founded a new station about 70 miles south of the kingdom called *Botshabelo*, or "place of refuge," where a number of converts relocated after 1865. There, next to the town lands of Middelburg, Merensky built a thriving settlement protected by a small fort and focused on industry and schooling, with workshops and even a press; Botshabelo in fact grew nearly five times in size (and many more in population) before Merensky's departure in 1882 and became the showpiece for the BMS in Africa (Delius 1983, xii, 161; Merensky 1996, 336–343, 497–498).

Alexander Merensky however was not solely a missionary or an educator, but also had some medical training from his mission preparation starting during his teens, and apparently some artistic leanings that may have been connected to his parents' early childhood efforts to groom him as a military cadet (Petrich 1919, 25–26, 36–37, 10–13). Before he arrived in South Africa in 1859, Merensky apparently already enjoyed riding and hunting and had a curiosity about the natural world, though his knowledge of geography has less certain origins (Petrich 1919,

12–13; Kayser 1937, 385). After his approval for Society mission work in 1861, Merensky appears to have busied himself primarily with the survival of the mission stations, but he took time to engage in exploration aimed at seeking Biblical sites (particularly Ophir of Solomonic fame) as early as 1862 based on information told him by African informants, and he collected such information that he would pass on to the explorer Karl Mauch in the late 1860s (Carroll 1988, 238–39; Hall and Stefoff 2006, 13).[6] As an educated individual, Merensky took on various local duties and also built networks with other scientifically-minded people, especially Germans, in the Transvaal. In mid-1866, Merensky became the postmaster for the ward of Olifantsrivier, a position that probably brought him into contact with Friedrich Jeppe, who succeeded his brother as Postmaster in Potchefstroom later that same year and published the *Transvaal Argus*, the territory's only newspaper.[7] Both came to know Mauch, whose travels were far more extensive than either of theirs, and they were in a good position to publicize his data.

It appears that Jeppe was interested in compiling a map of the Transvaal from 1866 onward, both for purposes of postal efficiency and scientific exposition, and Mauch's travels and sketches provided him with the opportunity for collaboration on a map the government hoped to send to the Paris Exhibition of 1867 (Carruthers 2007, 110; *Transvaal Argus*, May 29 and July 24, 1866). Indeed he appears to have produced a map with Mauch's collaboration which he sent to Cape Town for printing by Saul Solomon & Co., but Jeppe was dissatisfied with the results and announced both the impending publication and his unhappiness with it (and a consequent delay owing to unspecified new printing arrangements in England) on the front of the July 10 and November 6, 1866 editions of his *Transvaal Argus*. Jeppe and Merensky by this time had a correspondence about the latter's better knowledge of the far east and north of the region, which made him a natural ally in improvement of the map (Petrich 1919, 144). The pair recompiled Mauch's map, with expanded information from the region's few land surveyors, and sent it to the renowned German geographer August Petermann in 1867. This map in 1:1,850,000 scale was published in what Jeppe later described as "re-constructed and augmented" form by Petermann in 1868 as the "Original Map of the Transvaal or South African Republic" (in English) with an extended description by Jeppe in *Petermanns Geographischen Mittheilungen* (Jeppe 1877, 218; Jeppe and Merensky 1868; Jeppe 1868). Merensky's direct contributions are hard to divine but his consideration of the expulsion of the BMS from the boPedi may explain why the map's wash color was not extended over that region (Fig. 16.4). Jeppe worked hard to explain in following years, particularly between 1877 and 1881 when Britain governed the Transvaal as a colony, that this lack of coloring did not indicate that the Boers did not have a claim over those lands (Jeppe 1877, 221–22).

[6]For Merensky's own account of this *Untersuchungsreise*, and his encounters during it, see Merensky (1996), 186–205.

[7]See *Staatscourant der Zuid-Afrikaansche Republiek* nos. 151 (29 May 1866) and 181 (26 Dec 1866).

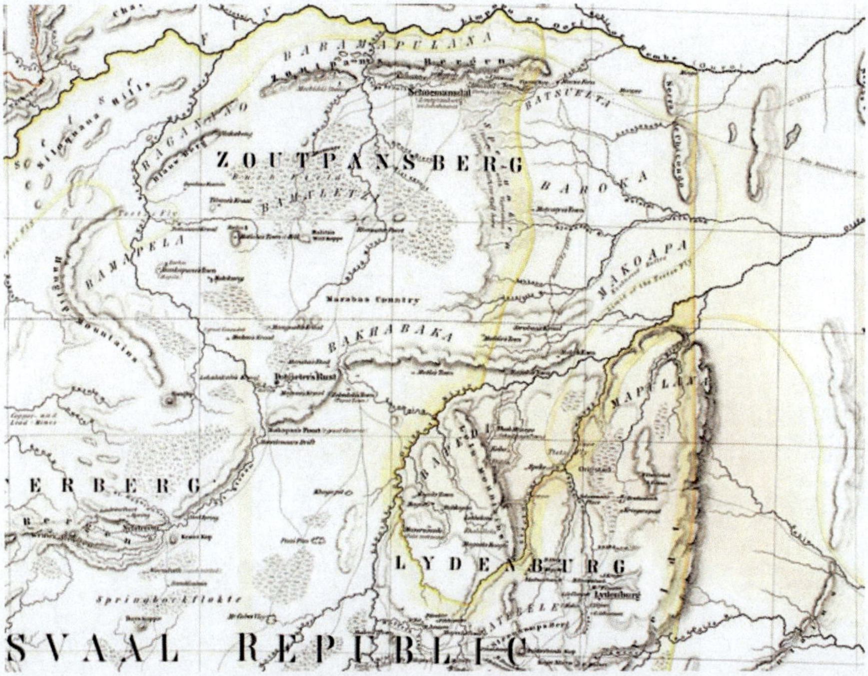


**Fig. 16.4** Detail from Jeppe and Merensky (1868), showing area excluded by the Rev. Merensky

After publication of the 1868 map, Merensky and Jeppe appear to have gone their separate ways. Jeppe involved himself in governance and other ventures; Merensky was busy at Botshabelo, although he received from the Republic the grant of a farm (which he named "Glück") in thanks for his cartographic work (Petrich 1919, 152). However, Merensky did not stop compiling cartographic information, and in 1875 he published a new map in the *Zeitschrift der Gesellschaft für Erdkunde zu Berlin* that included new data primarily from published sources but also his own information, collected with admittedly "deficient" (*mangelhafte*) instruments, and noting that an uncertain relationship existed between the government in Pretoria and the various kingdoms it claimed power over (Merensky 1875a, 268, 370). The map (Fig. 16.5) was still entitled "Original Map of the Transvaal or South African Republic," was still at 1:1,850,000 scale, and was labeled in English, to assure its appeal to the market for printed maps within South Africa and the British Empire.[8] Merensky however changed many of the toponyms' transliterations or names, likely in response to the improved knowledge of geographical names from the increasing number of residents and visitors at Botshabelo. This map was eclipsed within a year and a half by Jeppe's new map of the

[8]This is underscored by the printing of the map's stand-alone cover in English, even when sold by German firms. See the example at 912.682 MER, Map Collection, UNISA Library, South Africa.

**Fig. 16.5** Rev. Merensky's 1875 "Original Map"

Transvaal, which was published even more widely and in a number of editions, but both seem to have been aware of the other's work. One notable detail of Merensky's 1875 map is that it shows the position of the mission stations abandoned by the BMS without noting that fact; this may have been a bit of self-interested disinformation regarding the success of the Berlin society.

This kind of broad map marketing of small-scale maps was something Merensky continued after leaving South Africa; his later maps, four-sheet charts of all of southern Africa on an even smaller scale of 1:2,500,000, were apparently more successful or at least were printed more widely and passed through four known editions between 1884 and 1889. All of these maps still bore the "Original Map of..." moniker, in English, although it was printed in Germany and specified that it was all of South Africa including "native territories." During that time Merensky was in Germany and so his sources were thoroughly European. Although Merensky returned to Africa in 1890, he went to Nyasaland (today Malawi) to found stations there and does not appear to have produced further maps. At some point during that decade, he however contributed information to a broader thematic atlas of missionary activity but the details are not clear (Van der Heyden 1996, 17).

It is however Merensky's departure from South Africa, and the reasons for it, that reveals most starkly this missionary's cartographic connection to colonialism. While his work with Jeppe and in 1875 certainly helped the Boer state in the Transvaal define itself as a geographical entity, his relationship with the British who annexed the Transvaal in 1877 became more active. Merensky had been an intermediary between the Boers and Sekhukhune when they fought the war that bankrupted the Republic in 1876, and held the opinion that the Boers were hardly moral in their dealings with the baPedi. The British, on the other hand, shared mutual respect and even admiration with Merensky. Hoping to unseat the hostile Maroteng paramount and return to mission work within the boPedi, Merensky was an invaluable source of geographical data about Sekhukhune's kingdom for the British when the matter came again to war. An attempt to simply crush the kingdom turned into sieges and a failed attack on the capital in October 1878, and without good information on the area, the British turned to Merensky. Merensky in June 1879 provided not only an assessment but also a sketch map, which was gladly accepted and kept at Government House, where Sir Garnet Wolseley obtained it before he established his headquarters at Botshabelo in October.[9] Later, the British lithographed this map for purposes of the printed reports after Sekhukhune's capture in December; this lithographed version (Fig. 16.6), erroneously dated October, is the only extant copy.[10]

The map differs from Merensky's 1875 published map in level of detail, though the lithographed copy suggests it was "drawn from memory." Why, then, was Merensky consulted at all? The clue is probably in the personal names among the toponyms, indicating where people with potential loyalty to Sekhukhune were. Little other information appears on the map, but the topology of people who were transhumant or displaced by the unsettled conditions from 1875 to 1879 and important local landmarks were important for the British to know, and would have come from people at Botshabelo with personal connections within the kingdom. Merensky's own personal knowledge was probably inferior and outdated, but his mediation and provision of that information to Wolseley was critical.[11] The military importance of that timely spatial information was essential in planning the offensive against Sekhukhune's capital, but likely not to anyone else. Indeed, Friedrich Jeppe at the time was chief compiler for the British colonial administration in Pretoria, but although it is likely he had access to this map he did not revise the topography of his own 1877 map in any way.

---

[9]Merensky to Osborne (Colonial Secretary), 24 Jun 1879, R2165/79, SS 348, Transvaal Archives Depot, Pretoria (hereafter TAB).

[10]See "Chief of Staff's Journal of the Military Operations in the Transvaal, 1879," f. 86, CO 291/9, National Archives, Kew, UK (hereafter UK-NA).

[11]See "Native Locations and Strongholds of the Transvaal" (1907), A.1198, 75, WO 33/2753, UK-NA.

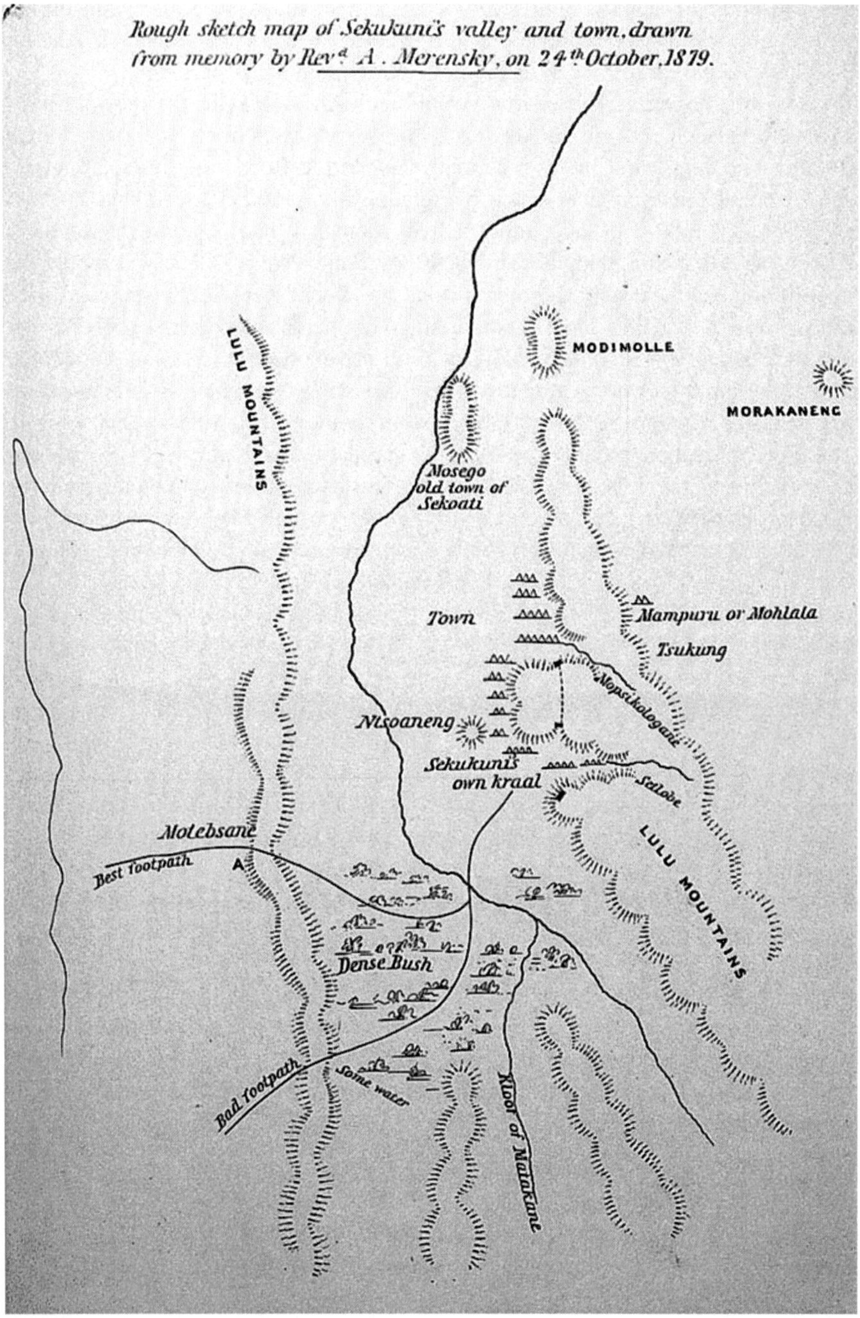


**Fig. 16.6** Lithographed version of Merensky's sketch for the British military, 1879

When the Transvaal Boers went into rebellion against British rule in 1880 over the imposition of new taxes and laws, Merensky was in favor of continued British administration; however, he was conscripted by the Boers for his medical capacity, and as such witnessed a number of key engagements in that conflict (Petrich 1919). The Boer irregulars probably had little use for Merensky's cartographic skills, and no further sketches appear from him. With retrocession it was clear that his pro-British sentiment was a liability, and he elected to return to Germany at the start of 1882.

Merensky's cartography is notable because information passed through him in a number of ways. Information from baPedi informants, passing travelers, and his own journeys appeared on maps he compiled or helped to compile. At the same time, he served to tailor the provision of information toward specific policy aims, such as the removal of Sekhukhune, the aggrandizement of mission activity in the region, or the publicization of the Transvaal as a conceptual unit. Although there is nothing about mission activity that restricts these activities to missionaries, the missionary's geographical and social positions made them privy to certain geographical knowledge. In providing his sketch to Wolseley and the Transvaal colonial government, however, Merensky served a similar, though much more focused, role to that of Thomas in the Transkei relative to Colonel Maclean. The difference is that Merensky's maps reached publication under his own name.

## Regionalizing Geography: Paul and Henri Berthoud in Eastern Limpopo

If the cartographic work of missionaries in the Transkei fed into a broader archive of formal cartography, and that of Merensky reached publication and embodied that archive, the cartography of the Swiss missionary brothers Paul and Henri Berthoud falls in between. They both produced their own cartography for publication and they provided crucial information in direct correspondence with people who had no information network in the areas of southwestern Mozambique and northeastern Limpopo where the Swiss mission stations were.

The Swiss mission stations were located to the north and northeast of Merensky's station. Both were trained in theology, Paul in Edinburgh (in medicine as well) and Henri at Lausanne (also in classics). Neither appears to have any formal training in geography or cartography. They arrived in southern Africa in the 1870s and set up mission stations first at Valdezia (1875) and Elim (1879), but agreed not to encroach on the Berlin missionaries who claimed the powerful Venda kingdoms as their own field of activity (Kirkaldy 2005, 30). The pair published maps and data narrowly, primarily in Swiss journals (or even more narrowly for bulletins of the Mission Society) and in French although this work did not go unnoticed by other geographical societies. Their primary work was in linguistics and corresponding translations of the Gospel among the xiTsonga-speaking people of northeastern South Africa and southern Mozambique, together with ethnological commentary, as one might expect of missionaries (Van Butselaar 1984, 40–41, 62–63; Harries 2007, 96–122). However they also collected and transmitted geographical data in the process.

Paul Berthoud made the first efforts to depict his mission field graphically. He produced a small (perhaps quarto in size) map of the mountainous Spelonken region in 1880, and a second edition in 1883 that cut down on the blank space at the map's top. These maps showed the deeded ground of the missions and some of the areas of the chiefs with whom they dealt in the far northeastern corner of the South African Republic's nominal Zoutpansberg district. The Germans' difficulty in making inroads with Venda society (and thus the limited transmittal of information) is also clear, because away from the roads between the mission stations there is little hard positional data and the Germans did not produce charts of their own. The Swiss Missions were kept at arm's length by Tsonga chiefs as well, but enjoyed a warm reception from the local "Native Commissioner," the Portuguese João Albasini, who also had a chiefly position and near whose farm "Goedewensch" appears a topological array of individual names on the early Berthoud maps (Van Butselaar 1984, 24). Most African political entities however saw little to be gained in changing the creeds on which kingship and social status were based so long as their societies were secure and reasonably prosperous, so risks outweighed rewards. This reluctance to engage the missionaries closely could nevertheless not efface lines of sight or broad descriptive information, so although geography away from the stations was poorly known to Berthoud, he simply added nameless kraals indicating an established but only vaguely identified population all over the area.

There is no record of these maps ever reaching British or Boer authorities, and no copies exist in the archives of the Republic, but the 1880 edition of the map (Fig. 16.7) at least was published with a separately-authored five-page description of the area (Berthoud 1880; Anon. 1881) by the young Swiss journal *L'Afrique Explorée et Civilisée*. In fact Berthoud clearly felt he needed the South African Republic's suspect geographical information, and obtained a compilation of the very dubious cadastres that made up the area near the Zoutpansberg mountains in December 1883. That compilation, based on an original at the Deeds Registry no longer extant, represented an effort either to identify lands available to purchase for mission stations or to chart the geography in a fundamentally flawed effort to differentiate vhaVenda-dominated areas from those of xiTsonga-speaking Shangaan people (Kirkaldy 2005, 15) and at the same time correct his earlier maps. Although Paul Berthoud did produce a revision (1883), there is no evidence that he either published that map in a geographical journal or undertook any revision beyond it. Both maps did appear (or were meant to appear) in Swiss mission society bulletins, based on text in their upper left corners, but their reach in terms of readership is unclear.

Difficulties in making inroads with established peoples of the region meant that personal journeys were the only means to make inroads, and that task fell to Paul Berthoud's brother Henri. Henri Berthoud made three extended journeys, in 1881, 1883, and 1885, and produced a map in 1886 from his extensive notes on those

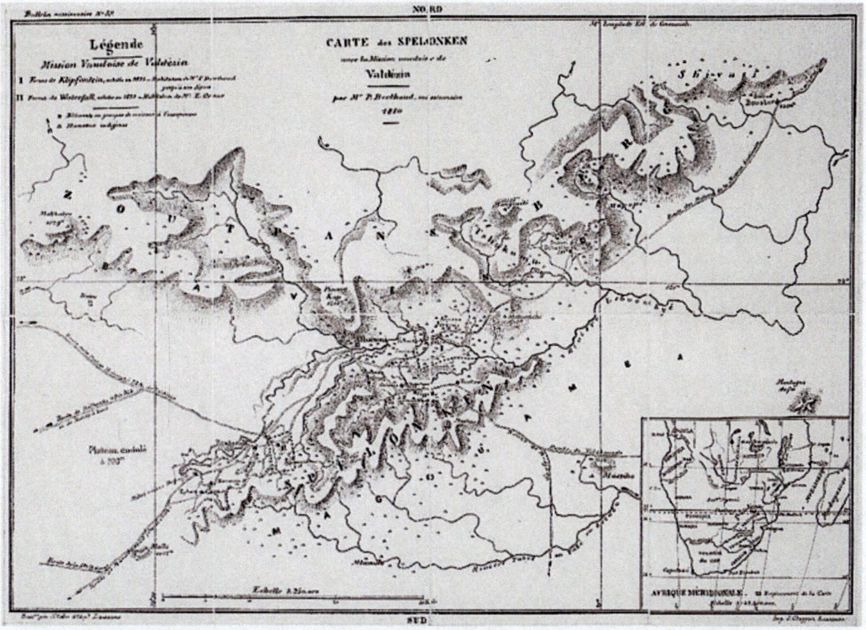


**Fig. 16.7** Rev. Paul Berthoud's (1880) "Carte des Spelonken"

journeys. Some of his later notes survive in copy form (Fig. 16.8), and they attest to Berthoud's careful observations and measurements which were precise (even if not always accurate or complete) and extended beyond creating topologies of space and into rationalizing language, politics, and society as he encountered them.[12] Berthoud certainly knew or learned enough of the local languages to ingratiate himself, and he or his companions could have made the necessary enquiries about land, society, and politics. His compilation route map is one of the first to show the transborder area between modern Mozambique and South Africa, and Berthoud presumably filled in details using Merensky's and Jeppe's maps of 1875 and 1877 as well as sketches by land surveyor St. Vincent Erskine who had published his work and a personal account in the *Journal of the Royal Geographical Society of London* (1869) but also drew another limited-circulation map of his travel in the transborder region for the use of the Swiss missionaries in the early 1880s.[13] Rather than restrict

[12]See for example, "Diary of a Journey 1889," Henri Berthoud Collection, A 1632, Cullen Library. The disposition of his earlier journals is not known.

[13]St. V. Erskine, "Route Map of the Gasa Country … Prepared for the Council of the Mission Romande," n.d. [1883-85?], Swiss Mission Collection, AC 1084/10/1/2.h, Cullen Library. This map is similar to Paul Berthoud's in printing and stock, and so was probably also part of a mission bulletin. The publication range is conjecture based on the period between Paul Berthoud's map and Henri Berthoud's 1886 work which draws in part from it; no publisher is credited.

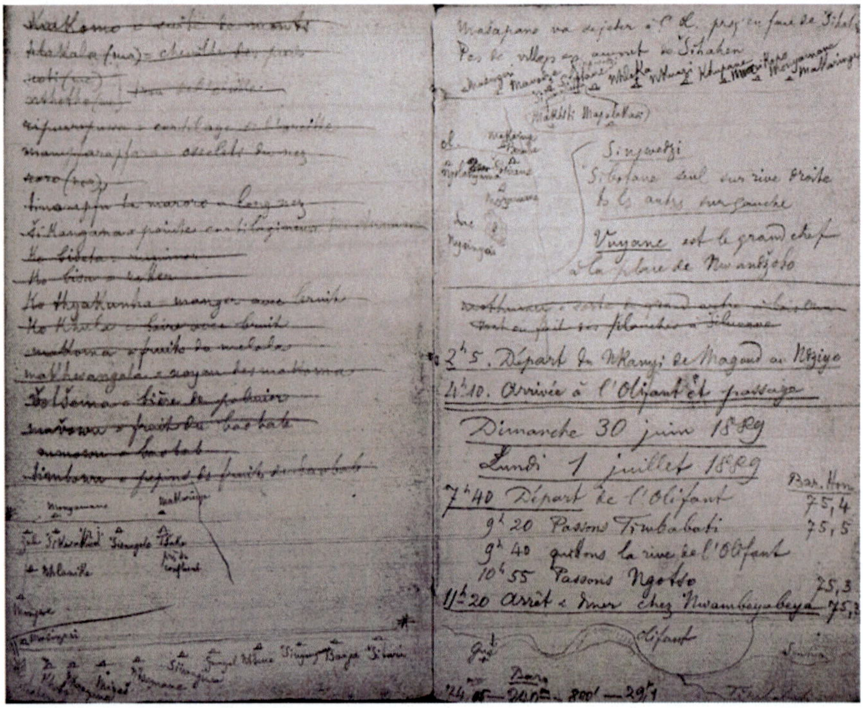


**Fig. 16.8** A page from Rev. Henri Berthoud's 1889 journal, showing the acquisition and schematizing of linguistic and meteorological (barometric) knowledge along with physical and human geographical data (A 1632, Cullen Library)

his map and notes to the mission field or seek a new and wider outlet, Berthoud kept with the practice of his brother and had the map and a descriptive article of unclear authorship published in *Afrique Explorée et Civilisée* that year (Berthoud 1886; Anon. 1886). The article explains Berthoud's encounters with local chiefs and kings, and it is clear even from the finished map's detail (Fig. 16.9) that many of the toponyms, ethnonyms, watercourses, and settlement locations he shows are inferred from informants and not direct visitation. Whether the information was accurate was less important than its existence.

The map is interesting for its effort to show districts of particular chiefs or kings with boundaries, but its relatively cursory indication of the ostensible Transvaal-Mozambique boundary. The latter is greatly erroneous and was probably taken from Merensky's 1875 map, which shows a similar kink in the northeastern corner. That the chiefly boundaries are colored solid lines while the colonial ones appear porous (and based on inaccurate topography) indicates just how irrelevant and geographically uncertain colonial boundaries were on the ground in the far northeast. Henri Berthoud's map of the Zoutpansberg and nearby region was still focused closely on his own movements and addressed within Francophone society, which

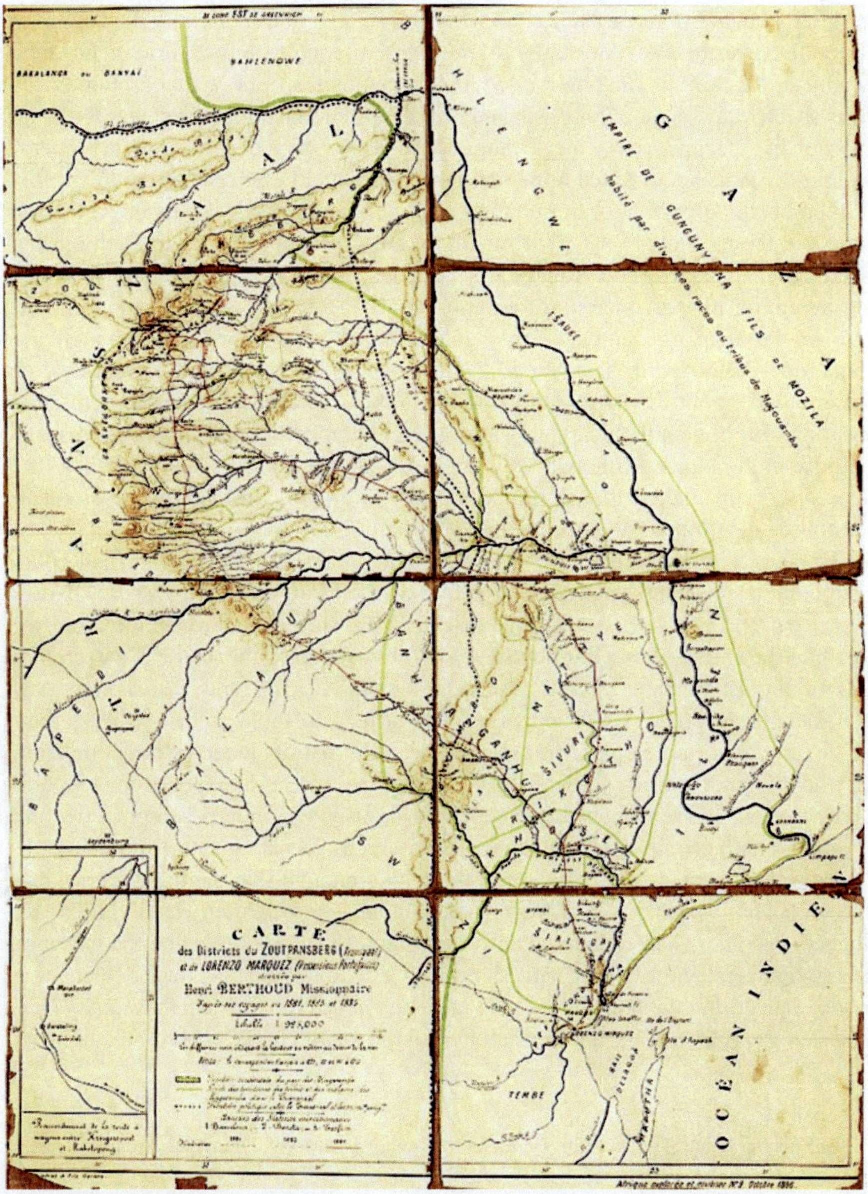


Fig. 16.9 Henri Berthoud's (1886) map of the northeastern trans-boundary area of the Transvaal; note the absence of detail in the Venda heartland (in the *upper left*) and the Gaza heartland (in the *upper right*), indicating their relatively stronger political stances. Copyright for this image by the Cullen Library, University of the Witwatersrand, Johannesburg

assured a fairly limited audience compared to publication in a major metropolitan journal. Nevertheless, Alexander Merensky's old collaborator Friedrich Jeppe finally saw a copy in 1888, and it led him to completely redraw that section of his four-sheet Map of the Transvaal and strike up a long correspondence with the missionary.[14] Jeppe, now Chief Draughtsman at the office of the Surveyor General in Pretoria, was an admitted armchair geographer and he prized Berthoud's information about "a portion of our country relatively unknown" – so much so that he involved the missionary as an informant on human and physical geography in the standing dispute over the Republic's boundary with Portuguese East Africa.[15] Berthoud, for his part, shared information quite openly and Jeppe named a mountain on his map of Zoutpansberg's goldfields (1893b) after him (although the name does not seem to have stuck).[16] Berthoud and Jeppe shared a great deal of information about African and colonial politics, methods of metrication, map printing, recent geographical publications, ethnography, language, and missionary activity in the North. Although we only have the side of the correspondence that Jeppe wrote it is clear that Berthoud's information was profoundly important in compiling his ultimately posthumous 1899 map.

Jeppe's 1899 work, in turn, was crucial to providing topography and route information to Henri Berthoud for his own "Map of Zoutpansberg, North Transvaal" (Fig. 16.10) which appeared in 1903, shortly before Berthoud's own death at the end of 1904 (Berthoud 1903). The map was published in Switzerland, but in English, which assured a broader South African (and British imperial) commercial audience. It was engraved and printed in three colors, so its quality set it apart from more ephemeral maps. The 1903 map included information from Berthoud's later travels, through 1898, as well as that he obtained from Jeppe to show routes and rivercourses. The cattle-destroying rinderpest epizootic and droughts that accompanied it between 1895 and 1898 also presumably helped to weaken independent chiefly power in the region enough that the missionaries and the colonial state alike had a freer hand to collect information (Dreier 2005, 8). Indeed, even the vhaVenda reluctantly submitted to colonial rule and the presence of missionaries, and when the British military attempted to assay the Zoutpansberg region late during the South African War (1899–1902) the German missionary Carl Beuster was able to offer them a small map of the major chiefs' spheres of authority,

[14]Jeppe to Berthoud, 27 Jan 1890, Berthoud-versameling, A 1529, Transvaal Archives Depot, Pretoria (hereafter TAB).

[15]Jeppe to Berthoud, 13 Oct 1890, Berthoud-versameling, A 1529, 1–2, TAB. The two sent sketches back and forth, and ultimately Jeppe sent Berthoud an advance copy of his proposed boundary map for comment; see Jeppe to Berthoud, 17 Nov 1890, Berthoud-versameling, A 1529, 1, TAB. A photocopy of the printed map derived from that, with Berthoud's writing, is extant; see copy of F. Jeppe, "Die Neue Grenze zwischen der Südafrikanischen Republik und den Portugiesischen Besitzungen," Swiss Mission Collection, AC 1084/10/1/1, Cullen Library.

[16]Jeppe to Berthoud, 19 Sep 1890, 4, Berthoud-versameling, A 1529, TAB. On Jeppe's 1893 map, "Mt. Berthoud" is at 23° 37' S, 30° 53' E. The official Mining Department and draft versions of this map, also drawn by Jeppe, do not include the toponym.

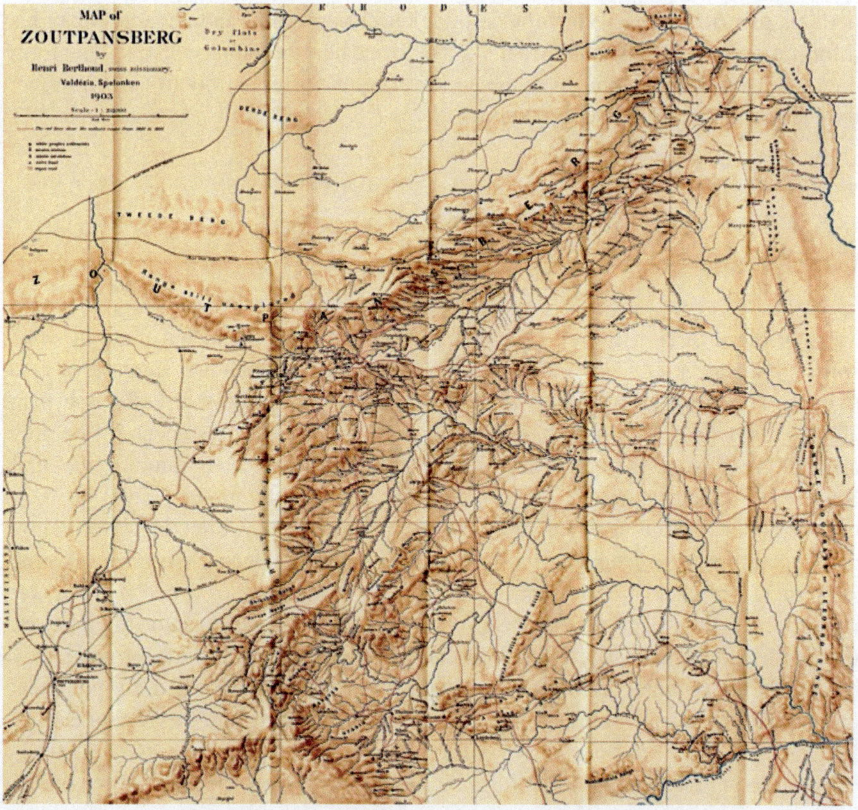


**Fig. 16.10** Henri Berthoud's (1903) map, revising his 1886 work with his later travel notes and adding much more impressive shading plus information from other maps. The basic form is however still topological, but the added features' apparent authenticity obscures gaps and subjectivities. Copyright for this image by the Cullen Library, University of the Witwatersrand, Johannesburg

with many localities indicated, that he had drawn in 1897 using topography from one of Merensky's earlier maps.[17]

British reconnaissance and other wartime information however does not appear to have figured heavily in Berthoud's 1903 compilation, and he indicates the mountains of the western vhaVenda as "still unexplored" and the area quite empty despite Beuster's inroads, military reconnaissance in 1898, and survey work beginning in 1902. Again, Berthoud was primarily interested in topological schematic, showing the topography and positions he himself collected or encountered, with only enough additional detail to aid the reader in orienting the map and

[17]C. Beuster to G. Y. Lagden (Minister for Native Affairs), 7 Sep 1901, N.A. Zout 6/01, SNA 2, TAB. The printer of the lithographed map is not indicated.

perhaps give it a greater appearance of scientific precision and accuracy than his information would likely warrant. As with earlier missionary maps, the form dictated by scientific rigor gives the appearance of positional fidelity when in fact the map remains primarily topological.

## Conclusion

Each of these three nodes of missionary mapping is different, but they all share some characteristics. The output of Paul and Henri Berthoud suggests a less expansive mindset than Merensky, but a more careful process of collection than Thomas or Dodd. This may have been a result of the circumstances in the area around present-day Kruger Park, where the Berthouds were among the first Europeans to live among xiTsonga-speakers and to attempt mediation of their culture and space into a structure of European knowledge. Collecting information and separating truth from fiction there were harder prospects than areas like the Eastern Cape where missionaries had been active longer, or when a particular ruler right on the edge of European settlement was the target. Nonetheless, the missionaries' positions as outsiders (both to African societies and the growing body of data in the colonial archive) meant their information always had a topological and partial quality, a combination of diligent route mapping and the sweeping in of less precise and even speculative information alike. Their maps concomitantly oriented around arranging that information in a usable form, sometimes in tandem with the transmission of other information also necessary for creating useful order out of what colonial states saw as chaotic or simply unintelligible. In all of these cases, the missionaries' route maps and observations found their way into the broader store of colonial geographical knowledge, so their position is not only between Stone's geographies of exploration and imperialism, but they also occupy a space of their own outside of it and within a broader framework of colonial information gathering.

## Biographical Note

**Lindsay Frederick Braun** earned his PhD in African and European History from Rutgers University in 2008. His work deals with cartography and surveying in the context of settler colonialism, primarily in South Africa. He is currently Assistant Professor of African History at the University of Oregon in Eugene, Oregon (USA).

## References

Anon. (1881) Les Spelounken (voir la carte jointe à cette livraison). L'Afrique Explorée et Civilisée 2(8):161–65

Anon. (1886) Exploration de M. H. Berthoud Entre les Spelonken et Lorenzo Marquez. Afrique Explorée et Civilisée 7(10):299–309

Arbousset T, Daumas F (1842) Relation d'un Voyage d'Exploration au Nord-Est de la Colonie du Cap de Bonne-Espérance. Arthus Bertand, Paris

Bassett TJ (1998) Indigenous Mapmaking in Intertropical Africa. In: Woodward D, Lewis GM (eds) Cartography in the Traditional African, American, Arctic, Australian, and Pacific Societies. Vol. 2, bk. 3 of the History of Cartography. University of Chicago Press, Chicago, pp 24–48

Berthoud P (1880) Carte des Spelonken avec la Mission vaudoise de Valdezia. 1:250,000. J. Chappuis, Lausanne

Berthoud P (1883) Carte des Spelonken avec la Mission Vaudoise de Valdezia. 1:250,000, 2nd edn. J. Reber, geometer, Lausanne

Berthoud H (1886) Carte des Districts du Zoutpansberg et de Lorenzo Marquez. 1:925,000. F. Noverraz & Fils, Geneva

Berthoud H (1903) Map of Zoutpansberg. 1:333000. H. Kümmerly & Frey, Bern

Beuster C (1897) Skizze des Bawenda-Landes entworfen von C. Beuster, gezeichnet von A. Merensky. Ca. 1:250000. n.p, Berlin

Braun LF (2009) The Colonial Archive and Maps of the Transkei, 1857–1898. In: Liebenberg EC, Demhardt IJ, Collier P (eds) Proceedings of the Symposium of the Commission on the History of Cartography in the 19th and 20th Centuries, Chap. 2. ICA, Pretoria

Butlin R (2009) Geographies of empire: European empires and colonies, c.1880-1960. Cambridge University Press, Cambridge

Butselaar V (1984) Africains, Missionaires, et Colonialistes. Brill, Leiden

Carroll S (1988) Solomonic Legend: the Muslims and the Great Zimbabwe. Int J Afr Hist Stud 21:233–247

Carruthers J (2007) Cartographic Rivalries: Friedrich Jeppe and the Transvaal. In: Etherington N (ed) Mapping Colonial conquest. University of Western Australia Press, Crawley, pp 102–122

Cartwright JF (1976) Maps of Southern Africa in Printed Books 1750–1856. UCT Libraries, Cape Town

Comaroff J, Comaroff J (1991) Colonialism and Consciousness in South Africa, vol 1, Of Revelation and Revolution. University of Chicago Press, Chicago

Delius P (1983) The Land Belongs to Us: the Pedi Polity, the Boers, and the British in the Nineteenth-century Transvaal. Ravan, Johannesburg

Van der Heyden U (1996) In: Merensky A (ed) Introduction to Mein Missionsleben in Transvaal. Edition Ost., Berlin

Dreier M (2005) Years of Terrible Drought: Surviving the 1895–1897 Supply-crisis in the Limpopo Area. Limpopo Research Project, University of Basel. http://pages.unibas.ch/afrika/limpopo/papers/limpopo_pdf/limpopo_dreier.pdf. Accessed 12 Feb 2011

Driver F (2001) Geography Militant: Cultures of Exploration and Empire. Blackwell, Oxford

Erskine St V (1869) Journey of Exploration to the Mouth of the River Limpopo. J R Geogr Soc 39:233–276

Etherington N (2004) A False Emptiness: How Historians May Have Been Misled by Early Nineteenth Century Maps of South-eastern Africa. Imago Mundi 56(1):67–86

Etherington N (2008) Missions and Empire. Oxford University Press, Oxford

Hall M, Stefoff R (2006) Great Zimbabwe. Oxford University Press, Oxford

Harries P (2007) Butterflies and Barbarians: Swiss Missionaries and Systems of Knowledge in South-east Africa. Ohio University Press, Athens

Hunt CG (1994) Some Notes on indigenous map-making in Africa. In: Stone JC (ed) Maps and Africa. Abderdeen University, Aberdeen, pp 32–35

Jeppe F (1877) Notes on some of the physical and geological features of the Transvaal, to accompany his new map of the Transvaal and surrounding territories. J R Geogr Soc Lond 47:218

Jeppe F (1893a) The Zoutpansberg Goldfields of the South African Republic. Geogr J 2(3):213–237

Jeppe F (1893b) Map of the Zoutpansberg Goldfields, compiled from latest information ... 1:600000. Turner and Shawe, London

Jeppe F, Jeppe, CFW (1899) Jeppe's Map of the S. A. Republic or Transvaal and Surrounding Territories. 1:476000. Edward Stanford, London

Jeppe F, Merensky A (1868) Original map of the Transvaal or South-African Republic. 1:850000. Justus Perthes, Gotha

Jeppe F (1868) Die Transvaal'sche oder Süd-Afrikanische Republik. Petermanns Geographischen Mittheilungen, ergänzingsheft no. 24

Kayser K (1937) Alexander Merensky als Geograph und deutscher Kolonialpionier. Koloniale Rundsch 6:385–393

Kirkaldy A (2005) Capturing the Soul: the Vhavenda and the Missionaries, 1870–1900. Protea Boekhuis, Pretoria

Kokkonen P (1993) Religious and colonial realities: cartography of the Finnish mission in Ovamboland, Namibia. Hist Afr 20:155–171

Lester A (2001) Imperial Networks: Creating Identities in Nineteenth-century South Africa and Britain. Routledge, London

Maclean J (1858) A Compendium of Kaffir Laws and Customs. Frank Cass, London

Merensky A (1875a) Eine Neue Karte der Süd-Afrikanische Republik. Z Ges Erdkunde Berlin 10:366–379

Merensky A (1875b) Original map of the Transvaal or South-African Republic including the gold and diamondfields. 1:850,000. J. Sulzer, Berlin

Merensky A (1996) Mein Missionsleben in Transvaal. Ed. and new introd. by Ulrich van der Heyden. Edition Ost., Berlin (Orig. pub. 1899)

Methodist Church of South Africa (1931) Minutes of the Fiftieth Annual Conference of the Methodist Church of South Africa. Methodist Publishing House, Cape Town

Moffat R (1842) Missionary Labours and Scenes in Southern Africa. J. Snow, London

Petrich H (1919) Alexander Merensky, Dr. theol; ein Lebensbild aus der deutschen evangelischen Mission des letzten Jahrhunderts. Verlag der Buchhandlung der Berliner Evangelischen Missiongesellschaft, Berlin

Price R (2008) Making Empire: Colonial Encounters and the Creation of Rule in Nineteenth century Africa. Cambridge University Press, Cambridge

Scriba G, Lislerud G (1997) Lutheran Missions and Churches in South Africa. In: Elphick R, Davenport TRH (eds) Christianity in South Africa: a Political, Social, and Cultural History. University of California Press, Berkeley, pp 173–94

Stone JC (1995) A Short History of the Cartography of Africa. Edwin Mellen, London

Transvaal Argus. (1866) Paris Exhibition 1867: Notice. July 10 and November 6

# Part VII
# General Cartography

# Chapter 17
# Earthcube: Christian Gottlieb Reichard's Point of View on the Earth

**Andreas Christoph**

**Abstract** Under the instruction of Franz Xaver von Zach (1754–1832) and Friedrich Justin Bertuch (1747–1822), Christian Gottlieb Reichard (1758–1837) designed an "Atlas des ganzen Erdkreises," which was first published by the "Landes-Industrie-Comptoir" at Weimar, Germany in 1803. Moreover, we find in a sales catalogue that it was possible to acquire the "Atlas" shaped on a cubical rack. My question is: Was the "Erdkubus" ever implemented? What was the full story behind this advertisement?

The aim of my paper is to analyze an unknown product of the cartographer Reichard. Looking at the correspondence between Reichard, the publisher Bertuch and the astronomer Zach should give us the opportunity to understand a special form of 'Cartography in the making'. Indeed, circa 1800 a synergetic cube-concept – applying the appeal of maps in combination with the representation-form of a globe – was mentioned for the first time by Bertuch and his staff. Reconstructing the development of the "Erdkubus" also offers miscellanea of travel literature and mapping culture in Germany in the early nineteenth century.

## Abbreviations

AGE Allgemeine Geographische Ephemeriden (1798–1815)
GSA Goethe and Schiller Archives in Weimar

A. Christoph (✉)
Friedrich Schiller University, Jena, Germany
e-mail: Andreas.Christoph@uni-jena.de

E. Liebenberg and I.J. Demhardt (eds.), *History of Cartography*,
Lecture Notes in Geoinformation and Cartography 6,
DOI 10.1007/978-3-642-19088-9_17, 

## Introduction

The following paper puts forward a micro-historical analysis regarding the non-academic context of geographic and cartographic scientific work around 1800. Well-developed travelogues, and particularly detailed maps and globes of this time, enable a scientific synthesis at the interface among scientific knowledge, practical technologies, and industrial arts. On the occasion of the "3rd International ICA Symposium on the History of Cartography," I will take a look at a special representation-form of the globe in the German classical era.

I found an article by Christian Gottlieb Reichard in the "Allgemeine Geographische Ephemeriden" (AGE), a famous geographic-cartographic periodical of Germany in the early nineteenth century. It focuses on the "Atlas des ganzen Erdkreises nach den besten astronomischen Bestimmungen neuesten Entdeckungen und eigenen Untersuchungen in der Central-Projection auf sechs Tafeln" – a hulking title for a sophisticated topic. I also had the chance to have a close look at a damage-free specimen in the "Anna Amalia Bibliothek" at Weimar. Behind the book cover laid six single sheets of maps. Four of them showed the area from the Earth's equatorial plane to the 35th parallel north and the 35th parallel south, the other ones showed the Polar Regions. Longitude and latitude were transformed to plane and hyperbola. Pieced together, the maps showed the globe as a cube. What was the background for this re-evaluation of entrenched forms of representation of cartographical knowledge?

## The "Erdatlas" – Project

### *From Initiation to Realization 1798–1803*

The leading lights of the Erdatlas-Project were three Germans interested in cartography. Friedrich Justin Bertuch (1747–1822) from Weimar was the coordinator. Bertuch was a famous author, scientific publisher, and salesman in Weimar, a small town in the duchy of Sachsen-Weimar-Eisenach in the centre of contemporary Germany. In 1791, he founded the "Landes-Industrie-Comptoir," a publishing and business house. He operated his geographical ventures in the "Geographisches Institut," created in 1804.[1] Maps, atlases, globes and travelogues contributed to the national and international recognition of his efforts to organize and diffuse scientific and economic knowledge about the physical world. The "Erdatlas"-project was initiated by Franz Xaver von Zach (1754–1832), an astronomer, surveyor, critical author and reviewer. He worked at the observatory of Seeberg

[1]For more details about the "Landes-Industrie-Comptoir" and the "Geographisches Institut," see for example Middell 2002, 2006 or Espenhorst 2003, 2008.

near Gotha. The draft versions of the "Erdatlas" were created by Christian Gottlieb Reichard (1758–1837), a town clerk and lawyer who had a great interest in cartography. This was the creative trio, working in the waning years of the eighteenth century on a cartographic innovation, a thought experiment of mathematical and astronomical geography.

Today, I will delineate the way the "Erdatlas" developed, showing the difficulties encountered in various phases of information procurement by correspondence, scientific underpinning of versatile position fixings, regional studies and ethnologies, and cartographical transformation. A plethora of original sources still awaits sustained examination. I found the most interesting material at the Goethe and Schiller Archives (GSA) in Weimar, the oldest literature archive in Germany, where the heritage of Bertuch and his firms is preserved. The background of the people was reconstructed by analysis of correspondence sent to Bertuch by Reichard between 1798 and 1803. Work in progress was also be determined by instructions of copperplate engravers and staff doing map colorations. Different stages of drawing, engraving and coloring were also followed up by the "Allgemeine Regeln für Landcharten-Zeichner" in a book about mathematical geography by the (military-) cartographer Friedrich Wilhelm Streit (1772–1839) that was published in 1837.[2]

## *The Methods of Operation Within the "Geographisches Institut"*

The question arises, how was Bertuch able to present a new representation-form of a globe belonging to the context of popularizing and diffusing knowledge within his limits of resources? To answer this question, the Bertuch's firm approach to work ought to be considered. During many years of dedicated work, academics, authors, and reviewers became integrated in his periodical projects.[3] For instance, in cooperation with Zach, Bertuch created the first German scientific periodical dedicated to geography, the "Allgemeine Geographische Ephemeriden." It was the figurehead of the "Geographisches Institut." The first volume appeared in January 1798.[4] This monthly periodical required a scientific approach, with articles from geographers and astronomers. The categories included treatises, book- and map-reviews, correspondence, and miscellanies. Not only national and international

[2]Streit, 1837.

[3]The tendency to turn the holistic approach towards empirical studies with commentary on images of nature can be mentioned in the periodicals "Allgemeine Literatur-Zeitung," "London and Paris," the "Journal des Luxus und der Moden," and especially in the "Bilderbuch für Kinder." For more details on the publications of Bertuch's firms, see Kaiser and Seifert 2000.

[4]Because of varying editors, the magazine's name changes three times. The last "Ephemeride" was issued in 1831, after a successful run of 30 years.

newspapers were a reliable source, but also correspondence with scientific societies and with the academic communities of researchers and explorers. Furthermore, the "Allgemeine Geographischen Ephemeriden" was an accumulation of advertising news.

The groundwork of Bertuch's 'Geographica' and 'Cartographica' was rarely original documents or the results of original surveying. However, only the latest foreign and domestic reviews, as well as miscellaneous issues and editions, were handled by Bertuch and his staff. This also applied to maps, atlases and globes. On this basis, Bertuch was able to generate a loyal readership. Purchasers were assured by a comprehensive list of subscribers, which continuously increased, thanks to complex self-advertisement in a multitude of in-house productions and the journals belonging to the Bertuch firm or by friendly publishers.

## *From Adaption and First Work Streams*

The cartographical self-development of the "Erdatlas" was initiated by the "Landes-Industrie-Comptoir" by the announcement of a celestial-atlas in 1798. The "Neuester Himmels-Atlas" by Christian Friedrich Goldbach (1763–1811), published in 1799, was polarized by its make-up: white stars on a black background. By Abraham Gotthelf Kästner (1719–1800), an esteemed mathematician and teacher of Carl Friedrich Gauß (1777–1855), the *Mathematicorum Principi*, the underlying central projection of the celestial atlas, attracted attention. A man imagined in the center of the earth looks at a selected section of the starry firmament. Kästner's complaisant review made Zach projecting this type of graticule on a special map.[5]

The association of Zach with Bertuch caused a swift implementation of this idea. Bertuch chose Reichard as a drawer, inspired by his cartographic precision of a self-contrived globe for educational use at school few years ago. At the beginning of the realization of the plan there stood several preliminary considerations. Which maps should be chosen and which dropped out? What was the best scale for preserving clarity of cartographic information while simultaneously allowing for manageability of the final product? The graticule of the map had to be transferred to the blank drawing paper and also should be applied to maps from other publishers. Were the astronomical position fixings reliable? Was there a further need for triangulation and surveying data? Information gaps had to be filled by compilation of itineraries and making excerpts of periodicals, which assured a continuous flow of data. Plausibility and accuracy of maps was only reached by adjustment of domestic

[5]AGE 2(1798), "Etwas zur Empfehlung der Central-Projection," p. 401–411. Here he stated: "Indessen, da man die Erdfläche auf so mancherley Arten vorgestellt hat, so könnte wol eine Darstellung der ganzen Erdfläche auf sechs Tafeln, der Sonderbarkeit wegen, Liebhaber finden."

and foreign maps, atlases and globes, not by own expeditions.[6] Zach contributed the projection, which was calculated in a more detailed way by Reichard.[7] The cartouche and the graphical scale were positioned carefully on the map sheets. Reichard decided to forgo a legend, too.

Obstructions were caused by the solitude of Reichard's workstation in the small town of Lobenstein in the Thuringia mountain range, a daytrip away from Weimar. For instance, through lack of drawing pens, he awaited eagerly an ample range of products of Bertuch's firm.[8] In Reichard's opinion, Lobenstein was a "damp hole of the earth,"[9] with a library and a literary club of moderate means. Reichard's map collection was still in the process of being established; he was not able to assess its quality. Therefore, he was captive to Bertuch's recurrent consignments of itineraries and common geographical and astronomical periodicals, like the "Allgemeine Geographische Ephemeriden," "Connaissance des temps," and "Monatliche Correspondenz zur Beförderung der Erd- und Himmelskunde." From these he gathered astronomically-specified longitudes and latitudes from single municipalities and local conditions of great significance, because absolute accuracy was supposed to be an outstanding feature of the "Erdatlas." Also, Bertuch reached his limit of information procurement, and Reichard learned to live with this fact.[10] Then again, information overkill at odd times caused redrafts of maps.[11] The long time on the drawing board caused wood and paper ruptures.[12]

[6]Selection of maps showing Europe used by Christian Gottlieb Reichard: Götze, J. A. F. 1800. *Charte des Schwarzen Meeres*. Weimar: Verlag des Industrie-Comptoirs; Güssefeld, F. L. 1785. *Charte von der Moldau und Walachey*. Nürnberg: Homaennische Erben; Güssefeld, F. L. 1799. *Charte des Mittelländischen Meeres*. Weimar: Verlag des Industrie-Comptoirs; Kitchin, T. 1803. *Charte von England und Wallis*. Augsburg: Johannes Walch; Mannert, C. 1800. *Charte von Frankreich*. Nürnberg: Adam Gottlieb Schneider und Weigel; Reinecke, J. C. M. 1800. *Charte des ganzen russischen Reichs in Europa und Asien*. Weimar: Verlag des Industrie-Comptoirs; Reinecke, J. C. M. 1800. *Charte des ganzen russischen Reichs in Europa und Asien*. Weimar: Verlag des Industrie-Comptoirs; Reinecke, J. C. M. 1800. *Charte von Island*. Weimar: Verlag des Industrie-Comptoirs; Reinecke, J. C. M. 1800. *Charte von Schweden und Norwegen*. Weimar: Verlag des Industrie-Comptoirs; Sotzmann, D. F. 1797. *Generalcharte von Ost-, West-, Süd- und Neu-Ost-Preussen nach dem Gränztractate von 1797*. Berlin: Simon Schropp et Comp.; Stieler, A. 1798. *Charte von Irland*. Weimar: Verlag des Industrie-Comptoirs; Stieler, A. 1800. *Charte von Spanien und Portugal*. Weimar: Verlag des Industrie-Comptoirs.

[7]"Von Grad zu Grad rechne ich, und es soll das Netz so scharf, als es nur möglich ist, werden [...]." Letter from Reichard addressed to Bertuch, 8 August 1799, GSA 06/1471.

[8]"[...] wegen des gänzlichen Mangels an solchen kleinen obwohl absolut nöthigen Hülfsmitteln in hiesiger Gegend [...]". Letter from Reichard addressed to Bertuch, 14 July 1799, GSA 06/1471.

[9]Letter from Reichard addressed to Bertuch, 8 August 1799, GSA 06/1471.

[10]"Da wird nun freilich manches willkürlich [...]." Letter from Reichard addressed to Bertuch, 10 February 1800, GSA 06/1471.

[11]Different quality of drawing paper restricted corrections. In addition to the original drafts, map add-ons had to be handled by Reichard and Bertuchs staff.

[12]"[...] was mir bey unsrer schwindenden durch Mark und Bein dringenden Luft schon einige mal passirt ist." Letter from Reichard addressed to Bertuch, 8 December 1800, GSA 06/1471.

## *Work in Progress*

As a foregone conclusion, the atlas size was limited for clarity's sake. The final maps measured 16.7 square inches. Also, globes in this time had similar sizes to warrant manageability. But size predefines the scale and limited information. Reichard felt impelled to make concessions and complained: "Wollte der Himmel, ich dürfte ihn noch größer machen."[13]

The graphical scale was portrayed on the map No. 1 of the "Erdatlas." Distances could be gathered by compass and then read off the scale. What Reichard and Bertuch kept back from the buyers was the fact that the drawings had been shrinking.[14] They compromised to swell the Africa drawing in a dank cellar or to dehumidify the other maps. Reichard trusted to Bertuch's experience in this field of map manipulation.

The long duration of "Erdatlas" map making, from the first announcement in 1798 until its publication in 1803, became a big problem. The estimated 4 weeks per map was unrealistic. Reichard's main job as a lawyer prevented him from doing drawings. In winter, limited daylight disabled working. Into the bargain, there was a fire which devastated parts of Lobenstein in summer 1800. Reichard fled helter-skelter with all of his drawing materials and maps. After that he felt ill with fever attacks for a couple of weeks.

Working gratuitously for Bertuch for 3 years, financial distress constrained Reichard to ask Bertuch for occasional advance of wages. Furthermore, Bertuch sent free copies of maps and books to Reichard. Also, draft versions and corrections circulated between Lobenstein, Weimar and Gotha by mail. This was oftentimes accompanied by difficulties like damaged maps or losses of books.[15]

Bertuch and Zach had a falling out in 1799 because of conflict about the scientific concept of the "Allgemeine Geographische Ephemeriden." Bertuch consulted Zach only in exceptional cases – even though Reichard wanted Zach's scientific assistance and map revisioning. In 1800, Zach tended his "Monatliche Correspondenz zur Beförderung der Erd- und Himmels-Kunde" on the astronomical-geographical science market – and Reichard adapted localizations and extracted other helpful information.

In summer 1802, the end of the "Erdatlas"-project was in sight. Drafts had been finished, the copperplate engraver was already working, and first prints appeared. Reichard did further work inscribing "Schnickschnack" like the antipodes, the diametrical opposites of the observatories of Vienna, Gotha (Seeberg), Greenwich

[13]Letter from Reichard addressed to Bertuch, 8 August 1799, GSA 06/1471.

[14]Letter from Reichard addressed to Bertuch, 8 March 1801, GSA 06/1471.

[15]"Bei der Reichspost muß es etwas confus hergehen." Letter from Reichard addressed to Bertuch, 13 January 1800, GSA 06/1471. Carriage maps were rolled on wooden rolls. See the "Post-Bericht" in GSA 06/1471.

**Fig. 17.1** The Marquesas Islands on map No. 3 of the "Erdatlas"

and Paris.[16] Some outstanding corrections were considered necessary by Reichard and Bertuch.[17] Also, 4,000 position fixings had to be kept track of.[18] The eventual launch of the "Erdatlas" reflected the level of information from February 1803, and Reichard was assuaged to send all drafts to Bertuch. In exhaustion, he wrote: "So wäre denn diese Arbeit vollendet, und die neueste Reform der Erde, so weit sie zur Zeit möglich ist, darinnen enthalten."[19]

The alphabets used for copperplate engraving were enclosed by Reichard. He had noticed the advantage of British maps improving "der Schärfe, dem Ebenmaß und dem Parallelismus" of the typeface and would rather not give up this in the "Erdatlas."[20] Different sizes and alignment of the alphabets were useful for clearness. In the process of copperplate engraving, intersection of words and overlaps were avoided. In context of the "Erdatlas," this acquired agreements (Fig. 17.1).

[16]Letter from Reichard addressed to Bertuch, 1 January 1801, GSA 06/1471.

[17]Bertuch could use his correspondence network for the critical review of stages of "Erdatlas" progress. Theophil Friedrich Ehrmann (1762–1811), who worked for Bertuch on the "Bibliothek der neuesten und wichtigsten Reisebeschreibungen," corrected the mapping of Africa. See the letter from Reichard addressed to Bertuch, 10 November 1803, GSA 06/1471.

[18]As an established procedure, the map was sectioned into quadrants. Beginning in the upper left corner, the correct words were crossed out, and the words to correct were underlined. If it was necessary, they were also numbered. Every quadrant was handled in this way.

[19]Letter from Reichard addressed to Bertuch, 8 March 1801, GSA 06/1471.

[20]Letter from Reichard addressed to Bertuch, 23 January 1800, GSA 06/1471.

The coloring of the atlas was considered carefully. The continents were differentiated by color, and the inner continents became washed. “Ich dächte es wäre nöthig, die Welttheile zu unterscheiden weil sie durch die Projection so sehr zerrissen werden, und gäbe den Karten ein muntereres Ansehen,” Reichard wrote to Bertuch.[21] The color application was effected in a matte finish. Bertuch’s painters abstained brown color, because of its “etwas widriges Ansehen.”[22] Details of the coastlines were distinguished from the ambient earth surface. The “Illuminierer” had also to look after exact compliance, according to Reichard’s instruction. The colors might not change to lighter or darker than instructed. Curvatures had to be colored exactly. To dry maps and atlases was the rule, because humidity could have fatal consequences for the embedded maps and atlases. All the artisans did not keep such standards; resourceful ones were modifying the original instructions concerning their mistakes.

## *The Launch of the “Erdatlas” – on the Way to the “Erdkubus”*

The appearance of the “Erdatlas” was announced in August 1803 in the “Allgemeine Geographische Ephemeriden”; to satisfy the expectations of potential buyers, the issue included a minimized trial proof of the Africa map (Fig. 17.2). Reichard authored the related commentary, with a detailed catalogue of used auxiliary means and also practical exercises. It was possible to find longitudes and latitudes of places all over the world and relate them mathematically.

In November, 1803, Reichard got his copy of the “Erdatlas” hot off the press – and he was highly pleased by the final product. A few days later, he wrote a letter to Bertuch to send him an unbound copy to bind. He suggested fixing the maps on a rack from wood and board. The idea of the “Erdkubus” was born (Fig. 17.3), and it was announced in the publishing catalogue of 1806. In 1805, the cost of the “Erdatlas” amounted to four Reichstaler; in 1806, the “Erdkubus” was available for six Reichstaler. In comparison, an 18in. globe cost at least 60 Reichstaler at Bertuch’s firm. The “Erdkubus” had its last sales catalog entry in 1821.

Why did the “Erdkubus,” a unique attempt of a small German publishing house, become effective in advertising cartographical products? After its release, the “Erdatlas” met with criticism. The earth not as a spheroid, but as a cube concept was abstract. Neither the “Erdatlas” nor the “Erdkubus” was offered in the context of the geographic-cartographical “Lehrmittelpaket” of the Bertuch firms. Reviewers noticed the avoidance of “altväterliche perspectivische Gebirgszeichnung” by Reichard, but in the end he was not able to capture the terrain by draft. Low mountain ranges in Thuringia look like the Himalayas.

[21]Letter from Reichard addressed to Bertuch, 11 August 1800, GSA 06/1471.

[22]Letter from Reichard addressed to Bertuch, 1 January 1801, GSA 06/1471.

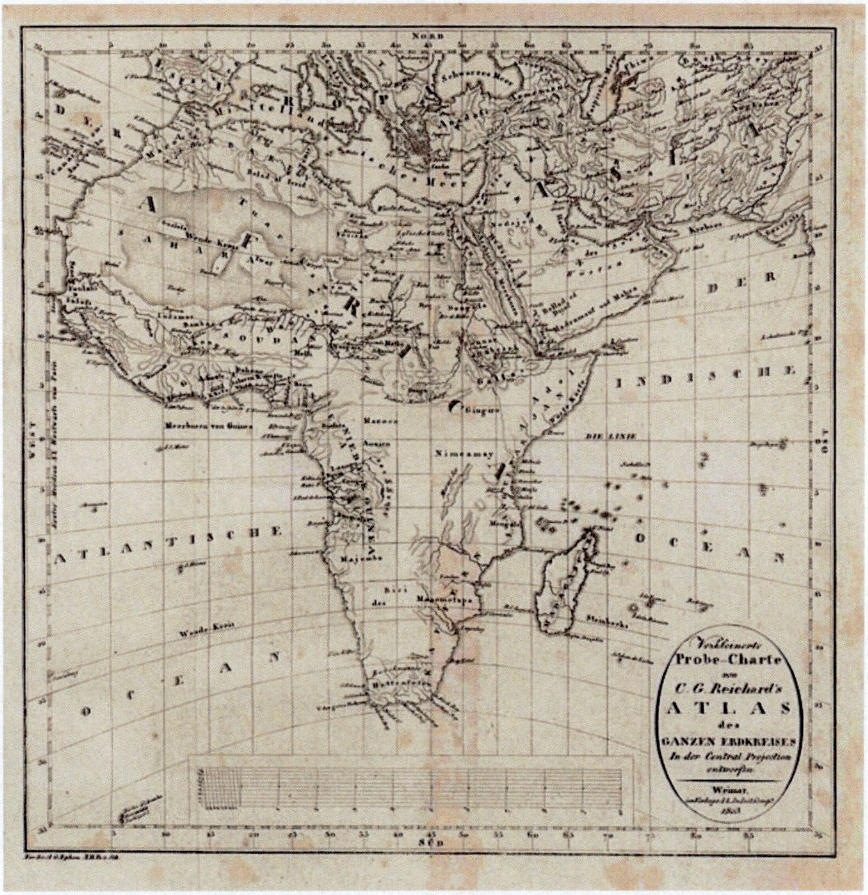


**Fig. 17.2** Minimized trial proof of map No.1 from the "Erdatlas" (AGE 1803)

The "Erdatlas" and "Erdkubus" had not been topographically useful in all respects. Also, Bertuch's and Reichard's efforts in compiling information seemed unbalanced in regard to cartographic distribution of information. Manifold data from courses of South American rivers was due to Alexander von Humboldt (1769–1859), but they had to dare gaps regarding to the West Australian coastline.[23]

When the "Erdatlas" project was finished, Reichard solicited orders from the publishing houses of Friedrich Campe (1777–1846) in Nürnberg and Justus Perthes

[23]"Und wahrhaftig, nichts als die Flüsse machen den Lärm, die wird doch nicht besser haben können, es müsste denn H[err] v[on] Humboldt etwas schicken." Letter from Reichard addressed to Bertuch, 25 December 1800, GSA 06/1471.

**Fig. 17.3** Reconstruction of the "Erdkubus"

(1749–1816) in Gotha. Furthermore, he provided Bertuch with maps, reviews and articles. Bertuch wanted Reichard to move his home and workplace to Weimar; he wanted to engage him as an expert map drawer. Notwithstanding the "Mangels einer öffentlichen Bibliothek, des Mangels von Freunden der Wissenschaften," Reichard abided to his home country.[24] In Lobenstein, he established his social position and officiated as town clerk. He worried about providing for his family as a full time cartographer, therefore, he carried on cartography only as a profitable auxiliary income.

[24] Letter from Reichard addressed to Bertuch, 25 November 1802, GSA 06/1471.

## Concluding Remarks

Friedrich Justin Bertuch's working methods and the sphere of the "Geographisches Institut" dominated and shaped the geographical knowledge produced in Germany from the early 1790s until 1820. The practical exploitation of geographical and cartographical knowledge by Bertuch and his scientific staff in the "Landes-Industrie-Comptoir" and the "Geographisches Institut" was a promising procedure for creating and re-creating maps, atlases, and globes for the German science market. The operating methods were translation and re-translation, review and revision of knowledge. In this period, cartography also passed through a process of establishing itself scientifically. It could depict how a non-academic organization supported the rise of geography as a scientific discipline, particularly with regard to a special focus on cartography. Bertuch's working methods and the "Geographisches Institut" dominated and shaped the geographical knowledge produced in Germany from the early 1790s until 1820. By the use of available materials and social structures, Bertuch was able to combine innovation and tradition in order to renew German geography and cartography. He exploited the market potential of his maps, while adding to contemporary knowledge about the physical world through reviews and compilations. The enhancement of maps, atlases and globes of topical interest was performed by their draft quality, stylistic elements and stringent symbolism. Prospective clients were drawn by advertising, exceptional offers and preprints of forthcoming projects.

At the same time, it is reasonable to ask why the Weimar publishing company fell out of favor after Bertuch's death in 1822. On the one hand, we can find in the history of the "Geographisches Institut" all the famous actors who have marked the pre-disciplinary period of German geography and cartography. As in a "Who's Who," you find the names of Christian Gottlieb Reichard, Heinrich Berghaus or Adolf Stieler, all of whom received technical training in Weimar. On the other hand, most of them had considered and used the Bertuch business empire only as stepping stone for future opportunities within the firm of Justus Perthes in Gotha or publishing companies in Leipzig or Berlin.

Aside from staff migration and the associated transfer of knowledge, the initial, highly original, concept stagnated. 'Pluripotent' periodicals had to make room for topically oriented professional journals linked with their scientific fields. Nevertheless, in the context of the industrial revolution, production-related proceedings were not stressed by Bertuch's firm, which, indeed, never went beyond the status of a workshop. Moreover, the multiple changes in ownership contributed to the demise of the name of the "Geographisches Institut" within the scientific world. And also, the "Erdatlas" and the "Erdkubus" were buried in oblivion.

As we have seen, my paper shows the bumpy road from a first idea to a final cartographical product at the beginning of the nineteenth century. We saw why the "Erdatlas" and the "Erdkubus" failed. The "Erdatlas" and "Erdkubus" had the character of "collector's items" and not of scientific teaching material. Lack of success prevented reissue, unlike other maps, atlases and globes from the Bertuch firm. The "Erdatlas" was accounted in worldwide libraries and map collections. However, the "Erdkubus" could not be located until today.

## Biographical Note

**Andreas Christoph** studied Biology, Physical Anthropology and History of Science at the University of Jena. His Ph.D.-study directs his attention to the history of cartography around 1800 in the context of the "Landes-Industrie-Comptoir" and the "Geographisches Institut" from Friedrich Justin Bertuch at Weimar.

## References

Espenhorst J (2003) Petermann's planet: the great handatlasses. Pangaea, Schwerte

Espenhorst Jürgen (2008) Petermann's planet: a guide to German handatlases and their siblings throughout the World 1800–1950. Pangaea, Schwerte

Middell K (2002) "Die Bertuchs müssen doch in dieser Welt überall Glück haben" Der Verleger Friedrich Justin Bertuch und sein Landes-Industrie-Comptoir um 1800. Universitätsverlag, Leipzig

Middell K (2006) "Dann wird es wiederum ein Popanz für Otto ..." Das Weimarer Landes-Industrie-Comptoir als Familienbetrieb (1800–1830). Universitätsverlag, Leipzig

Kaiser GR, Seifert S (eds) (2000) Friedrich Justin Bertuch (1747–1822): Verleger, Schriftsteller und Unternehmer im klassischen Weimar. Max Niemeyer, Tübingen

Streit FW (1837) Die mathematische Geographie in Verbindung mit dem Gebrauch des Globus und der Entwerfung geographischer Netze. Schroeder, Berlin

# Chapter 18
# Cartography's "Scientific Reformation" and the Study of Topographical Mapping in the Modern Era

**Matthew H. Edney**

**Abstract** This paper uses a historiographical paradox to explore the conviction that cartography "became a science" during the Enlightenment, when its primary locus moved from the office into the field. The systematic topographical surveys pursued by Western states since the mid-1700s lie at the heart of this conviction yet they remain little studied by most map historians. How have map historians studied topographical surveys and placed them in the larger narrative of cartographic history? Two sets of historians were interested before 1940 in systematic topographical surveys and their products. First, social historians and historical geographers were variously interested in the evidentiary worth of topographical maps for reconstructing past societies and landscapes. Second, leading members of official surveys wrote historical accounts of their institutions, in the process creating a larger narrative sense of mapping history. These narratives merged with those of general map historians to create the modern myth. Significantly, the new, idealized narrative collapsed the several revolutions previously identified by map historians into a single moment of reform, which helps explain the persistent confusion of just when cartography became a science.

## Introduction

The conviction that "cartography became a science" during the Enlightenment and that this "reformation" was particularly accomplished by the members of the Académie des sciences is firmly entrenched in the history of cartography. It is a recurring feature of post-WW2 discussions of the character of eighteenth-century

M.H. Edney (✉)
History of Cartography Project, University of Wisconsin-Madison, Madison, WI 53706, USA

also, Osher Professor of the History of Cartography, University of Southern Maine, Portland, ME, USA
e-mail: edney@wisc.edu

E. Liebenberg and I.J. Demhardt (eds.), *History of Cartography*,
Lecture Notes in Geoinformation and Cartography 6,
DOI 10.1007/978-3-642-19088-9_18, 

mapping and it is integral to the common, grand narratives of cartographic history (of the steady transformation of an art into a science and also of the sequence of epochs of national "leadership" between the sixteenth and twentieth centuries: Italian, Dutch, French, British, German, then American). However, historians have disagreed over a precise chronology for this scientific reformation, dating it anytime from the late seventeenth to the late eighteenth century. Coupled with the complete failure to define what is meant by "art," "science," and "leadership," this vagueness strongly suggests that the conviction is an ideological creation rather than empirical induction.[1]

This paper traces the development of this conviction by addressing an historical paradox. A key element of the concept of cartography's Enlightenment reformation is the shift of the primary locus of cartographic activity from the office to the field (Pelletier 1986). That is to say, the critical act of geographical compilation, exemplified by the work of Guillaume Delisle and J.B.B. d'Anville, was replaced as cartography's primary intellectual activity by the geodetic and systematic topographical works of four generations of Cassinis and other Academicians. (More broadly, this was apparently part of a shift in emphasis among mapmakers from conceptualizing to measuring the world.) Yet historians of cartography have largely ignored the systematic state surveys of the eighteenth and nineteenth century (see Edney 2011). For example, of the 600 substantive essays published in the first 75 years of *Imago Mundi*, from 1935 to 2010, 62% specifically addressed world and regional maps and another 16% marine charts. Only 7.5% of the essays addressed the detailed mapping of place and landscape; of these, only 3% overall (i.e., 18 essays) were concerned in any way with modern systematic surveys.

One obvious source for the concept of cartography's Enlightenment reformation is Christian Sandler's essay and facsimile atlas, *Die Reformation der Kartographie um 1700* (1905). Sandler argued that geographical mapping was "reformed" by Jean Dominique Cassini's implementation of Galileo's idea to determine longitude by timing the eclipses of Jupiter's moons. In the first application of this technique, members of the Académie des sciences determined the longitude of several points around France's coast. Jean Picard and Philippe de la Hire's 1693 map comparing the coastlines of France before and after this exercise was, for Sandler, the key

[1]Much of the "new" scientificness of eighteenth-century "cartography" (itself a nineteenth-century concept: Edney 2009b) stemmed from several significant trends – cosmology's realignment as astronomy; the new (but inconsistently applied) rhetoric of plain description; the "Enlightenment Project" to promote the disenchantment of the world (Horkheimer and Adorno 1994); the increasing precision of high-end instruments; and the new endeavor to measure the earth's shape – that together contributed to the Enlightenment ideal of "mathematical cosmography" (Edney 1993, 1994). It also depended on the bureaucratic reforms in both civil and military administration engendered by the so-called military revolution and the economic growth of Europe after 30 Years' War. That is, my rejection of "cartography's scientific reformation" is not a rejection of the increasingly scientific character of certain aspects of mapping endeavors, but the ineptness of the concept for historical investigation: it is just too underdetermined to be meaningful.

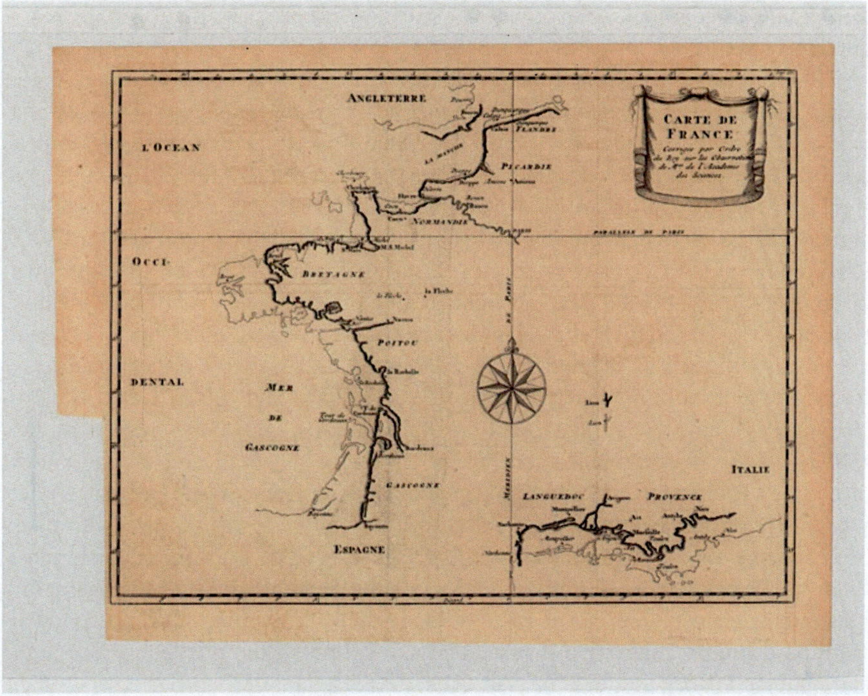


**Fig. 18.1** A late (1729) copy of Picard and de la Hire's corrected map of the coastlines of France, originally published by the Académie des sciences in 1693 (Courtesy of the Osher Map Library, University of Southern Maine)

symbol of this reformed approach to mapping (Fig. 18.1). Historians of cartography would eventually adopt Sandler's language and date of "ca.1700" for the "reformation of cartography" (Brown and Lloyd 1949, 241; Crone 1953, 128; Dorling and Fairbairn 1997, 18–21), enshrine the conviction that cartography "became a science" in the Enlightenment (e.g., Bagrow 1964, 22; Thrower 1991; even Crampton 2010, 68–70), and routinely reproduce the Picard-de la Hire map.[2] But Sandler was concerned only for the determination of geographical positions by latitude and longitude and so with the significant improvement after 1700 of the accuracy and comprehensiveness of the depiction of the continents. Sandler was not concerned with, and said nothing about, detailed topographical mapping. Nor was he suggesting that cartography shifted from an "art" to a "science." Rather, he sought to identify a particular moment in a progressive and ongoing mapping effort. So, how did topographical mapping get wedded to Sandler's

[2]Andrews (2009, 26–30) provides an interesting twist: his overview of the history of cartography recasts the structure of Sandler's reformation with empirical honesty: contemporaries and successors to the academicians were not exactly appreciative of the new scientificness.

astronomical-geographical reformation, and how did this become expressed as a qualitative shift from an art to a science?

The established historiography of map history has little to say about any of these issues. That literature developed within the heady atmosphere of post-war academia when many new academic disciplines flourished. In particular, cartography developed its own journals and degree programs and seemed, by the 1970s, about to "emerge" as a discipline in its own right (Wolter 1975). Indeed, some scholars argued that the "extradisciplinary" field of the history of cartography (Skelton 1972, 92) could emulate the success of the history of science and become its own discipline, if only it could develop a coherent intellectual and institutional identity (Edney 2005a, 51–56). The key historiographies – R.A. Skelton's 1966 Nebenzahl Lectures (published as Skelton 1972); Blakemore and Harley's (1980) manifesto for a new, culturally minded approach to map history; and Harley's introductory chapter for volume one of the ground-breaking *The History of Cartography* (Harley 1987) – were all written from a desire to reform the field and create a new discipline. Their story of redemption, of an unreformed past and present giving way to a reformed and disciplined future, was not entirely wrong. After all, it has been affirmed in part by the success enjoyed since 1990 by critical map historians. Even so, the established narrative is substantially incomplete. Key historiographic issues remain unstudied, not least how topographical mapping was incorporated within a communal narrative that defined and delimited the intellectual scope of cartographic history as a field of study, a narrative against which Harley and other critical historians would otherwise so profitably react.

## Early Studies of Topographical Maps by Historians

Agricultural and economic historians began in the later nineteenth century to use early topographical maps – and also early property maps – as evidentiary sources. Frederic Seebohm's (1833–1912) classic study, *The English Village Community* (1883), contained over 20 maps and plans. For example, he traced certain "tribal arrangements" of early Irish settlement in County Galway by means of a 1607 written survey and a sheet of the nineteenth-century six-inch Ordnance Survey of Ireland which recorded the old Irish toponyms and the remains of a couple of ancient circular enclosures (Seebohm 1883, 224–25). The primary proponent of such map usage was August Meitzen (1822–1910) who made extensive use of early plans of villages and landscapes in his monumental comparative analysis of early settlement forms across northern Europe, *Siedelung und Agrarwesen der Westgermanen und Ostgermanen, der Kelten, Römer, Finnen und Slawen* (1895). The three text volumes featured over 250 maps and other images, the atlas volume a further 150. Reviewers were amazed by the sheer scale of Meitzen's reliance on maps (Cheyney 1897, 124; Bateson 1897, 315; Ashley 1898, 143).

Meitzen was in particular a devotee of the Ordnance Survey; yet because its work mostly post-dated the massive landscape changes wrought by the agricultural and early industrial revolutions, he thought it "that most wonderful of all palimpsests" (Bateson 1897, 320). Just 2 years later, Frederic W. Maitland (1850–1906) deployed the OS in accordance with Meitzen's primary argument that settlement form manifested racial spirit (Seebohm 1897; Ashley 1898). Juxtaposing different sheets from the Ordnance Survey, Maitland distinguished between Anglo-Saxon settlement form, the "true village or nucleated village," and the scattered hamlets of "old Celtic arrangements." In doing so, he referred to the "ordnance map" as "that marvellous palimpsest, which under Dr. Meitzen's guidance we are beginning to decipher" (Maitland 1897, 15–16; Meitzen 1895, 2:119ff) (Fig. 18.2).

However, these historians did not reflect upon the nature and history of the maps they used. They took them at face value, selectively traced from them, and blended them with field observations. They mixed reproductions of original maps with historical reconstructions and did not distinguish between them. One comment by Meitzen (1895) is most revealing in this respect:

> In every village we wander, in a certain sense, among the ruins of the past; ruins which, in age, leave the romantic remains of the castles and city walls of the Middle Ages far behind. At every step, everywhere, in barnyard and field, we may meet the traces of the oldest conditions. The plan of ownership of the village lands is a veritable manuscript which conveys to us, legible as though in hieroglyphs, the ideas and aims of the early founders. (Quoted by Cheyney 1897, 125–26)

Is Meitzen's "plan of ownership" an early map, a modern reconstruction, or something more syncretic prepared in accordance with his preconceptions? Maitland (1897, 362 and 381–83) had certainly appreciated that the landscape represented in modern maps had undergone significant change since the classical and medieval eras, but he himself did not pursue the careful dating of each landscape component that was necessary for a full appreciation of early maps.

Scholarly reaction to Meitzen's and Maitland's cartographic arguments was grounded in this specific failure to understand the origins and nature of these evidentiary sources. As Paul Vinogradoff (1854–1925) would soon argue, maps showing pre-enclosure field systems manifested the agricultural and economic practices underpinning the field systems and had nothing to say about the social structure of village communities or the ethnic origins of their originators (Vinogradoff 1911, 182–83). Howard Gray (1875–1946) further outlined the difficulties inherent to interpreting cadastral plans from the eighteenth and nineteenth centuries for revealing field systems in use a millennium earlier, encouraged the use of early estate plans, and completely ignored topographical survey plans as having evidentiary value for the study of field systems and economic history (Gray 1915, 14–15; see also Bloch 1929).

Yet for geographers and historians interested in the evolution of landscapes, Meitzen stands as the "founder of historical geography in the modern sense" (Carl Sauer quoted by Block 1980, 34) and the examination of topographical maps remained an important tool among historical geographers. While some substantial projects were undertaken – such as the reconstruction, in the 1920s for the *Polish Historical Atlas*, by Karol Buczek (1902–1983) of "the map" of Poland ca.1790

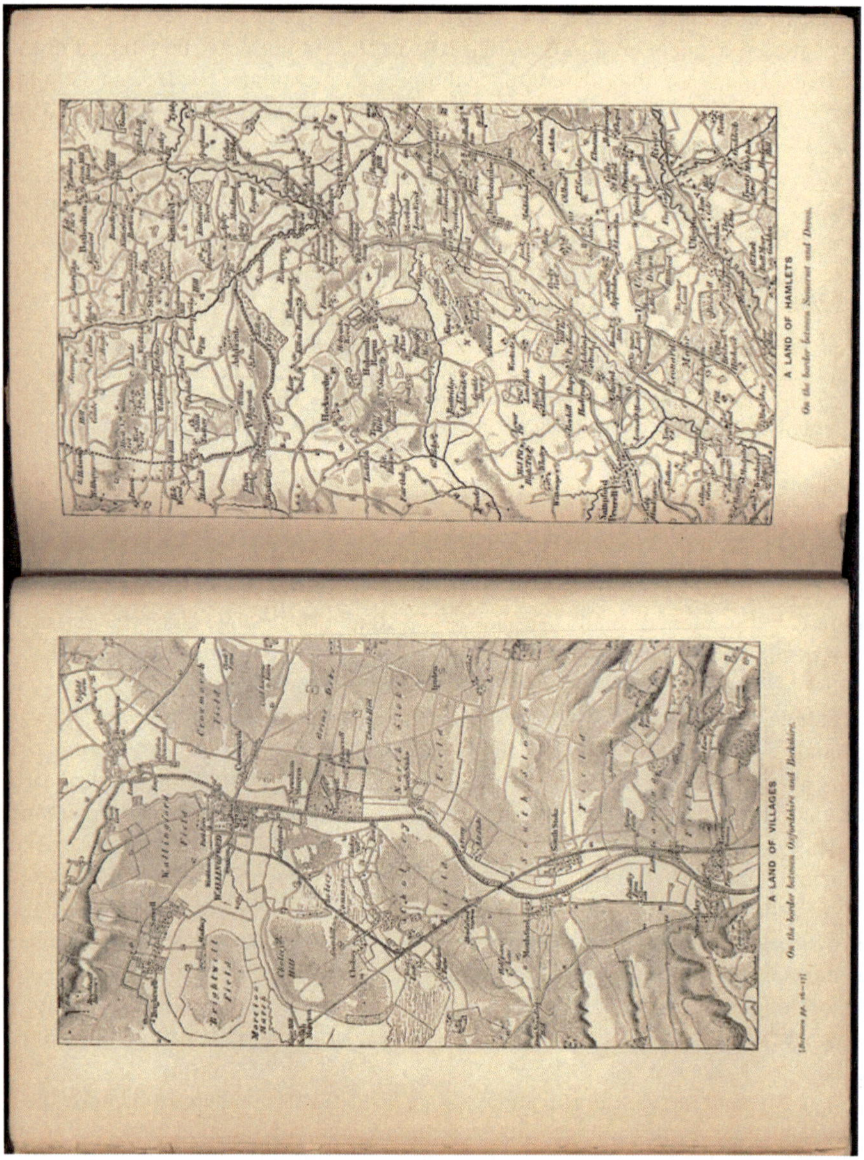


**Fig. 18.2** Maitland's 1897 reproduction of portions of two first edition OS maps (See Edney (2005a, 21–23) for more information (Author's collection)

from contemporary topographical maps by the French, Prussians, and Austrians as well as a variety of tax and census records (Buczek 1982, i) – interest was nonetheless sporadic. In 1924, J.K. Wright (1891–1969) could thus present a research agenda for the study of the products of the systematic topographical surveys of eighteenth- and nineteenth-century Europe as if the subject were quite novel. Wright explained the worth in terms of landscape history: tracing urban and infrastructural growth by comparison to modern maps; comprehending military history; and discerning landscape change through natural and social forces (Wright 1924, 9–13).

The most sustained early statement of the study of early topographical maps was made in conjunction with the seventh international historical congress, held in Warsaw in 1933. There, the commission for historical geography – formed only in 1930 – organized a large exhibition dedicated to the historical vision of the Austrian geographer Eduard Richter (1847–1905) (Anonymous 1933, esp. x-xi). The exhibition featured facsimiles of three types of map pertinent to historians: early maps, including topographical maps; urban maps; and historical maps depicting past events. The accompanying summary notes were largely bibliographical in nature, identifying the maps on display and describing their content. I should note that interest in systematic state survey maps seems in 1933 to have been largely limited to continental Europeans; the British component included nothing later than the 1680s while the U.S. component featured only early regional maps (Anonymous 1933, 99–101 and 141–42). One consequence of this exhibition was the publication in *Imago Mundi* of that journal's only essay before 1960 on early systematic surveys, specifically Ernst Nischer-Falkenhof's (1937) detailed listing of the many surveys undertaken across Europe by the Austrian Monarchy significantly expanding his brief exhibition in Warsaw (Nischer-Falkenhof in Anonymous 1933, 124–29).

Nischer's work exemplifies the manner in which geographically minded historians and historically minded geographers then approached the subject of early topographical maps. Their concern was to establish an empirical foundation for landscape studies by identifying surviving maps, the areas they covered, and perhaps the relative quality of their content. In this, they followed the broad concern of early historians of cartography to present old maps and to elucidate their content for the benefit of other historians. In style, Nischer's listing, Wright's (1924, 4–9) listing of early survey maps in the AGS library, and the bibliographical entries in the Warsaw catalog (Anonymous 1933, 1–142) all differ little from most of the essays identifying and describing world and regional maps that filled *Imago Mundi's* pages.

## Professional Histories and Narratives of Cartographic Progress

The primary historiographies (Skelton 1972; Blakemore and Harley 1980; Harley 1987) identified three primary communities of map historians in the nineteenth and early twentieth century: academic geographers and historians; map librarians, especially those in the great national libraries; and dealers and collectors of antiquarian maps. But a fourth community has also been active since the later nineteenth century, comprised of professional cartographers who wrote histories of their surveying and mapping institutions. Accounts of surveyors' field experiences had been a crucial element of survey reports since the eighteenth century: in describing how surveyors had functioned, and had often suffered for their craft, they established the trustworthiness and consistency of the survey not only for other cartographic and scientific experts but also for politicians and bureaucratic paymasters. It was natural for surveyors to take an historical approach to their profession, to trace the work of their predecessors, in order to understand the nature and need of contemporary operations. The result was accordingly a large number of professional histories of particular national surveys: Clements Markham (1878) and Reginald Phillimore (1945–1958) on British India; Pilkington White (1886) and Charles Close (1926) on the Ordnance Survey; Henri Berthaut (1898–1899; 1898–1902) on France; Attilio Mori (1903) on Italy; Ernst Nischer (1925) on Austria; Émile Hennequin (1891) on the Austrian Netherlands and Belgium; and so on. In addition, two works provided comparative analyses of the historical progress of national surveys across Europe: George M. Wheeler (1885) sought to demonstrate that base mapping is most effectively accomplished by the military so that US base mapping should be transferred from the newly created USGS to the corps of engineers;[3] Willibrod Stavenhagen (1904) undertook his comparative review of topographical surveys outside of Germany[4] as part of an argument for international cooperation in topographical mapping at scales larger than the international map of the world.

While professional map historians addressed topics other than those that interested other map historians, they were not completely segregated. In particular, professional map historians adapted the narrative of cartographic development already developed by other scholars. That narrative identified two key revolutionary moments: first, the nearly simultaneous discoveries of the New World and Ptolemy's *Geography*; second, Gerard Mercator's apparent unification in 1569 of geographical mapping and marine charting with his famous map projection.[5] The narrative solidified in the 1870s with the formation of "the history of cartography" as a field of study. For example, one of the first essays specifically to address "the history of cartography" laid out a clear narrative of progress in the endeavor of reducing the world to paper, from the classical era through the Renaissance (Daly 1879; see also Breusing 1883; Wolkenhauer 1904). The catch-phrase in such work

[3]See also Comstock's (1876) a historical review of the various European surveys then under way, undertaken to ensure that the U.S. Lake Survey followed the established best practices.

[4]Stavenhagen promised that a parallel study on surveys within Germany was in preparation; I would be pleased to know if he or another scholar ever published it.

[5]Some early texts referred to Mercator's work as comprising the "reformation" of cartography, in turning the older artistic practices into a science. See Gelcich et al. (1909, 75–89).

was, in fact, "the *early* history of cartography" (as also, e.g., Nordenskiöld 1889), indicating an implicit contrast to modern cartography.[6]

Professional map historians simply extended the narrative to encompass the modern era of topographical mapping. Stavenhagen (1904, 1–11) thus began his comparative study of modern European surveys with a broad overview of the highlights of European cartography, starting in Antiquity, that he had derived from Wilhelm Wolkenhauer's (1895) bibliographical history of cartography. Stavenhagen used lists of key atlases and maps to illustrate how cartographers had progressively produced ever more detailed maps of European territories, from one-sheet provincial maps in Ortelius's *Theatrum orbis terrarum* (1570), to the 20-plus sheet provincial maps of the late seventeenth and early eighteenth century, to the inevitable adoption of triangulation-based surveys for still more detailed coverage at topographical scales.[7] Similarly, John G. Bartholomew (head of the commercial mapmaking firm that was significantly invested in topographical mapping) introduced his bibliographical lists of current maps of the world, from continental, general maps to detailed topographical and cadastral maps, with brief historical commentaries emphasizing the significant progress made by Europeans during the nineteenth century in base mapping (Bartholomew 1891; see also Laussedat 1892).

None of these narratives imputed any special *qualitative* change to cartography in the eighteenth century. Bartholomew (1902, 36), for example, had noted that France was the "first country to undertake a general survey of the country at the expense of the state" and that others followed her lead, but he did not understand either the Cassinis' surveys or the Académie's astronomical and geodetic work as redefining the character of modern cartography. Instead, Bartholomew argued that the British had taken the lead in maritime charting and hydrography and that it was Alexander von Humboldt and subsequent German geographers who had attained the glories of "scientific mapping" in the nineteenth century. Wilhelm Wolkenhauer (1895, 70–88) even claimed that "modern cartography" began only in 1855! Aimé Laussadat (1892), as a colonel of engineers and the principal French topographical surveyor, understood Europe's armies as having had the key role in the development of systematic surveys and therefore downplayed the role of scientists and of cartographic "science." That is, these narratives simply traced the progressive development of an endeavor understood to have always been a science.

---

6 Bagrow (1951, 7) thus contrasted early cartography, the subject of his *Geschichte der Kartographie*, with technologically distinct modern cartography. The reference to "specialized science" in the much-expanded foreword to the English edition was presumably added by the editor, R.A. Skelton (Bagrow 1964, 22).

7 Stavenhagen's work apparently influenced only one other scholar. It inspired in J.K. Wright a (fleeting) interest in the history of the modern surveys as a crucial component of modern cartography. Wright (1924, 14–36) outlined a research agenda for a detailed, internal history of modern topographical mapping. Unfortunately, Wright's statement was unnoticed by other map historians and Wright did not sustain his interest and his work had no discernible effect. Stavenhagen's and Wright's work might therefore be dismissed as historiographically insignificant, except that together they suggest that there was a broader interest in the early twentieth century for extending the history of cartography to encompass modern topographical surveys.

This perspective persisted throughout the early twentieth century. Fritz Curschmann, in his preface to the German contribution to the Warsaw exhibition, understood the development of systematic surveys in the eighteenth century as a function strictly of the needs of Europe's absolute monarchies and showed no interest at all in their scientific nature; at the same time, H. Almquist suggested that the Swedish surveys of the second half of the eighteenth century were functional rather than scientific (Anonymous 1933, 22–23 and 133). The clearest statements of the necessarily complex nature of cartography's history and so the inability to reduce its progress to a simple narrative were provided by Max Eckert's (1921–1925) thorough analysis, in large part structured historically, of the entire field of map making and by Erwin Raisz's (1937, 1938) graphic timelines of cartography from antiquity to the present prepared for the introduction to his innovative manual of cartographic practice. Tracing an ever-broadening front of mapping activities, Raisz's images empirically preclude any simplification of the historical narrative. We see in these narratives prepared by professional historians the origins of the progressive model of the accretion of knowledge. A corollary to the modern cartographic ideal that solidified after 1870 (Edney 2011), it was inevitably incorporated into the broader stream of map history.[8]

The grand narrative was sufficiently flexible to permit some authors to insert a causal role for "science." Bartholomew (1902, 34) introduced yet another potted history of cartography by stating that professional histories were motivated by the belief that "Maps and Atlases may be said to be the epitomised record of our earth-knowledge, and, in their evolution, ought to reflect the progress of civilization." W.W. Jervis (d. 1959), an academic geologist, surveyor, and geographer, thus sought to integrate cartography with modern narratives of the history of western science through heavy-handed comments about cartography's increasingly "scientific tendencies" in his otherwise highly idiosyncratic *The World in Maps* (Jervis 1936, esp. 42–43 and 88–89).[9]

## Reconfiguring Cartography as the Art That Became a Science

Only in France was Sandler's vision of the "reformation" of cartography widely applied to *all* aspects of cartography before 1940. Sandler's moment of geographical reform was explicitly transformed into the general pivot between early and

---

[8]Buczek (1982, 7), writing originally in 1963, presented an abnormal sequence of cartographic development, in which socio-economic factors prevented the eighteenth-century mapping of Poland from living up to the ideals of reformed (as per Sandler) cartography.

[9]Reviewers (*Imago Mundi* 2 [1937]: 98; Gerald Crone ["G.R.C."] in *Geographical Journal* 90, no.1 [1937]: 85–86) noted Jervis's idiosyncrasies, e.g., by starting his chapters on "great map-makers" and "modern map production" in the modern era before working backwards in time to the Renaissance.

modern mapping by Lucien Gallois. Gallois's (1909) study of the origins of the Cassini family's great *Carte de France*, which of course drew on Berthaut's (1898–1899) professional history, placed the determination of longitude within the larger pattern of the Académie's mapping activities and in particular its initial geodetic measurement of the Paris meridian. His argument that cartography became a science in France in the late seventeenth and eighteenth century was thereafter promulgated by a variety of works, including the key bibliography of French history (Bourgeois and André 1913–1926, 1:1–7), Yves Marie Goblet's summary of modern French territorial surveys in the Warsaw exhibition catalog (Anonymous 1933, 69–99, esp. 76), and Georges Perrier's (1939) short history of geodesy.

The Sandler-Gallois reinterpretation of cartography's progressive history found its clearest expression in the work of Lloyd Brown (1907–1966). Brown first conflated the new longitude measurements with geodetic triangulations, drawing partly on Sandler, in his study of Jean Dominique Cassini's 1696 world map (Brown 1941). World War Two had brought home how the world had yet been mapped in detail only "after a fashion," even as the capacity to undertake topographical surveys had greatly increased; Brown set out to explain this situation as an introduction to the new wave of detailed, systematic world mapping that he now anticipated (Brown and Lloyd 1949, 10–11; also Brown 1953). His overarching theme, first stated in his best-selling *The Story of Maps* (Brown 1949) and subsequently reiterated, was the manner in which the "ancient art" of cartography "grew to be a science" (Brown 1952, xi; also Brown 1941, 10; Brown 1960). *The Story of Maps* defined a normative sequence of mapping, of ever-larger maps of each area that were permitted by the scientific spirit inculcated by the first Cassini and his fellow academicians. Drawing extensively Gallois (1909), Brown blurred the significant differences between astronomical techniques for determining longitude, geodetic triangulations, and detailed topographical surveys. With a purposefully unclear chronology, he implied that the scientific elements of non-French surveys in the eighteenth century stemmed from French influence even as he glossed over the surveys' more persistently traditional elements, as can be seen for example in how he treated the Austrian surveys (Brown 1949, 241–79, esp. 271–72). For the nineteenth-century surveys, Brown (1949, 280–309) relied on Wheeler (1885). His work remains the only general, scholarly text within the history of cartography to examine the modern systematic surveys at length and has had great influence. In particular, it served as the primary source for Emil Bachmann's *Wer hat Himmel und Erde gemessen?* (1965) and John Noble Wilford's bestselling *The Mapmakers* (1981); Wilford's work in turn strongly influenced Simon Berthon and Andrew Robinson's *The Shape of the World* (1991).

Significantly, Brown's work seems not to have been used by Gerald Crone (1899–1982) in the preparation of his concise study, *Maps and their Makers* (1953). Crone's chapter (128–40) on "The Reformation of Cartography in France" was clearly indebted to Sandler (1905) and also drew on Goblet (Anonymous 1933, 69–99) and Perrier (1939). National sentiment required Crone to add a chapter on eighteenth-century British contributions to cartography before he discussed the "regular topographical surveys" that had "characterized ... cartography since the early decades of the nineteenth century" (Crone 1953, 151). Like Brown, Crone drew extensively on recent historical literature by professional surveyors. The parallel gestation of Brown's and Crone's ideas again suggests a common concern.[10] Indeed, Crone concluded that it "is true to say, therefore, that a new 'reformation of cartography' is now in progress" (Crone 1953, 171). For both Brown and Crone, the war had brought cartography to a new institutional and technological plateau that invited not nostalgia but a celebration of cartography's past progress as prelude to its future perfection.

The imminent new wave of post-war global base mapping can also be seen to have influenced those geographers, notably Arthur H. Robinson (1947, 1952), who created the new field of academic cartography specifically to focus on the design of small-scale maps and to reclaim a space for cartography-as-art. In the process, they only reaffirmed the new post-war sense that the fundamental character of modern cartography is "scientific." The subsequent definition of cartography as "both the art and science" of making maps (see Maling 1991) has only obscured any attempt at clarity in understanding the nature of maps (Delano Smith 1990; Krygier 1995).

## Detailed Histories of Topographical Surveys After 1960

Despite the repeated allusions to the overall sweep of map history – of the ever-increasing quantity and quality of geographical information stored in increasingly detailed regional maps and eventually topographical surveys (Crone 1953, xi) – and despite Brown's prominent discussion of topographical surveys, most map historians continued after 1950 to pay little attention to modern topographical surveys. The field remained committed to publicizing sources of historical information and did not worry too deeply about how the information had been gathered. This situation only began to change in the 1960s through two developments.

First, led by H.C. Darby, a fan of Maitland and Meitzen, British historical geographers began actively to supplement fieldwork with documentary evidence of past landscapes. A variety of historical geographers – most notably Harley and J.H. Andrews – began to study large-scale surveys of the eighteenth and nineteenth century. Thus, in organizing the second international conference on the history of cartography in 1967, Harley dedicated it specifically to the analysis of early maps as sources of evidence; the resultant volume of *Imago Mundi* is rather anomalous (Harley 1968; see Sims and Krogt 1995). What distinguished this new scholarship was a concern for the detailed history of the surveys in order to assess the trustworthiness of the evidence provided by early topographical maps: why were

[10]It is perhaps possible that both authors might have been influenced by a common source, specifically Jervis's (1936) *World in Maps* which Brown cited and which Crone had reviewed. Jervis's work certainly guided at least one subsequent interpretation of the artistic qualities of cartography before "science claimed" it (Rees 1980).

the maps made, how were the surveys undertaken; how were features represented? The result was a wide array of studies that brought systematic surveys firmly within the ambit of the history of cartography in Britain (explored in detail by Edney 2005a, 19–31). A similar phenomenon can be identified among U.S. historical geographers (e.g., Bartlett 1962; Friis 1965; Marshall 1976; Ehrenberg 1983).

The second development was the formation by Arthur Robinson and other academic cartographers of a new, internal paradigm of cartographic history that explored the history of cartography as a profession and that paid particular attention to mapping techniques and technologies (Edney 2005b; Edney 2011). This promoted, to some degree, a more complex treatment of systematic surveys as an integral element of the field, as evident for example in Eduard Imhof's (1965) classic analysis of the depiction of relief or in the *Lexikon zur Geschichte der Kartographie* with its lengthy discussions of modern topographical mapping (Kretschmer et al. 1986).

Together, these two movements led to some serious considerations of large-scale, systematic surveys: the motives behind them; the techniques and technologies they deployed; the maps they produced; and their successes and their failures. In exploring these surveys on their own terms, rather than simply listing their maps as unproblematic documents, the new studies accepted that the idealized progressive narrative of cartography is indeed an ideal and that, when historians pay attention to the empirical evidence, the narrative is exposed as a myth that has little relevance for explaining the actual history of maps and mapping. In this respect, post-war studies of topographical mapping – still few by comparison to the rest of the studies in map history – formed a significant reason for the shift after 1980 to a new, socio-cultural paradigm of map history. A main theme of that paradigm has been the analysis of state surveys, especially in colonial and imperial settings but also in domestic contexts (see Edney 2009b). In this last trend, scholars have sought to augment professional histories with careful analyses of institutional structures and technological practices. It is here that we can see a future for research that focuses on modern topographical surveys and maps. By addressing not only the processes that surveyors have variously followed in interacting with the land and its inhabitants, but also how their resultant maps have circulated and been used, we will be able to submit the modern cartographic ideal to critical scrutiny. In breaking free of the ideal, we will be able to attain a more complete understanding of the complexity of past and present cartography.

## Biographical Note

**Matthew H. Edney** was born and raised in London, UK, and read geography at University College London (B.Sc. Hons, 1983) before moving to the USA to pursue graduate studies with David Woodward at the University of Wisconsin-Madison (M.S., Cartography, 1985; Ph.D., Geography, 1990). In his first job he taught computer mapping and GIS at SUNY-Binghamton (1990–1995). He moved to the

University of Southern Maine in 1995 to be the faculty scholar in its newly founded Osher Map Library; since 2007 he has held the Osher Chair in the History of Cartography. Already coeditor of *Cartography in the European Enlightenment*, Volume Four of *The History of Cartography*, in 2005 he also became director of the History of Cartography Project in Madison. Edney's principal scholarly interest lies in the history of surveying (from property to regions), especially in state and imperial contexts. This work is exemplified in his *Mapping an Empire: The Geographic Construction of British India, 1765–1843* (Chicago, 1997; New Delhi, 1999) and "The Irony of Imperial Mapping" (Edney 2009b). This interest has led to broader work on mapping practices in what might be called a processual approach to map history (Edney 1993, 2011). This theoretical work has entailed reflections upon the nature of Enlightenment cartography (e.g., Edney 1994). It also combined with Edney's ongoing study of the mapping of colonial New England into a project to understand the processes whereby regional maps were produced, circulated, and consumed within the British transatlantic empire before 1775. Essays on this topic have appeared in *William & Mary Quarterly* (2004), *Imago Mundi* (2008), and *Word & Image* (2010). This conceptual and programmatic work has also extended into some recent historiographical studies (Edney 2005a, b, 2013).A list of better publications and a fuller intellectual biography is online at http://www.usm.maine.edu/~edney/research.html, in something like an old-style *curriculum vitae*.

## References


Andrews JH (2009) Maps in those days: cartographic methods before 1850. Four Courts Press, Dublin

Anonymous (1933) Catalogus mapparum geographicarum ad historiam pertinentium quae curante collegio historico-geographorum adiuvantibus viris congressi ordinando in polytechnico Varsoviensis exponentur. International Committee of Historical Sciences, Commission for Historical Geography, Warsaw

Ashley WJ (1898) Meitzen's Siedelung und Agrarwesen der Germanen. Political Sci Q 13 (1):143–155

Bachmann E (1965) Wer hat Himmel und Erde gemessen? Von Erdmessungen, Landkarten, Polschwankungen, Schollenbewegungen, Forschungsreisen und Satelliten. Ott Verlag, Munich

Bagrow L (1951) Die Geschichte der Kartographie. Safari-Verlag, Berlin

Bagrow L (1964) The history of cartography (trans. DL Paisley; ed. RA Skelton). CA Watts, London

Bartholomew JG (1891) The mapping of the world. Scottish Geogr Mag 6:293–305, 6:575–597, 7:124–152, 7:586–611

Bartholomew JG (1902) The philosophy of map-making and the evolution of a great German atlas. Scottish Geogr Mag 18(1):34–39

Bartlett RA (1962) Great surveys of the American West. University of Oklahoma Press, Norman

Bateson M (1897) Review of August Meitzen, Siedelung und Agrarwesen der Westgermanen und Ostgermanen, der Kelten, Römer, Finnen und Slawen, 4 vols. (Berlin: Hertz, 1895). English Hist Rev 12(46):313–323

Berthaut HMA (1898–1899) La Carte de France, 1750–1898: Étude historique, 2 vols. Imprimerie du service géographique de l'armée, Paris

Berthaut HMA (1898–1902) Les ingénieurs géographes militaires, 1624–1831: Étude historique, 2 vols. Service géographique de l'armée, Paris

Berthon S, Robinson A (1991) The shape of the world: the mapping and discovery of the earth. Rand McNally for Granada TV, Chicago

Blakemore, MJ, Harley, JB (1980) Concepts in the history of cartography: a review and perspective. Cartographica 17(4): Monograph 26

Bloch M (1929) Les plans parcellaires. Annales d'Histoire Economique et Sociale 1(1):60–70 and 1(3):390–398

Block RH (1980) Frederick Jackson Turner and American Geography. Ann Assoc Am Geogr 70(1):31–42

Bourgeois É, André L (eds) (1913–1926) Les sources de l'histoire de France: XVIIe siècle (1610–1715), 5 vols. Auguste Picard, Paris

Breusing A (1883) Leitfaden durch das Wieganalter der Kartographie bis zum Jahre 1600 mit besonderer berücksichtigung Deutschlands. Beitrage zum Kataloge der Ausstellung des Dritten Deutschen Geographentages zu Frankfurt a.M., am 29., 30. und 31. März 1883. Verlag von Mahlau & Waldschmidt, Frankfurt

Brown LA (1941) Jean Dominique Cassini and his world map of 1696. University of Michigan Press, Ann Arbor

Brown LA (1949) The story of maps. Little, Brown & Co, Boston. Reprinted New York: Dover, 1979

Brown LA (1952) The world encompassed: an exhibition of the history of maps held at the Baltimore Museum of Art, October 7 to November 23, 1952. Trustees of the Walters Art Gallery, Baltimore

Brown LA (1953) Maps: the necessary medium to world progress. Surveying Mapp 13:277–285

Brown LA (1960) Map making: the art that became a science. Little, Brown and Co., Boston

Buczek K (1982) The History of Polish Cartography from the 15th to the 18th Century (trans. Andrzej Potocki), 2nd edn. Meridian Publishing Co, Amsterdam. Reprinted from Dzieje kartografii polskiej od XV do XVIII wieku (Krakow, 1963)

Cheyney EP (1897) Review of August Meitzen, Siedelung und Agrarwesen der Westgermanen und Ostgermanen, der Kelten, Römer, Finnen und Slawen, 4 vols. (Berlin: Hertz, 1895). Ann Am Acad Political Soc Sci 9:124–128

Close C (1926) The early years of the ordnance survey. Institution of Royal Engineers, London

Comstock CB (1876) Notes on European surveys. In Annual Report of Major C. B. Comstock, Corps of Engineers, for the Fiscal Year Ending June 30, 1876, 126–217. Vol. 2.3 of Report of the Secretary of War . . . at the Beginning of the Second Session of the Forty-Fifth Congress, Washington, DC

Crampton JW (2010) Mapping: a critical introduction to cartography and GIS. Wiley-Blackwell, Oxford

Crone GR (1953) Maps and their makers: an introduction to the history of cartography. Hutchinson, London

Daly CP (1879) The early history of cartography, or what we know of maps and map-making before the time of Mercator. J Am Geogr Soc 11:1–40

Delano Smith C (1990) Art or cartography? The wrong question. Hist Hum Sci 2(1):89–93

Dorling D, Fairbairn D (1997) Mapping: ways of representing the world. Addison-Wesley Longman, Harlow, Essex

Eckert M (1921–1925) Die Kartenwissenschaft: Forschungen und Grundlagen zu einer Kartographie als Wissenschaft, 2 vols. Walter de Gruyter, Berlin

Edney MH (1993) Cartography without 'Progress': reinterpreting the nature and historical development of mapmaking. Cartographica 30(2–3):54–68

Edney MH (1994) Mathematical cosmography and the social ideology of British cartography, 1780–1820. Imago Mundi 46:101–116

Edney MH (2005a) The origins and development of J. B. Harley's cartographic theories. Cartographica 40(1-2): Monograph 54

Edney MH (2005b) Putting 'Cartography' into the history of cartography: Arthur H. Robinson, David Woodward, and the Creation of a Discipline. Cartographic Perspectives, no. 51:14–29. Reprinted, with corrections in A Reader in Critical Geographies, ed. Salvatore Engel-Di Mauro and Harald Bauder (Praxis (e)Press www.praxis-epress.org, 2008), 711–728

Edney MH (2009a) Cartography, history of. In: Derek Gregory RJ, Johnston GP, Watts M, Whatmore S (eds) The dictionary of human geography. Blackwell, Oxford, pp 69–72

Edney MH (2009b) The irony of imperial mapping. In: Akerman JR (ed) The imperial map: cartography and the mastery of empire. University of Chicago Press, Chicago, pp 11–45

Edney MH (2011) Field/map: an historiographic review and reconsideration. In: Christopher JR, Kristian HN, Michael H (eds) Ways of knowing the field: studies in the history and sociology of scientific field work and expeditions. Aarhus University Press, Aarhus

Edney MH (2013) Histories of cartography. In: Monmonier M (ed) Cartography in the twentieth century. Vol. 6 of The history of cartography. University of Chicago Press, Chicago

Ehrenberg RE (1983) Mapping land, sea, and sky: federal cartography from 1775 to 1950. Government Publications Rev 10:361–374

Friis HR (1965) A brief review of the development and status of geographical and cartographical activities of the United States Government: 1776–1818. Imago Mundi 19:68–80

Gallois L (1909) L'Académie des sciences et les origines de la carte de Cassini. Annales de géographie 18(99):193–204 and 18(100):289–310

Gelcich E, Sauter F, Dinse P, Groll M (1909) Kartenkunde, geschichtlich dargestellt, 3rd edn. G. J. Göschen, Leipzig

Gray HL (1915) English field systems. Harvard University Press, Cambridge

Harley JB (1968) Conference on the history of cartography, London, September 1967. Imago Mundi 22:9

Harley JB (1987) The map and the development of the history of cartography. In Harley JB, Woodward D (eds) Cartography in prehistoric, ancient, and medieval Europe and the Mediterranean, vol 1, The history of cartography. University of Chicago Press, Chicago, 1–42

Hennequin É (1891) Étude historique sur l'exécution de la carte de Ferraris et l'évolution de la cartographie topographique en Belgique. Bulletin de la Société Royale Belge de Géographie 15:177–296

Horkheimer M, Adorno TW (1994) The dialectic of enlightenment (trans. John Cumming). Continuum, New York

Imhof E (1965) Kartographische Geländedarstellung. Walter de Gruyter, Berlin

Jervis WW (1936) The world in maps: a study in map evolution. George Philip & Son, London

Kretschmer I, Dörflinger J, Wawrik F (eds) (1986) Lexicon zur Geschichte der Kartographie von den Anfängen bis zum ersten Weltkrieg, 2 vols. Franz Deuticke, Vienna

Krygier JB (1995) Cartography as an art and a science? Cartographic J 32:3–10

Laussedat A (1892) Histoire de la cartographie. Revue scientifique 49(23, 4 June): 705–719 and 49 (24, 11 June): 742–761

Maitland FW (1897) Domesday book and beyond: three essays in the early history of England. Cambridge University Press, Cambridge

Maling DH (1991) The origins of that definition. Cartographic J 28(2):221–223

Markham CR (1878) A memoir on the Indian surveys, 2nd edn. Allen & Co, London

Marshall DW (1976) The British Military Engineers, 1741–1783: a study of organization, social origin, and cartography. Ph.D. dissertation, University of Michigan (History)

Meitzen A (1895) Siedelung und Agrarwesen der Westgermanen und Ostgermanen, der Kelten, Römer, Finnen und Slawen, 4 vols. Wilhelm Hartz, Berlin

Mori A (1903) Cenni storici sui lavori geodetici e topografici e sulle principali cartografiche eseguite in Italia della metà del secolo XVIII ai nostri giorni. Florence.

Nischer E (1925) Österreichische Kartographen: Ihr Leben, Lehren und Wirken. Österreichischer Bundesverlag für Unterricht, Wissenschaft und Kunst, Vienna

Nischer-Falkenhof, E von (1937) The survey by the Austrian General Staff under the Empress Maria Theresa and the Emperor Joseph II, and the subsequent initial surveys of neighbouring territories during the years 1749–1854. Imago Mundi 2:83–88
Nordenskiöld AE (1889) Facsimile-atlas to the early history of cartography with reproductions of the most important maps printed in the XV and XVI centuries (trans Ekelöf JA). Stockholm
Pelletier M (1986) La France mesurée. Mappemonde 86(3):26–32
Perrier G (1939) Petite histoire de la géodesie: Comment l'homme a mesuré et pesé la Terre. Presses Universitaires de France, Paris
Phillimore RH (1945–1958) Historical records of the survey of India, 4 vols. Survey of India, Dehra Dun
Raisz E (1937) Time charts of historical cartography. Imago Mundi 2:9–16
Raisz E (1938) General cartography. McGraw-Hill, New York
Rees R (1980) Historical links between cartography and art. Geogr Rev 70(1):60–78
Robinson AH (1947) Foundations of cartographic methodology. Ph.D. dissertation, Ohio State University (Geography)
Robinson AH (1952) The look of maps: an examination of cartographic design. University of Wisconsin Press, Madison
Sandler C (1905) Die Reformation der Kartographie um 1700. R. Oldenbourg, Munich
Seebohm F (1883) The english village community examined in its relations to the manorial and tribal systems and to the common or open field system of husbandry: an essay in economic history. Longman, Green & Co., London
Seebohm F (1897) Review of August Meitzen, Siedelung und Agrarwesen der Westgermanen und Ostgermanen, der Kelten, Römer, Finnen und Slawen, 4 vols. (Berlin: Hertz, 1895). Econ J 7(25):71–79
Sims DW, Krogt P van der (1995) [2nd] Conference on the history of cartography, London, 21–22 Sept 1967. http://cartography.geog.uu.nl/ichc/1967.html. Last accessed 6 Mar 2005.
Skelton RA (1972) Maps: a historical survey of their study and collecting. University of Chicago Press, Chicago
Stavenhagen W (1904) Skizze der Entwicklung und des Standes des Kartenwesens des ausserdeutschen Europa. Justus Perthes, Gotha
Thrower NJW (1991) When mapping became a science. UNESCO Courier, June 1991:31–34
Vinogradoff P (1911) The growth of the manor, 2nd edn. George Allen, London
Wheeler GM (1885) Government land and marine surveys (Origin, Organization, Administration, Functions, History, and Progress). In: Wheeler GM (ed) Report upon the third international geographical congress and exhibition at Venice Italy, 1881, Accompanied by Data Concerning the Principal Government Land and Marine Surveys of the World, Washington, DC, 76–569
White TP (1886) The ordnance survey of the United Kingdom. William Blackwood & Sons, Edinburgh
Wilford JN (1981) The mapmakers: the story of the great pioneers in cartography from antiquity to the space age. Vintage Books, New York
Wolkenhauer A (1904) Beiträge zur Geschichte der Kartographie und Nautik des 15. bis 17. Jahrhunderts. Mitteilungen der geographische Gesellschaft zu München 1(2):161–260
Wolkenhauer W (1895) Leitfaden zur Geschichte der Kartographie in tabellarischer Darstellung. Breslau: Ferdinand Hirt, Königliche Universitäts- und Verlags-Buchhandlung. Reprinted, with many additions, as Aus der Geschichte der Kartographie, Deutsche geographische Blätter 27 (1904):95–116, 33 (1910):239–64, 34 (1911):120–29, 35 (1912):29–47, 36 (1913):136–158, and 38 (1916/17):101–128 and 157–201
Wolter JA (1975) The emerging discipline of cartography. Ph.D. dissertation, University of Minnesota (Geography)
Wright JK (1924) Early topographical maps: their geographical and historical value as illustrated by the Harrison Collection of the American Geographical Society. American Geographical Society, New York